XXV International Conference and Meeting of Departments (CONSTRUMAT 2019)

IOP Conference Series: Materials Science and Engineering
Volume 549

Zuberec, Slovakia
29 – 31 May 2019

ISBN: 978-1-5108-9089-3
ISSN: 1757-8981

Printed from e-media with permission by:

Curran Associates, Inc.
57 Morehouse Lane
Red Hook, NY 12571

proceedings.com

Some format issues inherent in the e-media version may also appear in this print version.

Printed by Curran Associates, Inc. (2019)

For permission requests, please contact the Institute of Physics at the address below.

Institute of Physics
Dirac House, Temple Back
Bristol BS1 6BE UK

Phone: 44 1 17 929 7481
Fax: 44 1 17 920 0979

techtracking@iop.org

Additional copies of this publication are available from:

Curran Associates, Inc.
57 Morehouse Lane
Red Hook, NY 12571 USA
Phone: 845-758-0400
Fax: 845-758-2633
Email: curran@proceedings.com
Web: www.proceedings.com

Table of contents

Volume 549

XXV International Conference and Meeting of Departments, CONSTRUMAT 2019

29–31 May 2019, Zuberec, Slovakia

Accepted papers received: 13 May 2019

Preface

Papers

IOP Conf. Series: Materials Science and Engineering **549** (2019) 011001 doi:10.1088/1757-899X/549/1/011001

PREFACE

The XXV. International Conference „CONSTRUMAT 2019" has traditionally provided a forum for participants from industry, universities, research and testing laboratories. The conference will take place in Zuberec, Slovakia on May 29 -31, 2019. It has been organised by Faculty of Civil Engineering, University of Žilina, Slovakia. The conference presents the latest knowledge of science and research activities in the field of construction materials, performance of concrete, steel and composite structures and discuss the manufacturing process, testing and assessment methods in laboratory and service conditions. The CONSTRUMAT series of conferences focuses on new trends and provides guidance for future developments within Central Europe context.

First name	Surname	Institution	E-mail address
Miroslav	Brodňan	University of Žilina, Slovakia	brodnan@fstav.uniza.sk
Peter	Koteš	University of Žilina, Slovakia	kotes@fstav.uniza.sk
Petra	Bujňáková	University of Žilina, Slovakia	petra.bujnakova@fstav.uniza.sk
Patrik	Kotula	University of Žilina, Slovakia	kotula@fstav.uniza.sk
Richard	Hlinka	University of Žilina, Slovakia	hlinka@fstav.uniza.sk

First name	Surname	Institution	E-mail address
Vlastimil	Bílek	Technical University of Ostrava	vlastimil.bilek@vsb.cz
Ján	Bujňák	University of Žilina	jan.bujnak@fstav.uniza.sk
Robert	Černý	Czech Technical University in Prague	cernyr@fsv.cvut.cz
Rostislav	Drochytka	Brno University of Technology	drochytka.r@fce.vutbr.cz
Adriana	Eštoková	Technical University of Košice	adriana.estokova@tuke.sk
Jiří	Kolísko	Czech Technical University in Prague	jiri.kolisko@cvut.cz
Petr	Konvalinka	Czech Technical University in Prague	petr.konvalinka@fsv.cvut.cz
Jitka	Malá	Brno University of Technology	mala.j@fce.vutbr.cz
Pavel	Schmid	Brno University of Technology	schmid.p@fce.vutbr.cz
Stanislav	Unčík	Slovak University of Technology in Bratislava	stanislav.uncik@stuba.sk
Josef	Vičan	University of Žilina	josef.vican@fstav.uniza.sk

Peer review statement

All papers published in this volume of *IOP Conference Series: Materials Science and Engineering* have been peer reviewed through processes administered by the proceedings Editors. Reviews were conducted by expert referees to the professional and scientific standards expected of a proceedings journal published by IOP Publishing.

Quality control of the concrete and corrosion damage of cables of prestressing reinforcement of KA beams, bridge in Chomutov

S Rehacek[1], D Citek[1], M Krystov[1] and J Kolisko[1]

[1] Klokner Institute, Czech Technical University in Prague, Solinova 7, Prague 6, Czech Republic

E-mail: stanislav.rehacek@cvut.cz

Abstract. The aim of this paper is to assess the condition of the diagnosed bridge structure, at which previous bridge inspections identified corrosion damage of cables of prestressing reinforcement of KA beams, to assess the extent of corrosion damage of prestressing reinforcement, the condition of chloride ions in concrete of the bearing structure, and to draw up a detailed corrosion assessment of the prestressing reinforcement, then to propose a possible method of the bridge renovation with partial use of new materials (UHPC).

1. Description of the structure

The bridge with the reg. no. 7i-052.1 [1,2,3,4] is a bridge over the Chomutovka Spring. It is a single-span bridge carrying an access ramp to the 1st class road (no. 7) [1].
Bridge abutments and the Caps are made of reinforced concrete. The bearing structure consists of 9 reinforced-concrete prefabricated KA-73 elements. The superstructure is placed on elastomeric bearings.

2. Concrete

2.1. Destructive tests of concrete compression strength

Cored specimens (Ø about 78 and 100 mm) were taken for purposes of destructive tests of concrete compression strength (abutment web, abutment cap and Superstructure).
Summary results of the destructive tests of concrete compression strength and corresponding strength classes, or concrete classes, respectively, are stated in the following tables 1 and 2.

Table 1. Summary of results of the concrete compression strength.

Diagnosed structural elements		Concrete compression strength (MPa)		Variation
		Test average value	Characteristic	coefficient v*
Abutments - web	destructively	51.2	44.2	4.7
Abutments – cap	destructively	44.8	37.8	6.4
Superstructure	destructively	81.4	-	2.7

* ČSN 73 2011 [3] states a limit value of the variation coefficient for homogeneous concrete of the class C 30/37 and higher: v = 12 % (homogeneity from the viewpoint of strengths).

Table 2. Concrete strength class or concrete class, respectively, on the basis of the tests carried out.

Diagnosed structural elements		Concrete class / concrete strength class	
		ČSN EN 1992	Documentation [2, 3] requirement
Abutments - web	destructively	C 40/50	B 250, C 16/20
Abutments – cap	destructively	C 30/37	
Superstructure	destructively	At the level C 55/67 – C 60/75	B 330, C 25/30

On the basis of destructive tests of concrete compression strength in a very limited range we recommend considering the following concrete classes for the monitored reinforced-concrete structures of the bridge, according to ČSN EN 1992 [5]:

- Abutment web: C 30/37
- Abutment cap: C 25/30
- Superstructure: C 45/55

2.2. Comparison of the concrete carbonation and thickness of the reinforcement cover layer

The concrete carbonation depth was determined evenly along the entire length of the bridge structure. It is possible to estimate the measurement uncertainty within ± 2 mm. Steel reinforcement is exposed to corrosive processes which are affected by a number of factors. The following factors can be considered as the most essential ones:

- Ambient humidity
- Filling of the porous system of concrete with water,
- Depth of placement of the reinforcement under the surface,
- Thickness of the carbonated concrete layer,
- Content of certain pollutants in concrete (chloride ions, presence of acids and other aggressive substances),

If the reinforcement loses its passivation protection due to the concrete alkalinity (carbonation), the presence of humidity calls out corrosion processes leading to known failures, such as droping out surface layers and reinforcement cross section decrease.

Determination of the thickness of the reinforcement cover layer and concrete carbonation depth was carried out on these elements with the following result:

Substructure

- Concrete carbonation depth: 8 to 20 mm
- Concrete cover of reinforcement: 30 to 45 mm

IOP Conf. Series: Materials Science and Engineering **549** (2019) 012001 doi:10.1088/1757-899X/549/1/012001

Superstructure

- Concrete carbonation depth: 1 to 5 mm
- Concrete cover of prestressing reinforcement at the lower surface: 15 to 50 mm
- Concrete cover of stirrups at the lower surface: 0 to 20 mm

2.3. Determination of the content of chlorides in concrete

The content of chloride ions above a certain limit significantly increases the risk of corrosion of the reinforcement. For this reason, a chemical analysis of concrete was carried out within the framework of diagnostic works for the purpose of identification of the content of chloride ions in concrete.

Sampling was distributed evenly along the structure of the bridge. Altogether 30 samples of concrete was taken in 10 sites (always three samples from different heights in one site).

An informative calculation review was carried out under these presumptions and qualified estimations and conditions:

- The quantity of cement used for production of 1 m^3 of concrete is 350 kg for the substructure concrete and 420 kg for the concrete of the beams.
- The concrete density was identified from cored specimens and is at a level of 2,340 kg/m^3 for concrete used for abutment web and cap, while for the concrete superstructures it is 2,510 kg/m^3.

Limit content of Cl^- [% of weight] related to the cement weight is, according to ČSN EN 206+A1 [6], 1 % (of weight) for simple concrete; 0.4 % (of weight) for reinforced concrete; 0.2 % (of weight) for prestressed concrete.

The following can be stated from the carried out chloride content analysis:

- The average content of Cl^- [% of weight] found out by the laboratory analysis for concrete of abutments is 2.24 % for the sampling depth 0-15, 1.97 % for the sampling depth 15-30 and 1.40 % for the sampling depth 30-45 mm. The limit content of Cl^- [% of weight] related to the cement weight is, according to ČSN EN 206+A1, 0.4 % of weight for reinforced concrete.
- The average content of Cl^- [% of weight] found out by the laboratory analysis for concrete of sills is 1.32 % for the sampling depth 0-15, 0.56 % for the sampling depth 15-30 and 0.29 % for the sampling depth 30-45 mm. The limit content of Cl^- [% of weight] related to the cement weight is, according to ČSN EN 206+A1, 0.4 % of weight for reinforced concrete.
- The average content of Cl^- [% of weight] found out by the laboratory analysis for concrete of the superstructure is 0.83 % for the sampling depth 0-15, 0.57 % for the sampling depth 15-30 and 0.46 % for the sampling depth 30-45 mm. The limit content of Cl^- [% of weight] related to the cement weight is, according to ČSN EN 206+A1, 0.2 % of weight for prestressed concrete.

3. Reinforcement of structural elements

3.1. Destructive probes

Altogether 20 destructive probes into the bearing structure and 2 destructive probes into the substructure were made for identification of the reinforcement position.

On the basis of destructive probes carried out and other facts found out it is possible to state in general:

- The probes made into the prestressed beams "KA-73" with a length of 15 m identified prestressing reinforcement in the forms of cables with individual wire diameters 4.5 mm.
- Individual cables with surface corrosion were found in most probes. At the three probes, individual wires were weakened to 3.5 – 4.2 mm. This means weakening of the wire cross section area by more than 40 %. A fully over-corroded cable of prestressing reinforcement was found in one probe.

- Sleeves were mostly filled in with an injection mixture. At the two probes the sleeves were not filled in.
- The probes made into abutments identified a rebar reinforcing system with a diameter of 24 mm on average. Surface corrosion was found in the place of the probes on the reinforcement.
- The probes details are shown in figure 1 and 2.

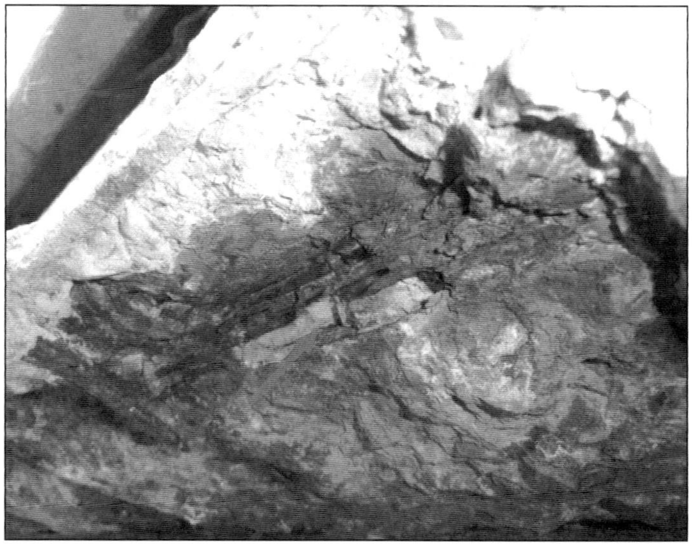

Figure 1. Probe detail, prestressing reinforcement, KA beam, lower face approx. 3.2 metres from the abutment. General decomposition of the prestressing reinforcement.

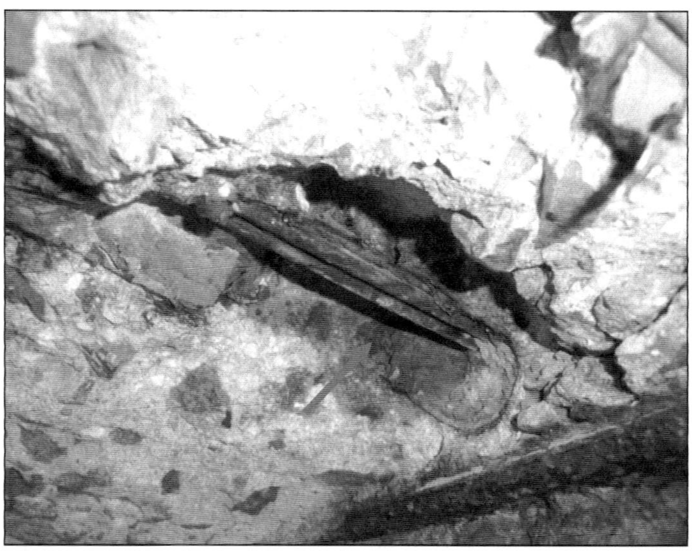

Figure 2. Probe detail, prestressing reinforcement, KA beam, lower face approx. 3.5 metres from the abutment. Heavy or even depth corrosion was found on the prestressing reinforcement. The sleeve was party filled in with injection mixture. Prestressing reinforcement coverage is about 15 mm.

3.2. Corrosion assessment of the prestressing reinforcement

The corrosion survey on the bridge, reg. no. 7i-052.1, included the sampling of corrosion products (very often with fine-grain fragments of the cement injection) and their element analysis by using the XRF method (X-ray fluorescence analysis) and the phase representation analysis by using the XRD method (X-ray diffraction analysis).

On the basis of the probes carried out and other facts found out it is possible to state in general:

- From the results it is clear that corrosion damage was induced and especially stimulated by chloride ions (Cl^-), which are components of defrosting salts (NaCl). The content of chloride anions in corrosion products was very important and oscillated between values 1.19 – 3.24 weight %. At the same time, akaganeite (β-FeO(OH)) was found in the corrosion products, containing embedded chlorine atoms, and also generally significant quantity of hydrocalumite (hydration phase of cement containing chlorine atoms) was found in injection residuals.
- From the results it is clear that at cables it is possible to register highly localised and unpredictable corrosion damage stimulated by chloride anions. Localisation is given mainly by penetration of humidity into the area of the routing of prestressing reinforcement (leakage, condensing). The injection system was originally modified by using finely ground limestone in a rheological way.
- The result from one of the sample is shown in table 3 and figures 3 and 4.

Table 3. Results of the XRF element analysis of the corrosion products sampled from the A1 sampling site.

Element	Representation [weight %]
Na	1.28
Mg	1.38
Al	3.40
Si	11.50
S	1.11
P	0.04
Ca	41.69
Cl	2.71
Fe	35.89

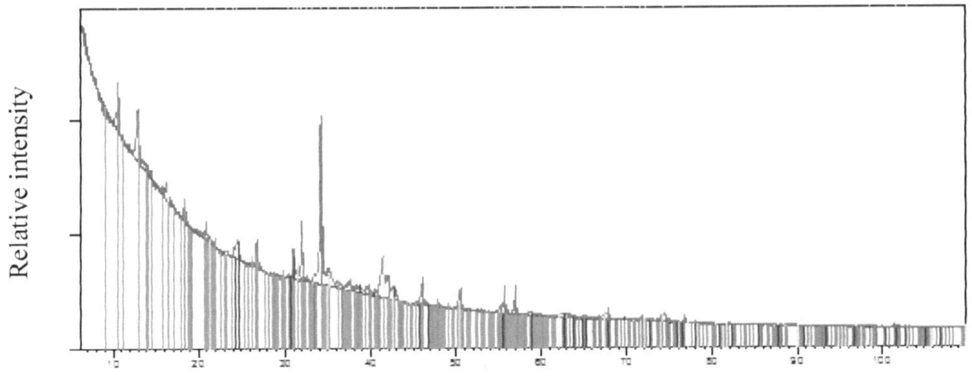

Figure 3. Diffraction pattern of corrosion products sampled from the A1 sampling site.

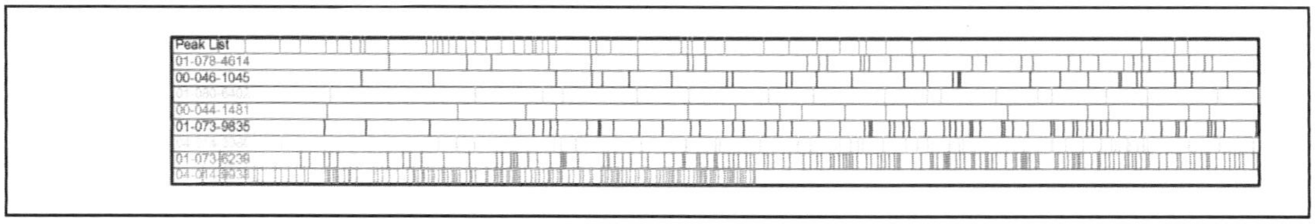

Phase name	Chemical formula	Representation [semi-quant %]
Calcite	$CaCO_3$	41
Ettringite	$Ca_xAl_yS_zO_oH_p$	22
Hydrocalumite	$Ca_xAl_yC_zO_oCl_kH_p$	13
Quartz	SiO_2	5
Magnetite	Fe_3O_4	7
Goethite	α-FeO(OH)	6
Lepidocrocite	B-FeO(OH)	5

Figure 4. Legend to the diffraction pattern from **Figure 3.**

4. Summary and conclusions

- Because of the limited number of destructive tests of concrete compression strength we recommend consideration of the concrete class:
  - Abutment web: C 30/37
  - Abutment cap: C 25/30
  - Superstructure: C 45/55
- While comparing the cover layer of concrete and the carbonation depth found out it is clear that a part of the diagnosed reinforcement (stirrups on the lower face of the beams) is already situated in the carbonated concrete layer and is not protected from corrosion with its natural alkalinity anymore. The beam stirrups are locally fully exposed and over-corroded. The same holds for cables of the prestressing reinforcement of the beams where the concrete cover layer has fallen off and the actual prestressing reinforcement has been exposed due to the influence of expansion corrosion pressures. Especially edge beams.
- The content of chloride ions in the substructure concrete is high in the entire range of the sampling depth (0 – 45 mm) and does not meet requirements of ČSN EN 206+A1. The content of chloride ions is within the standard limits only in the depth range of 30 – 45 mm at abutment sills.
- The content of chloride ions in the superstructure concrete in the range of the sampling depth (0 – 45 mm) is high and does not meet requirements of ČSN EN 206+A1.
- Individual cables with surface corrosion were found in the place of destructive probes to the prestressing reinforcement of the KA beams. At three probes, individual wires were weakened to 3.5 – 4.2 mm. This means the weakening of the wire cross section area by more than 40 %. A fully over-corroded cable of the prestressing reinforcement was found in one probe.
- The sleeves were mostly filled in with an injection mixture. At two probes the sleeve was not filled in.
- From the results of the corrosion survey of the prestressing reinforcement it is clear that corrosion damage was induced and especially stimulated by chloride anions (Cl^-), which are components of defrosting salts (NaCl). The content of chloride anions in corrosion products was very important and oscillated between values 1.19 – 3.24 weight %. At the same time, akaganeite (β-FeO(OH)) was found in the corrosion products, containing embedded chlorine atoms, and also generally significant quantity of hydrocalumite (hydration phase of cement containing chlorine atoms) was found in injection residuals.

- From the results it is clear that at cables it is possible to register highly localised and unpredictable corrosion damage stimulated by chloride anions. Localisation is given mainly by penetration of humidity into the area of the routing of prestressing reinforcement (leakage, condensing). The injection system was originally modified by using finely ground limestone in a rheological way.
- On the basis of the facts found out, the superstructure correspond to the building state V according to ČSN 73 6221 – poor condition = The defects and failures have a principal influence on reliability of the structure and are removable only through major interventions into the superstructure.
- It implies from the results of the corrosion survey that the beams are approaching the end of their lifetime due to the long-term leakage of salty solutions into the structure and effects of other corrosion factors from the viewpoint of durability. It is only possible to remove the identified defects of the superstructure by replacing the superstructure with a new one. With regard to the defects found, we recommend execution of replacement of the supertructure with the new one not later than within 5 years.

Acknowledgments
This article was written with kind contribution of the project SGS19/139/OHK1/2T/31.

References

[1] Listing from the BMS system (Bridge management system)
[2] Archive documentation, digital copies of longitudinal and cross sections, 5/1983. Supplied by the Client
[3] Type ground documents concerning the KA-73 prefabricated beams, Dopravoprojekt 11/1973, digital copies
[4] ČSN 73 6221 Inspection of Road Bridges
[5] ČSN EN 1992: Eurocode 2: Design of concrete structures
[6] ČSN EN 206+A1: Concrete - Specification, performance, production and conformity

IOP Conf. Series: Materials Science and Engineering **549** (2019) 012002 doi:10.1088/1757-899X/549/1/012002

The impact of aggregates from recycling, discharge time and mixing technique on the consistency of concrete

A Sicakova[1] and K Urban[2]

[1] Institute of Environmental Engineering, Faculty of Civil Engineering, Technical University of Kosice, Vysokoskolska 4, 042 00, Kosice, Slovakia
[2] Stavby Kosice s.r.o., 040 11, Kosice, Slovakia

E-mail: alena.sicakova@tuke.sk

Abstract. Concretes prepared by the application of the triple mixing approach and by the use of both recycled concrete aggregate (RCA) and recycled clay brick aggregate (RBA) were tested in terms of loss of consistency. Fly ash was used to improve the performance of recycled aggregates during mixing – this is a specific aspect of the triple-mixing method. The slump test was performed immediately after mixing, as well as after next 90 minutes. The results were evaluated from the point of view of prolonged discharge time. Concrete consistency changes with time, having the impact on technological processes at construction site. Over the discharge time, the RBA-based concrete achieved the worst result – by 87% slump loss. Discharge time influenced the slump loss more favourably in the case of normal mixing and in the case of using natural aggregate (NA). The substitution of NA with both the RCA and RBA caused the slump loss turned worse, while RCA provided better results than RBA. However, the substitution of NA with both RCA and RBA resulted in more favourable change in the slump loss when concrete was mixed by 3M method.

1. Introduction

The mixing and delivery process can influence the homogeneity and uniformity of fresh concrete, resulting in a negative effect on the technical parameters (workability and long-term performance characteristics of concrete). Concrete is a dynamic material and its consistency changes with time as mixing water is consumed through absorption by aggregates, cement hydration and evaporation. These factors are dependent on ambient conditions, types and combination of concrete components as well as the total available water. Prudent control of the factors that affect concrete workability requires an understanding of the factors, how they are interrelated, and what can be controlled by reasonable means. Mixture proportions, aggregate quantities, moisture contents, and admixtures are some of the factors that affect the rate and extent of slump loss that a ready mixed concrete producer can control.

The slump loss of concrete mixture at a construction site is one of the fundamental rheological properties considered to be responsible for the strength and durability aspects of concrete [1]. Concrete should achieve the required slump with consideration of the delivery time that can range from 45 to 90 minutes [2]. Prolonged mixing in a truck mixer accelerates stiffening of concrete so thus the rate of slump loss and the increased rate of the slump loss mostly brings about inconvenience, particularly when long hauling periods are involved as generally it is the case for ready-mixed concrete deliveries [1]. The duration that elapses in the course of mixing, delivering, placing, compacting, and finishing operations of concrete is considered to be the main parameter that affects the slump loss appreciably.

The depletion of water in fresh concrete increases as the length of delivery time extends since the hydration of cement and evaporation are directly related to elapsed time [1,3].

Several studies present the results of prolonged mixing time on the properties of concrete. They are not consistent and include the improvement, as well as the worsening of the results. Ravina [4] reported that the 7-, 28-, and 90-day compressive strength increased with increasing mixing time up to 135 min for various mixtures. The author also noted that the strength gain of these mixtures were linear. Kırca et al. [5] studied the 7- and 28-day compressive strength as a function of mixing time at a constant mixing speed. The authors recorded an increase in compressive strength as mixing time increased and reported that the strength gain was the result of the loss of water due to evaporation, which led to a decrease in w/c ratio. The researchers also stated that another possible reason for the increase in strength was that longer mixing times could have resulted in grinding of the cement particles, resulting in finer cement grains and greater degree of hydration. On the other hand, Trejo and Chen [6] reported the results of laboratory-mixed concrete as exhibiting no significant reduction in the properties (compressive strength, splitting tensile strength, modulus of elasticity, and modulus of rupture) when mixed up to 180 min at 8 rpm or less, while laboratory mixtures mixed at a mixing speed of 15 rpm exhibited reduced values. The field-mixed concrete exhibited significant reductions in compressive strength, splitting tensile strength, and modulus of elasticity after 120 min of mixing.

Based on volume, Construction and Demolition (C&D) waste is the largest waste stream in the EU – it represents about one third of all waste produced. Proper management of C&D waste and recycled materials – including the correct handling of hazardous waste - can have major benefits in terms of sustainability and the quality of life. But it can also provide significant benefits for the EU construction and recycling industry, as it boosts demand for C&D recycled materials [7]. Therefore, recycling of construction and demolition waste (C&DW) as a source of aggregates for the production of new concrete has attracted the growing interests of the construction industry.

Recycled aggregate (RA) is defined as aggregate resulting from the processing of inorganic or mineral material previously used in construction. When consisting mainly of crushed concrete it can be classified as recycled concrete aggregate (R_{CA}), similarly for crushed clay brick – recycled brick aggregate (RBA) [8]. Shortly, the main technical problems of RA may be specified as follows [9, 10, 11, 12]:

- high content of cement paste/mortar,
- weak interfacial transition zones between original mortar and aggregate,
- high porosity / water absorption,
- high level of sulphates,
- high level of chlorides,
- impurity,
- poor grading,
- high variation in quality.

One of the major problems with the use of recycled aggregate in structural concrete is its high water absorption capacity, which leads to difficulties in controlling the properties of fresh concrete [13]. Higher absorption causes the need for much more water for mixing and a high slump loss rate depending on the elapsed time. These characteristics of RA account for low workability, strength, and durability of concrete [14-16].

To improve the unfavourable properties of recycled aggregates, the following methods of surface treatments of RA are investigated worldwide.

- impregnation / coating in advance before mixing the concrete [17-22]
- coating during mixing using specific mixing procedures [23-27]. For this purpose, mainly various Supplementary Cementitious Materials are used.

By the impregnation / coating process, the surface properties of recycled aggregates could turn to that of being more suitable for concrete production.

In this study, the slump loss as a technological parameter of fresh concrete being prepared with two kinds of recycled aggregate, is presented and evaluated. The principle of the triple-mixing method was adopted. Samples with natural aggregates, as well as samples prepared by standard mixing were also tested, being considered as the reference ones. The slump test was performed immediately after mixing, as well as after next 90 minutes. The results were evaluated from the point of view of prolonged mixing time, application of recycled aggregates instead of natural one, and the use of triple mixing instead of normal mixing.

2. Materials and methods

The presented investigations were intended to assess the slump loss of concrete mixture. The following aspects of concrete composition, mixing and testing were included:

- application of the recycled aggregates (RCA and RBA) in comparison with natural aggregate (NA)
- application of the triple mixing (3M) approach in comparison with normal mixing (NM)
- application of fly ash (FA) as a coating material
- assessment of the loss of consistency due to prolonged mixing time, by testing the slump 90 minutes after mixing, in comparison with the slump being measured immediately after mixing

2.1. Materials

The characteristics of the materials used in the study are as follows:

Aggregates: natural aggregates NA and recycled aggregates RA (RCA and RBA) of D_{max} = 16mm. As for RA, 0/32 fraction was obtained from a company dealing with C&DW treatment (ENVIRONCENTRUM Ltd., Slovakia). Within the experiment, RA was crushed and sorted to standard fractions (4/8 and 8/16). As for fraction 0/4, only NA was used in the experiment. The physical properties of the aggregates are shown in Table 1.

Material for coating the coarse aggregates: fly ash (FA) from the energy segment of the steel-making factory in eastern Slovakia. The original grain size was d(0.9) = 95μm. The chemical composition was as follows: CaO - 4.1%; SiO_2 - 51.6%; Al_2O_3 - 23.1% and Fe_2O_3 - 7.8%.

Cement (CEM): CEM I 42.5 R

Admixture: polycarboxylate type of plasticizer.

Table 1. Properties of natural and recycled aggregates.

Type	Fraction	Density [kg/m^3]	Water absorption capacity [%]
NA	0/4	2650	1.2
	4/8	2650	1.0
	8/16	2650	1.0
RCA	4/8	2200	6.8
	8/16	2300	5.3
RBA	4/8	2050	9.6
	8/16	2150	7.0

2.2. Design of concrete composition and mixing technique

The compositions of concrete were designed keeping the limiting amounts of cement and water for the specific class of exposure, in accordance to the standard for concrete production (min 300 kg of cement and max w/c ratio = 0.5). The amounts of real mixing water (W_1 and W_2) were determined taking into account the actual absorption capacity of aggregates. The effective amount of water (W_{ef}) needed for the paste was increased by the amount of water corresponding to the absorption value of aggregate. For calculation of the additive amount, the thickness of a coating layer was considered as δ

= 0.150 mm. The coating layer was considered as a part of the concrete volume, so the amount of coating paste was taken into account when calculating the total volume of binder. Thus, the volume of binder paste consists of volume of coating layer and volume of paste filling the voids between grains. The compositions were calculated taking into account the density of individual components to keep a constant volume of $1 m^3$. For more details on the calculation methodology, see the articles [28, 29]. The compositions of concrete mixtures are given in Table 2.

The constant parameters of mixtures were as follows:

- the grain size distribution of aggregates (for both the NA-based samples and RA-based samples): 0/4: 50%; 4/8: 15% and 8/16: 35%,
- water/binder ratio (w/b) = 0.5, both for the coating layer paste and for the filling paste.

The triple mixing (3M) method was intended as a way to improve the quality of RA surface directly during mixing the concrete while the coarse part of the aggregate was only subjected to this treatment. Therefore, the order of adding components was specifically designed, see Fig. 1. For comparison, the same samples – recipes as given in Table 2 were mixed using the standard mixing procedure, see Fig. 2.

Table 2. Mix proportions of tested concretes.

Components [kg]		Samples according to the type of aggregate		
		NA	RCA	RBA
CEM I 42.5 R		336	310	323
NA	0/4	896	898	898
	4/8	269	-	
	8/16	627	-	
RA	4/8	-	224	208
	8/16	-	545	510
Coating material - Fly ash (FA)		47	68	56
Admixture		2.7	2.5	2.6
Effective water	W_{ef-1}	23.2	33.7	28.0
	W_{ef-2}	168	155	162

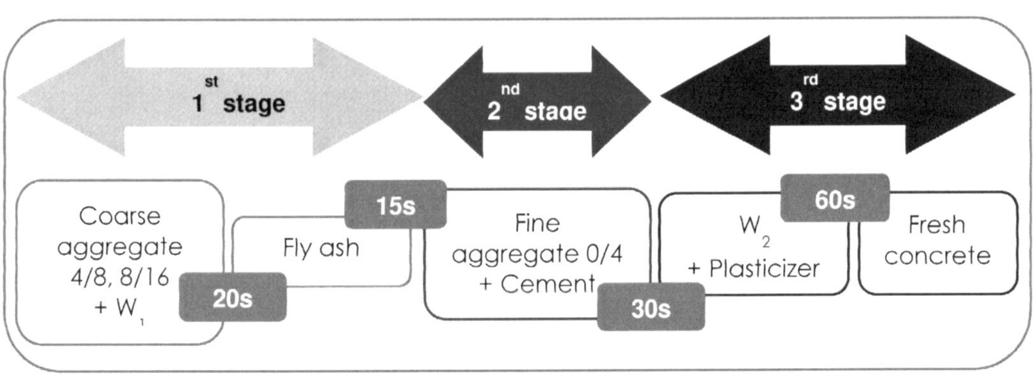

Figure 1. Scheme of triple mixing procedure (3M).

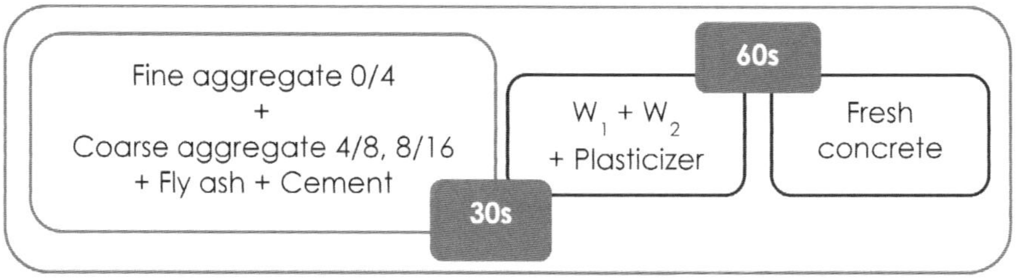

Figure 2. Scheme of normal mixing procedure (NM).

2.3. Testing the consistency

The standard slump test [30] was applied for testing the consistency of fresh concrete while it was tested in the two time sequences: immediately after mixing (S_0) and after next 90 minutes (S_{90}). Up to 90 minutes, the concrete mixture was re-mixing every 15 minutes. The results were obtained by repeating the test twice.

3. Results and discussion

The results of slump test of individual samples are given in 3. Here, the slump loss is also expressed as a percentage difference between the values found immediately after mixing (S_0) and 90 minutes after mixing (S_{90}), as well as it is expressed through the change in slump class according to [31]. To compare how the change of mixing method influences the consistency, the percentage differences are calculated and given in Table 3. The values are presented for mixtures according to the type of aggregate, as well as for both time sequences: 0 minutes and 90 minutes. Similarly, to compare, how the substitution of natural aggregate by RCA and RBA respectively influences the consistency, the percentage differences are calculated and given in Table 4. The values are presented for mixtures according to mixing method, as well as for both time sequences: 0 minutes and 90 minutes.

Table 3. Results of slump test and changes of slump in two time sequences.

Mixtures of concrete	Slump [mm] S_0	S_{90}	Slump loss [%] S_0 / S_{90}	Slump class change S_0 / S_{90}
NM-NA	220	180	-18	S5/S4
NM-RCA	210	70	-67	S4/S2
NM-RBA	80	50	-38	S2/S2
3M-NA	210	80	-62	S4/S2
3M-RCA	200	70	-65	S4/S2
3M-RBA	150	20	-87	S3/S1

Table 4. Change in consistency due to substitution of NA by RCA and RBA, respectively.

Samples according to mixing method	Change in consistency [%] NA / RCA		NA/RBA	
	S_0	S_{90}	S_0	S_{90}
NM	-5	-61	-64	-72
3M	-5	-13	-29	-75

The starting values of slump (S_0) are quite similar (200-220 mm) for both the NA and RCA-based samples, as well as for both mixing methods (NM and 3M). The slump values of samples having RBA are lower, while the difference is more significant when mixtures are mixed normally (NM - 80 mm). Different values for RBA were found despite the fact that the same principle in water dosage was applied for all kinds of aggregates: providing sufficient water to soak up the aggregate at the start of mixing.

The loss of consistency in time was noticeable in all cases; it is similar as presented for standard mixing and agrees with [32]. The researchers, presenting their results in [32], tested the influence of different moisture states of RCA on concrete workability and found that the replacement of the natural coarse aggregate with coarse RCA at different moisture states caused significant rapid slump loss in the mix with time - even the samples having RCA in saturated state. In the conducted tests, the highest slump loss from 0 to 90 minutes was achieved by the mixture with RBA (87%), while the lowest one (18%) was achieved by NA-based sample. As for mixing methods, the slump loss 90 minutes after mixing seems to be higher for 3M; only in the case of RCA the difference between NM and 3M is negligible.

Looking at Table 4, and evaluating the influence of the aggregate type immediately after mixing (S_0), it is clear that the use of RCA instead of NA worsens the slump only by 5%, while the use of RBA instead of NA worsens the slump by 64% in the case of NM and 29% in the case of 3M method. After 90 minutes, using of both the RCA and RBA instead of NA causes much higher change in consistency. This is in agreement with [33], presenting that the use of RCA decreases the slump and slump flow of concrete for both the normal and high workability concrete. They also achieved slump results revealing that higher RCA content caused lower workability, which is attributed to the physical characteristics of the RCA particles. This can be linked to our experiment, in which the full replacement of 4/8 and 8/16 fractions was applied. The difference between the effect of RCA and RBA was also described in [34] where authors present a tendency for slump to decrease as the mixing ratio of RBA increased.

In terms of mixing methods, changes in consistency in line 3M are smaller or insignificant, than that of the relevant values in line NM; this points to the advantage of 3M mixing in this sense.

4. Conclusion

Concretes prepared by the application of the triple mixing approach and by using both the recycled concrete aggregate and recycled brick aggregate were tested in terms of loss of consistency. The slump test was performed immediately after mixing, as well as after next 90 minutes. The results were evaluated from the point of view of prolonged mixing time, the application of recycled aggregates instead of the natural one, and the application of the triple mixing instead of the normal one.

The results obtained from the research can be summarized as follows:

- the slump loss was detected in all evaluated situations – by discharge time, as well as by the incorporation of recycled aggregates into the concrete mixture
- over the discharge time, the RBA-based concrete achieved the worst result – by 87% slump loss
- discharge time influenced slump loss more favourable in the case of normal mixing and when the NA was used
- the substitution of NA with both the RCA and RCB caused the slump loss turned worse, while RCA provided better results than RBA
- the substitution of NA with both the RCA and RBA resulted in more favourable change in slump loss when concrete was mixed by 3M method

Acknowledgements
This article has been carried out within the project supported by Slovak Scientific Grant Agency VEGA: Effective use of alternative aggregate of lower quality through modern concrete preparation techniques (grant number 1/0524/18).

References
[1] Erdogdu S 2005 Effect of retempering with superplasticizer admixtures on slump loss and compressive strength of concrete subjected to prolonged mixing. *Cem. and Con. Res.* 35 pp. 907– 912
[2] NRMCA."TIP 12 Slump test of concrete". [online]. [cit. 2017-10-25].Available on: <http://www.nrmca.org/aboutconcrete/downloads/TIP12.pdf>
[3] Ravina D, Soroka I 1994 Slump loss, compressive strength of concrete made with WRR and HRWR admixtures and subjected to prolonged mixing. *Cem. and Con. Res.* 24 pp 455–1462
[4] Ravina D 1996 Effect of Prolonged Mixing on Compressive Strength of Concrete with and without Fly Ash and/or Chemical Admixtures. *Mater. J. 93* pp 451–456
[5] Kırca Ö, Turanlı L, Erdoğan TY 2002 Effects of retempering on consistency and compressive strength of concrete subjected to prolonged mixing. *Cem. Concr. Res.* 32 pp 441–445
[6] Trejo D, Chen J 2014 *Effects of Extended Discharge Time and Revolution Counts for Ready-mixed Concrete (* Washington State Department of Transportation: Olympia, US)
[7] *EU Construction and demolition Waste Management Protocol*, September 2016, Available on: http://ec.europa.eu/growth/content/eu-construction-and-demolition-waste-protocol-0_en
[8] STN EN 12 620 + A1: 2008 Aggregates for concrete. Slovak Office of Standards, Metrology and Testing, Bratislava.
[9] Tam W Y V, Tam C M, Lea K N 2007 Removal of cement mortar remains from recycled aggregate using pre-soaking approaches. *Resources, Conservation and Recycling,* 50 pp 82 – 101
[10] Descarrega A 2011 *Quality improvement of the recycled aggregates through surface treatment*. Master theses, Barcelona.
[11] Sivakumar N et al. 2014 Experimental Studies on High Strength Concrete by using Recycled Coarse Aggregate. *Int. J. of Eng.and Sci.* 4/1, pp 27-36
[12] Kartini K, Rohaidah MN and Zuraini ZA Performance of Ground Clay Bricks as Partial Cement Replacement in Grade 30 Concrete. *World Academy of Science, Engineering and Technology*, 68 pp 312-315
[13] Poon C S, Kou S C, Lam L 2007 Influence of recycled aggregate on slump and bleedingof fresh concrete. *Mat. and Struct.* 40 pp 981–988
[14] Tabsh S W, Abdelfatah A S 2009 Influence of recycled concrete aggregates on strength properties of concrete". *Constr. and Build. Mat.* 23 pp 1163–1167
[15] Levy S M, Helene P 2004 Durability of recycled aggregates concrete: A safeway to sustainable development". *Cem. Concr. Res.* 34(11) pp 175–180
[16] Eguchi K, Teranishi K, Nakagome A, Kishimoto H, Shinozaki K and Narikawa M 2007 Application of recycled coarse aggregate by mixture to concrete construction *Constr. Build. Mater.* 21 pp 1542–1551
[17] Ryou J S and Lee Y S 2014 Characterization of Recycled Coarse Aggregate (RCA) via a Surface Coating Method *Int. J. Concr. Struct. Mater.* 8 pp 165–172
[18] Kim N W et al 2005 A study on the mechanical properties of concrete using the recycled aggregates by surface coating. *J. of the Korean society of Civ. Eng.* 25 pp 387-393
[19] Choi H et al 2016 Evaluation on the mechanical performance of low-quality recycled aggregate through interface enhancement between cement matrix and coarse aggregate by surface modification technology. *IJCSM* 10 pp 87-97
[20] Li J, Xiao H, Zhou Y 2009 Influence of coating recycled aggregate surface with pozzolanic powder on properties of recycled aggregate concrete". *Constr. and Build. Mat*, 23(3) pp

1287-1291

[21] Spaeth V and Djerbi Tegguer A 2013 Improvement of recycled concrete aggregate properties by polymer treatments *Int. J. Sustain. Built Environ.* 2 pp 143–152

[22] Ondova M, Sicakova A 2016 Evaluation of the Influence of Specific Surface Treatments of RBA on a Set of Properties of Concrete. *Materials*. 9/3 pp 1-12

[23] Kong D, Lei T, Zheng J, Ma C, Jiang J and Jiang J 2010 Effect and mechanism of surface-coating pozzolanic materials around aggregate on properties and ITZ microstructure of recycled aggregate concrete *Constr. Build. Mater.* 24 pp 701–708

[24] Urban K, Sicakova A 2016 Impact of specific mixing method on the properties of concrete with BFS aggregate. *Proc SGEM (Albena), Sofia, Bulgaria*, pp 49-56

[25] Rejeb S K 1996 Improving compressive strength of concrete by a two-step mixing method. *Cem. and Con. Res.* 26 pp 585-592

[26] Corinaldesi V, Moriconi G, 2009 Influence of mineral additions on the performance of 100% recycled aggregate concrete. *Constr. and Build. Mat.* 23 pp 2869-2876

[27] Sunayana S, Barai S V 2017 Recycled aggregate concrete incorporating fly ash: Comparative study on particle packing and conventional method. *Constr. and Build. Mat.* 156 pp 376-386

[28] Sicakova A, Urban K 2018 The influence of discharge time, kind of additive, and kind of aggregate on the properties of three-stage mixed concrete. *Sustainability* 10(11) pp 1-13

[29] Sicakova A, Urban K, Kováč M 2018 Slump Loss of Concrete Based on RCA and Prepared by Specific Mixing Approach. *Periodica Polytechnica Civil Engineering*, 62(4) pp 1015–1020

[30] STN EN 12350-2: 2010Testing fresh concrete. Part 2 Slump test. Slovak Office of Standards, Metrology and Testing, Bratislava, 2016

[31] STN EN 206:2013 + A1:2016 Concrete. Specification, performance, production and conformity. Slovak Office of Standards, Metrology and Testing, Bratislava.

[32] Ismail S, Ramli M 2014 Effect of Different Moisture States of Surface-Treated Recycled Concrete Aggregate on Properties of Fresh and Hardened Concrete". *Int. J. of Civ. Envi. Struct. Constr. and Arch. Eng.* 8(1) pp 65-71

[33] Safiuddin M et al 2011 Properties of high-workability concrete with recycled concrete aggregate. *Mat. Res.* 14(2) pp 248–255

[34] Jian Y, Qiang D, Yiwang B 2011 Concrete with recycled concrete aggregate and crushed clay bricks. *Constr. and Build. Mat.* 25/4 pp 1935-1945

IOP Conf. Series: Materials Science and Engineering **549** (2019) 012003 doi:10.1088/1757-899X/549/1/012003

Moisture and its condensation in external construction of wooden house

D Bladová [1], A Struhárová[1], M Kostelecká[2]

[1] Slovak University of Technology in Bratislava, Department of Material Engineering, Slovak Republic
[2] Czech Technical University in Prague, Klokner Institute, Czech Republic

E-mail: diana.bladova@stuba.sk

Abstract. This work is focused on the assessment of moisture condensation in different external constructions of wooden house. The studied wooden construction is constructed from 3 different external wall compositions. There were sensors measuring relative humidity within the external constructions. Meanwhile, external and internal conditions were measured. These conditions and composition of external constructions were used in Teplo 2017 software to assess the moisture condensation. In this work, the influence of the imperfect realization of the vapor barrier on the moisture in external constructions is also theoretically evaluated.

1. Introduction

Wood as construction material was very popular in the past. In the building industry, it fulfilled many functions from load-bearing (e.g. cottages) to decorative. Wood was gradually replaced by reinforced concrete after the beginning the industrial revolution. Today, wood is getting back to its place in people's minds and is used in construction of modern wooden houses. The return to traditions or mature is not the only reason why people opt for the wooden houses. Main reasons for that are the fulfilment of much stricter energy requirements for reasonable price and larger floor area due to smaller thickness of external walls. Some also like the fact the wooden houses are mostly build from environment friendly materials that cause lesser ballast to the nature or have less complicated production and transport.

There are two possible external wall compositions. First is diffuse open and it is more preferable from the health aspect, because the house "breaths" through it. Materials allow water vapour transmission and at the same time make the construction airtight. Diffusion resistance of individual layers decreases towards the exterior so that vapour can be released from the structure. Permeable ecological insulation is the most used in this composition. Second is diffuse closed composition. This composition differs from the open one by a vapour barrier. This barrier inhibits water vapour transmission. Barrier tightness is secured by a proper execution without any gaps. It is a material with high diffuse resistance and installation fore-wall is in front of it to prevent it from damage [1].

Possible water vapor condensation within the construction has to be taken into consideration during design and execution. It is an unwanted phenomenon that appears when the partial pressure of water vapor is higher than the partial pressure of saturated water vapor at certain point within the construction [2]. Condensation can influence the functionality, durability and degradation of construction materials. Their thermo-insulating and mechanical characteristics are getting worse. It also leads to mould formation and loss of integrity of the structure due to canker [3].

2. Methods

2.1 Object description

Measured object is a prefabricated wooden house showed on figure 1. It is placed in Dubňany on the east of Czechia. The gross floor area is 148 m^2 and gross building volume is 645 m^3. The house meets the requirements for energy passive construction. It uses the technologies such as recuperation, wall cooling and solar panels.

The object consists of underground floor, floor and residential attic. The technical building equipment is situated in the underground floor. There are 2 types of external constructions. Diffuse open construction is located on the east and south side and diffuse closed construction is located on the north and west side.

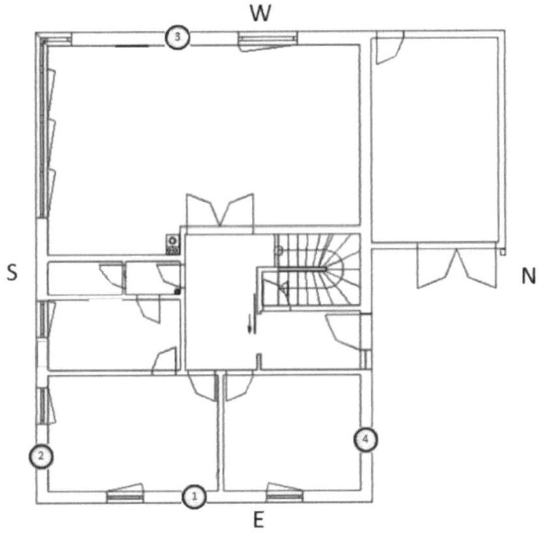

Figure 1. Object ground plan with sensor placement.

2.2 Specifically used compositions of external constructions

From inside out, each composition of external constructions consists of: silicone plaster, adhesive reinforced with glass mesh, fibreboard insulation, column construction from STEICO wall I profiles filled with fibreboard insulation, oriented strand boards OSB 3, adhesive tape/vapor barrier/permeable barrier, installation fore-wall with wooden grate construction and plasterboard. Composition of external walls is always same except from one layer placed on the OSB. It is the adhesive tape, vapor barrier or permeable barrier. Changing this layer makes individual compositions diffuse open or diffuse closed.

2.2.1 External construction Nr.1.

This is a diffuse open construction. The OSB joints are taped over with AIRSTOP FLEX 50mm wide. It is situated in the east side of the object. Measuring sensor Nr. 1 is placed here. Its complete composition can be seen in the figure 2.

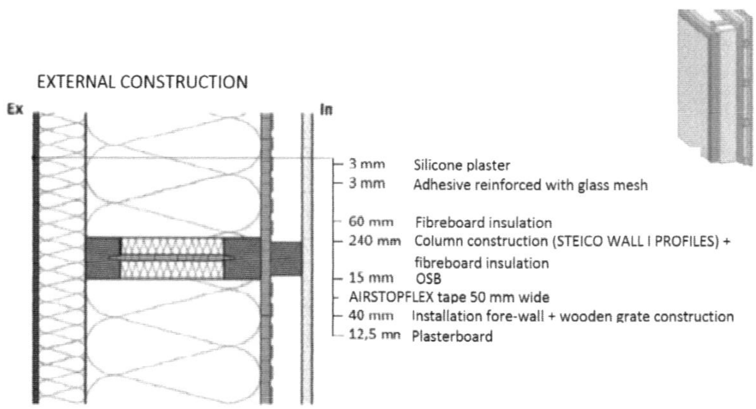

Figure 2. External sheet composition Nr.1.

2.2.2 External construction Nr.2

This is a diffuse open construction. Instead of adhesive tape the ISOCELL permeable fibre-reinforced barrier with diffuse resistance 9090 is applied. It is situated on the south side of the object. Measuring sensor Nr. 2 is placed here.

2.2.3 External construction Nr.3.

This is a diffuse closed construction due to vapor barrier, specifically Guttafol WB foil with diffuse resistance 210 154. It is situated on the north and west side of the object. Measuring sensors Nr. 3 and 4 are placed here. Sensor Nr. 3 is on the west side and sensor Nr. 4 is on the north side.

2.3 Measuring

During the measurement, the parameters such as temperature and relative air humidity of the external and internal environment were measured. The measurements took place every 27 minutes and they were done by the resistive measuring sensors WS-16 by Elbez which were installed in the external walls.

The company Úsporné bydlení Ltd. provided us with the measurements from July 2017 – March 2018. To evaluate the occurrence of condensation we have chosen the measurements from March 2018. After the measurement analysis, it was obvious that the lowest exterior temperature was reached in this month. Therefore we chose those.

Measurements were evaluated in the Teplo 2017 program. Teplo 2017 is program that serves to assess the compositions of building structures in terms of heat and water vapor distribution. The compositions of external constructions and conditions mentioned in table 1. were imported to the program. We also used the supplied data and diffuse resistance factor of vapor/permeable barrier to determine the amount of relative humidity in the structure when the barrier was improperly executed.

3. Results and analysis

In table 1, the measured humidity and temperature values of the exterior and interior are shown, as well as the relative humidity values measured by the sensors placed in the external walls of the house in the time of lowest exterior temperatures in March 2018.

Table 1. Measured humidity and temperature values in March 2018.

	Temperature external	Temperature internal	Humidity external	Humidity internal	Humidity – sensor [%]			
					1	2	3	4
March 2018	-10.9 °C	23.0 °C	29.7 %	30.0 %	3.1	4.0	6.9	7.8

3.1 Evaluation of condensation occurance inside external walls

Condensation inside external walls occurs when the partial pressure of water vapor (pink line) is higher than the partial pressure of saturated water vapor (red line) at certain point within the construction.

3.1.1 External construction Nr.1.

Measuring sensor Nr. 1 is placed here. According to results from Teplo 2017 program (figure 3), water vapour condensation does not occur in the external construction Nr.1 with adhesive tape on the OSB (diffuse open construction).

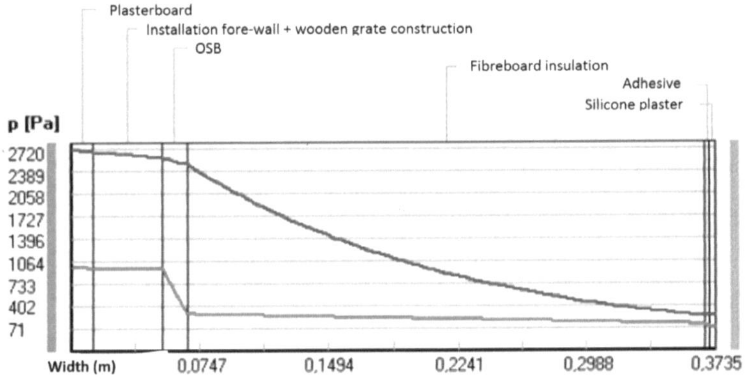

Figure 3. Condensation evaluation of external sheet Nr. 1.

3.1.2 External construction Nr.2.

Measuring sensor Nr. 2 is placed here. According to results from Teplo 2017 program (figure 4), water vapour condensation does not occur in the external construction Nr. 2 with the ISOCELL permeable fibre-reinforced barrier (diffuse open external construction).

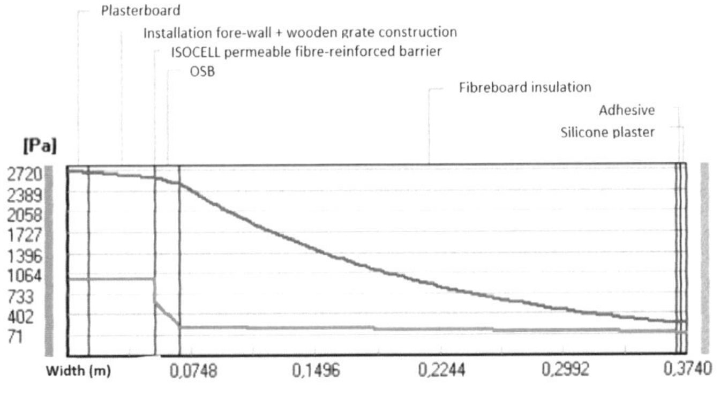

Figure 4. Condensation evaluation of external sheet Nr. 2.

3.1.3 External construction Nr.3.

Measuring sensor Nr. 3 and 4 are placed here. For the evaluation we used humidity measured by the sensor Nr. 4 as it recorded higher humidity on the north side. According to results from Teplo 2017 program (figure 5), water vapor condensation does not occur in the external construction Nr. 3 with Guttafol WB foil (diffuse closed external construction).

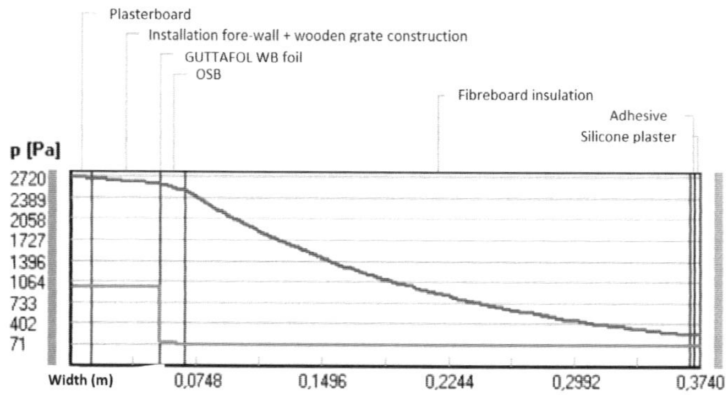

Figure 5. Condensation evaluation of external sheet Nr. 3.

3.2 Evaluation of the influence of an improper execution of vapour barrier

In the figure 6, the influence of improper execution of vapor/permeable barrier on the relative humidity in the external construction is shown. The ISOCEL permeable fibre-reinforced barrier that is in the external construction Nr. 2 was used for this analysis. This particular barrier has a diffuse resistance of 6902 and the proper execution causes the relative humidity of 7.3% at the measured point in the external sheath. If the barrier is damaged or is executed improperly, the relative humidity rises up to 10.06%. Not using the barrier at all, the relative humidity in the construction is 12.84% which is the value calculated in the external sheath Nr. 1.

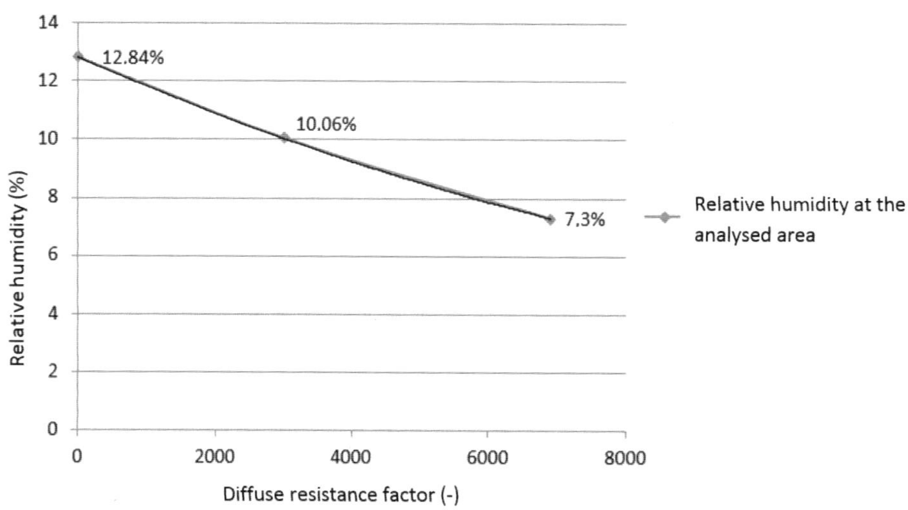

Figure 6. Evaluation of improper execution of vapour barrier.

4. Conclusion

The aim of this work is to evaluate various types of external constructions concerning the water vapor condensation within it. Evaluation was processed by Teplo 2017 program. All external wall compositions have proved up to the standards. Which means there is no water vapor condensation inside construction. That is why we consider the result positive.

For the future we would recommend installing 2 more sensors than are installed now. Specifically, one in the air layer resp. installation fore-wall, and one closer to the exterior. More sensors would significantly help in better analysis of water vapor transmission through construction.

In conclusion, we can state the wooden houses are very convenient alternative of traditional-build houses concerning the environmental impact, construction time, space options, costs and all-year-round construction possibilities.

Acknowledgments

This article was created with the support of the Ministry of Education, Science, Research and Sport of the Slovak Republic within the Research and Development Operational Programme for the project "University Science Park of STU Bratislava", ITMS 26240220084, co-funded by the European Regional Development Fund.

References

[1] Anderson L 2002 *Wood Frame House Construction* (The Minerva Group)
[2] Halahyja M, Chmúrny I and Sternová Z 1998 *Bulding thermal technology* (Bratislava: JAGA GROUP)
[3] Katunský D 2017 *Building physics* Lecture (Košice: TUKE)
[4] Kandráč P 2015 *Thermodynamics of the atmosphere. Air humidity* Lecture (Košice: TUKE)
[5] Kantor P 2012 *Chemical compostion of wood* Lecture (Brno: MENDELU)
[6] Makovíny I 1995 *Measurement of wood moisture* (Zvolen: TUZVO)
[7] Reinprecht L 2008 *Wood protection* (Zvolen: TUZVO)
[8] Nagy E 2009 *Low energy and passive house* (Bratislava: JAGA GROUP)
[9] Roberts N 1981 *How to build your own wood-frame house from scratch* (Tab Books)

Wood pulp as a potential raw material source for manufacturing bio-based building materials

N Stevulova[1], V Hospodarova[1], V Vaclavik[2] and T Dvorsky[2]

[1] Technical University of Kosice, Faculty of Civil Engineering, Kosice, Slovakia
[2] VSB-Technical University of Ostrava, Faculty of Mining and Geology, Ostrava, Czech Republic

E-mail: nadezda.stevulova@tuke.sk

Abstract The application of bio-based material such as cellulose fibres derived from renewable source to cement mortars was investigated and the properties of fresh mixtures and hardened mortars were characterized. The mixtures composition consisted of combined addition of the selected kind of cellulose fibres (bleached wood pulp) and superplasticizer (polycarboxylate ester). The aim was to produce a cement mortar with the improved properties for indoor application. The 0.5 wt.% fibre addition (from weight of filler and binder) and superplasticizer dosage of 1% in the mixtures were used. Reference cement mortar did not contain cellulose fibre and superplasticizer. The comparative study of the fresh fibre cement mixture properties has been showed worse flowability behaviour of mixture with plasticizer, as expected. The the beneficial effect of the superplasticizer was not proven even on the compressive and flexural strength of fibre cement mortars after 7, 28 and 90 days of hardening. The performances of the mortars were dependent not only on fibres and superplasticizer nature but also on the organic/inorganic interface.

1. Introduction

Nowadays, the use of raw material from renewable sources to bio-based building materials is significantly interested in terms of development and production of environmentally friendly materials. Generally, renewable sources are considered as option in their use in building materials to improve the sustainability of the construction industry [1]. To improve sustainability in building materials usage, the construction industry must include the reuse of industrial by-products and renewable materials in construction [2]. Natural fibres in cement composites are becoming increasingly popular. The use of cellulosic fibres as a renewable source coming from different sources such as wood, vegetables, agro-plants and waste from their processing in the production of cement composites has drawn an increasing interest [3, 4]. Cellulosic fibres offer a variety of advantages such as wide availability, biodegradability, fast recoverability at relatively low costs, low density, adequate stiffness and strength, variety of morphologies, controllable aspect ratio and surface roughness, an interfacial compatibility through appropriate modification of the fibre surface and are without no known health hazards [5]. Natural fibres obtained from renewable sources are readily available and a relatively cheap material compared to man-made fibres. The utilization of cellulose fibres into cement-based composites/mortars in European today´s construction industry is renewed since the last century due to environmental benefit, energy and resource conservation [2].

The incorporation of different type and size of cellulose fibres into cementitious matrix led to favourable effects regarding the reduction of matrix brittleness and increase in the durability, which is proportional to the resistance to crack propagation offered by the fibres bridging the matrix, thereby effectively transferring the load [6]. Many research works on the use of cellulose fibres coming from different renewable sources such as wood [7, 8, 9], sisal [10, 11], hemp [12-15], coconut [11], and bamboo [16, 17] in bio-based fibre cement materials have been performed and reported. The use of cellulose fibres and fibrous waste residues promotes bio-architecture and relates to low environmental impact at the end of the lifecycle.

In the present work, the cellulosic fibres coming from wood pulp as a potential source of raw material are incorporated to the cement mixture. This study is focused on the properties of fresh and hardened cement mortars containing wood pulp in dosage of 0.5 % (respect to binder and filler weight) and superplasticizer addition (1 % from cement weight). The flowability of fresh mixture, density (28 days hardening time) as well as mechanical properties (compressive and flexural strength) of hardened fibre cement composites after 7, 28 and 90 days of hardening were studied.

2. Materials and methods

2.1. Materials

CEM I 42.5 N, the ordinary Portland cement supplied by Cement Factory Ltd. (Povazska cementaren Ladce, Slovakia) was used for the preparation of the cement mortars. Standard natural silica sand from company Filtracni pisky Ltd (Chlum, Czech Republic) in accordance with standard STN EN 196-1 [18] was used as filler. The wood pulp Greencel was obtained from Slovak company Greencel Ltd (Hencovce, Slovakia). The white bleached wood pulp GW-500 with 99.11 % of holocellulose content was used as 0.5 % addition in relation to the total weight of the filler and binder. The physical properties of this sample of cellulose fibres are shown in Table 1. Water for the cement mixtures preparation was used in accordance with standard STN EN 1008 [19]. To adjust the workability of fibre cement mixtures a superplasticizer MasterGlenium Sky 665 (as polycarboxylate ester type) acquired by BASF - Chemical Company Ltd (Prague, Czech Republic) was taken in dosage of 1 % from weight of cement according to standard STN EN 934-2+A1 [20].

Table 1. Properties of wood pulp GW-500.

Density [10^3kg/m^3]	Bulk density [kg/m^3]	Average fibre width [μm]	Average fibre length [μm]	Aspect ratio	Specific surface area [m^2/g]	Thermal conductivity [W/m.K]
1.508	60	21.7	504	23.2	8.26	0.0674

2.2. Specimens preparation, curing end testing

Three different formulations were used for the preparation of the mortar samples. The weight ratio of components of cement: sand: water was kept 1: 3: 0.55 in the mix design. Mixture sample WPA contained wood pulp addition of 0.5 wt. % from weight of cement and sand. Unlike this sample, the WPAP mixture was also prepared with the addition of a plasticizer of 1 wt. % dosage from weight of cement. For comparison to mixtures WPA and WPAP, reference sample (RF) without wood pulp as well as superplasticizer was prepared.

Preparation of mortars started with soaking of wood pulp and manual mixing in approximately 50 wt. % of water. Remaining water, superplasticizer (only for WPAP sample), the required amount of sand and the cement was mixed by mechanical stirring in a mixer in accordance with standard [18] (to ensure the homogenous distribution of wood pulp in cement mixture). Mortars bodies were produced by placing the fresh mix to standard steel prism in the dimension of 40 x 40 x 160 mm. The samples of mixture WPA were consolidated using a jolting apparatus and for the samples WPAP (prepared with

IOP Conf. Series: Materials Science and Engineering **549** (2019) 012004 doi:10.1088/1757-899X/549/1/012004

wood pulp and superplasticizer) were used a vibrating table. Specimens were demoulded one day after mixing and subjected to the following curing regime. Specimens were kept in water under laboratory temperature (+ 20 °C) for 7, 28 and 90 days.

Flow behaviour of the fresh fibre cement mixtures (indicating its workability) was estimated by a flow table test in accordance with standard STN EN 1015-3 [21].

Dry bulk densities of fibre cement mortars were calculated after 28 days of hardening [22].

The thermal properties such as thermal conductivity, thermal diffusivity and volume heat capacity were experimentally determined. Before testing these parameters, all specimens (40x40x160 mm) were dried at 105°C until a constant mass is reached. The thermal parameters were measured by using an ISOMET 2114 with a planar sounder.

The strength tests of 7, 28 and 90 days hardened fibre cement composites were carried out by using a compression test machine (FORM+TEST Seidner & Co. GmbH, Riedlingen, Germany) according to standard [23]. In the case of the compressive strength testing, a loading rate was 2 400±200 N/s, while the three-point bending test was performed at a loading rate 50±10 N/s. To determine the fibre cement mortar physical and mechanical average values, series of three prisms with and without fibres were fabricated.

3. Results and discussion

3.1. Consistency of fresh mixtures

Consistency measurements of three of fresh wood pulp cement mortar samples with constant water/cement ratio of 0.55 are illustrated in Figure 1. Test results of their flow ability are presented in Figure 2. As known, the flow diameter value of the fibre mixture is proportional to the fluidity of the matrix. In the papers [24, 25], there was noticed the influence of cellulosic fibres on flow ability of fresh cement mixture. In other study [26], the superplasticizer was used for better flow ability as well as improvement the workability of fibre reinforced cement mortars and it is generally preferable for moldability and practical use. Polycarboxylate-based superplasticizers significantly affect the properties of fresh and hardened concrete/mortar by their physical and chemical properties. Polycarboxylates are formed by long side chain organic macromolecules that adsorb on the surface of the cement particles and due to the steric effect significantly affect the fresh mortars consistency. This leads to their dispersion, deflocculation and the stabilization of the cement mixture. This creates better conditions for hydration and for increasing the strength of cement composites.

Figure 1. Consistency of reference sample RF (a) and fiber cement fresh mixtures WPA (b) and WPAP (c) with addition of cellulose fibers (0.5 wt.%) and superplasticizer (1 wt.%).

Our experiments demonstrated that the flow values decreased with incorporation of cellulosic fibres and superplasticizer addition into fresh cement mixture in comparison to cement mixture RF

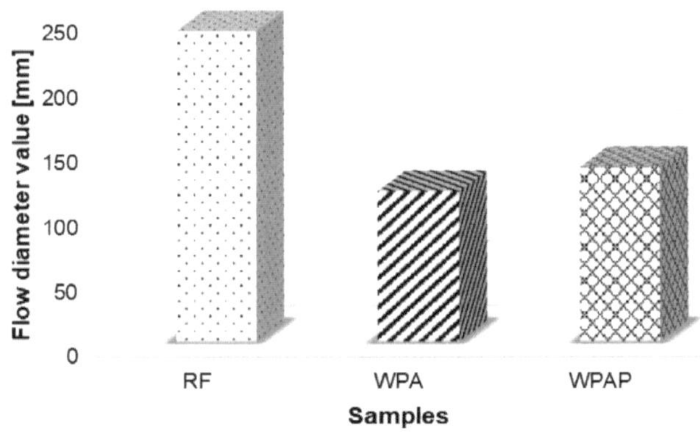

Figure 2: Flow diameter values of fresh fibre cement mixtures.

(without fibres and superplasticizer). This decrease was 51.5 % and 43.9 % for WPA and WPAP, respectively. WPA sample had the lowest value of flow diameter (116 mm). The reason for this behaviour of fresh fibre cement mixtures (without plasticizer) is that the natural fibres have a hydrophilic nature, causing absorption of larger amounts of water and it consequently significantly reduces the resulting consistency of the fresh mixture. WPAP sample with superplasticizer reached a slightly higher value (134 mm) than WPA mortar. The goal of using the superplasticizer into fibre cement mixture was to achieve or approach the consistency value of the reference sample. The experiment showed that the selected superplasticizer used in the given amount did not improved consistency of fibre cement mixture. As known, the main factors affecting the rate of workability loss of fresh concrete are the initial consistency, the type and amount of plasticizer, the type and amount of cement, the time of addition of the admixture, the laboratory conditions (temperature, humidity), the mixing method and the presence of other additives [27]. Authors assume that one of the main factors why the consistency of the reference sample is not achieved is the presence of organic fibers. Based on results of the flow behaviour study of recycled paper cellulosic fibre cement composites [28], the surface nature of fibres is the important factor determining the consistency of fresh fibre cement mixture.

3.2. Density and thermal conductivity of fibre cement based mortars

The density results of 28 days hardened mortars are shown in Figure 3. In accordance with published work [29], the addition of lightweight porous fibres occupies part of the volume of cement specimen resulting in a density reduction.

The decrease in density of the fibre cement mortar is recorded due to lower density of cellulosic fibres than density of cement [30]. Reduction in density is also caused by fibre incorporating air into the matrix during the mortar mixing. However, this occurrence of this phenomenon becomes more significant in the case of using high percentages of fibres [31]. The WPA mixture with a small amount of wood pulp (0.5 wt. %) led to relatively lower density value by 4% as compared to RF sample but on the other hand WPAP fibre cement sample with addition of superplasticizer showed very slightly gain in density (1.2 %) in comparison to RF sample. These differences in density values are at the measurement error level.

The average values of thermal conductivity, thermal diffusivity and volume heat capacity of 28 days hardened fibre cement mortar samples are given in Table 2. The values of thermal conductivity and thermal diffusivity are in line with the trend of density values. Whereas the decrease in the thermal

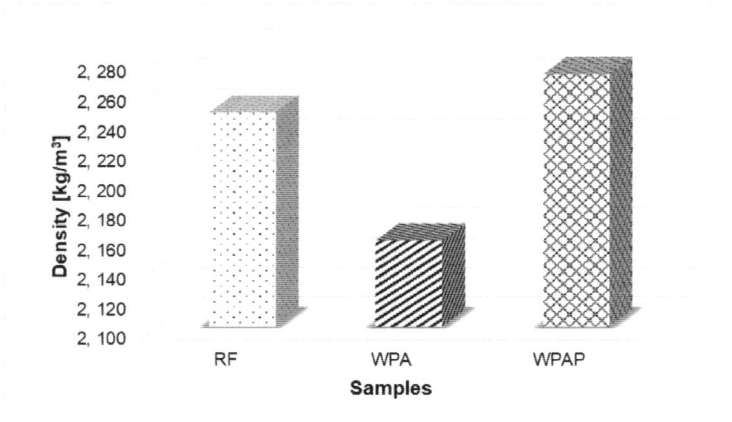

Figure 3. Density values of 28 days hardened fibre cement mortars and reference sample.

conductivity was 20% for sample WPA compared to RF sample, higher value of thermal conductivity coefficient was measured for WPAP sample than for WPA mortar sample. As shown in [32], a higher content of cellulose fibres in the mortars (4-12%) led to a lower thermal conductivity values, i.e. a greater thermal insulation capacity.

Table 2. Thermal conductivity, thermal diffusivity and volume heat capacity for 28 days hardened fibre cement mortars and reference sample.

Sample	Thermal conductivity λ [W/m.K]	Thermal diffusivity $\alpha \times 10^6$ [m^2/s]	Volume heat capacity cp x 10^{-6} [J/m^3.K]
RF	2.699	1.607	1.679
WPA	2.162	1.273	1.697
WPAP	2.45	1.299	1.885

3.3. Mechanical properties of fibre cement based mortars

The development of compressive and flexural strengths on hardening time (7-90 days) is illustrated in Figures 4 and 5. According to the experimental results, the strength values have been reduced by incorporating of the fibre content to the cement matrix. The WPA cement mixtures containing only wood pulp reached lower values of the compressive strength by 16.6 % in comparison to reference sample after hardening time of 90 days. The compressive strength values of fibre cement mortars with a superplasticizer addition (WPAP mortar) were very close to the RF sample in each hardening time. The compressive strength values of 28 day hardened fibre mortars are in the accepted range for plastering mortars [33].

As shown in Figure 4, the addition of cellulose fibers did not enhance the flexural strength of the fibre cement specimens (WPA sample) in the prolonged hardening time, the dosage of superplasticizer to the fibre cement mixture (WPAP sample) caused improvement in the tested strength parameter as compared to WPA mortars. These findings are in accordance with the published research results where a general reduction in mechanical properties of fibre reinforced cement composites was observed [34, 35]. This decrease in strength parameters of fibre cement samples is attributed to the creation of air voids due to the incorporation of porous cellulosic fibres in the cement matrix resulting in a reduction of bond strength of the matrix, causing overall reduction in strength [36]. As generally known, the application of a superplasticizer admixture in the cement mixture can improve workability, increase strength, conserve cement and help reduce shrinkage and thermal cracking due to maintains the fluidity and providing the mix stability.

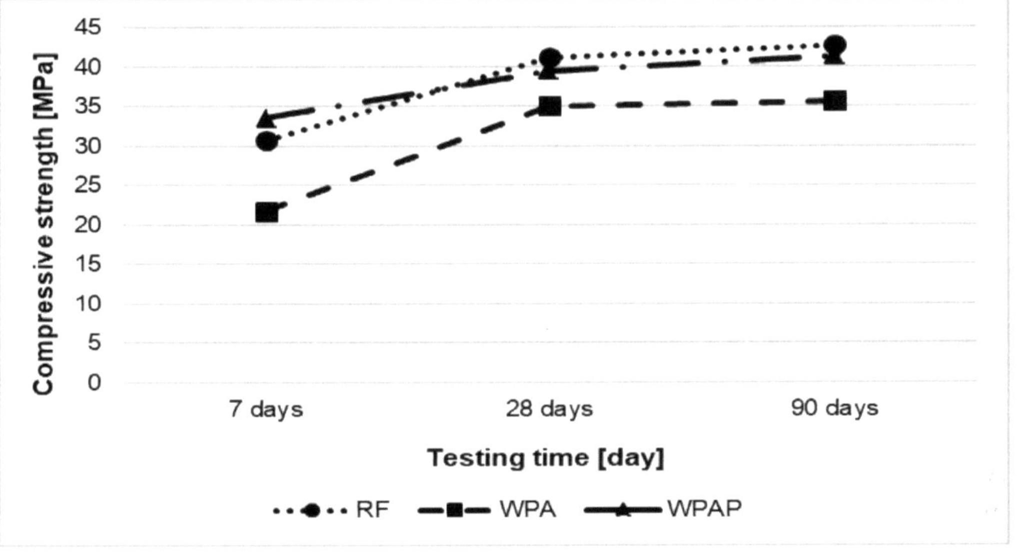

Figure 4. Compressive strength of hardened fibre cement mortars after 7, 28 and 90 days.

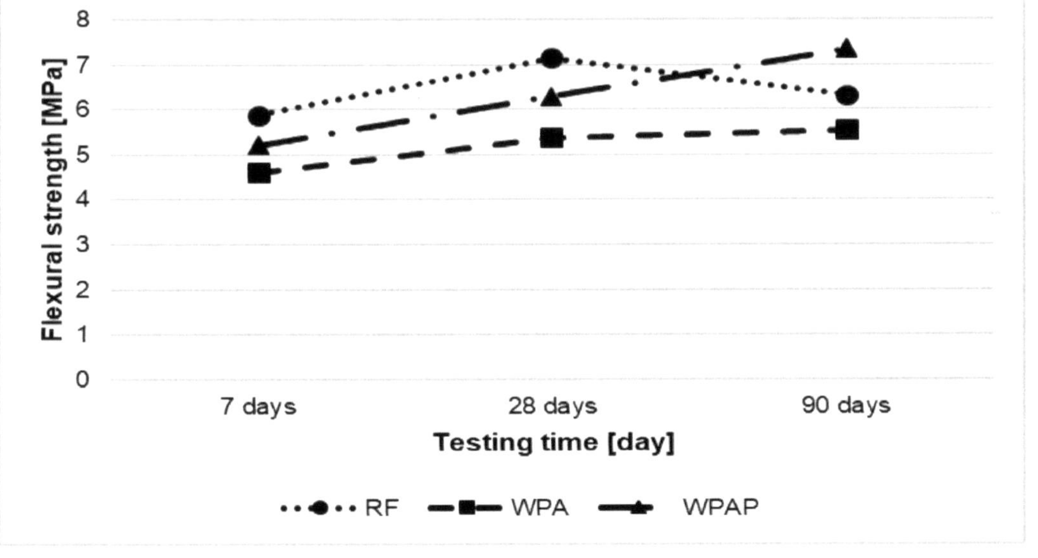

Figure 5. Flexural strength of hardened fibre cement mortars after 7, 28 and 90 days.

As reported above, the performance of the fresh fibre cement mixtures was not improved by a small addition of superplasticizer. The plasticizer efficiency control parameter is strength of hardened mortars. The beneficial effect of polycarboxylate superplasticizer on the compressive and flexural strengths of WPAP fibre cement mortar specimens at all hardening times has been not recorded, although the works [37, 38] demonstrated that the presence of a superplasticizer in the cement matrix has a favourable impact on the compactness of the structure; plasticizer reduces the pore size or numbers the pores and as a result the compressive strengths are increased. However, the introduction of fibre into the cement mixture did also not promote a significant increase in the strength of the mortars both in compression and in bend. It is likely that the cause of this reduction in mechanical properties of fibre cement based mortars is the presence of cellulosic fibers and their surface properties. The properties at the cement particles/cellulosic fibres interface play a key role in the mortars toughness [16]. The hardened cement specimens based on waste paper cellulosic fibres with lower holocellulose content and the addition of same superplasticizer achieved the highest values of compressive (48.4 MPa) and flexural (up to 7 MPa) strength due to improved bond strength between fibres and matrix [28].

4. Conclusions

Two fibre cement mortars containing cellulose fibres from bleached wood pulp in 0.5 wt.% addition (from weight of filler and binder) and superplasticizer based on polycarboxylate ester (dosage of 1% of cement weight) were prepared. The objective of this study was to determine the effect of superplasticizer addition on the workability of fibre fresh mixtures and on the physical and mechanical properties of hardened specimens in comparison to the parameters for reference sample. The results can be concluded as follows:

- the minimum differences in density values of fibre cement mortars with and without superplasticizer in comparison to reference sample were obtained. The variance of calculated density values was at the measurement error level.
- the flow values of fresh cement mortars, the thermal and mechanical properties (compressive and flexural strength) of hardened mortars were dependent on the presence of cellulosic fibres and superplasticizer in mixtures compared to reference cement mixture (without fibres and superplasticizer).
- the performances of the mortars at all hardening times were dependent not only fibres and superplasticizer nature but also on the organic/inorganic interface.
- the values of compressive strength of 28 days of hardened fibre cement mortars are in the accepted range for plastering mortars.

Preliminary investigation of the impact of superplasticizer on the properties of fresh and hardened fibre cement mortars under the studied conditions revealed that the selected superplasticizer used in the given amount is not appropriate for fibre cement mixture. Further research will be therefore focused on the selection plasticizer of another material base and optimization of cellulose fiber cement mortar composition to develop thermal insulating cement mortar.

Acknowledgment

The authors are grateful to the Slovak Grant Agency for financial support of the project VEGA 1/0222/19, Project Institute of Clean Technologies for Mining and Utilization of Raw Materials for Energy Use (No. LO1406) and the student grant competition project SGS SP2017/116.

References

[1] Xie X *et al* 2015 *Compos. Part B-Eng.* **78** 153
[2] Onuaguluchi O and Banthia N 2016 *Cem. Concr. Compos.* **68** 96
[3] Andiç-Çakir Ö *et al* 2014 *Compos. Part B-Eng.* **61** 49
[4] Petrella A *et al* 2018 *J. Sustainable Cem.-Based Mater.* **0** 1
[5] Tonoli G H D *et al* 2011 *Cem. Concr. Compos.* **33** 225
[6] Anju T R *et al* 2016 *Cem. Concr. Compos.* **74** 147
[7] Coutts R S P 2005 *Cem. Concr. Compos.* **27** 518
[8] Kidalova L *et al* 2012 *J. Clean. Prod.* **34** 116
[9] Quiroga A *et al* 2016 *Compos. Part B-Eng.* **84** 25
[10] Melo Filho J A 2013 *Cem. Concr. Compos.* **40** 30

[11] Toledo Filho R D *et al* 2000 *Cem. Concr. Compos.* **22** 127
[12] Cigasova J *et al* 2016 *Chem. Engineer. Trans.* **50** 373
[13] Jarabo R *et al* 2012 *Ind. Crop. Prod.* **39** 89
[14] Stevulova N *et al* 2016 *Chem. Engineer. Trans.* **50** 367
[15] Zhou X M *et al* 2016 *Key Eng. Mater.* **711** 163
[16] Ardanuy M *et al* 2015 *Constr. Build. Mater.* **79** 115
[17] Lima H C *et al* 2008 *Materials and Structures* **41** 981
[18] STN EN 196-1 2016 *Methods of testing cement. Part 1: Determination of strength.* (Bratislava: Office of Standards, Metrology and Testing)
[19] STN EN 1008 2003 *Mixing water concrete. Specification for sampling, testing and assessing the suitability of water, including water recovered from processes in the concrete industry, as mixing water for concrete.* (Bratislava: Office of Standards, Metrology and Testing)
[20] STN EN 934-2+A1 2013 *Admixtures for concrete, mortar and grout. Part 2: Concrete admixtures definitions, requirements, conformity, marking and labelling.* (Bratislava: Office of Standards, Metrology and Testing)
[21] STN EN 1015-3 2000 *Methods of test for mortar for masonry. Part 3: Determination of consistence of fresh mortar (by flow table).* (Bratislava: Office of Standards, Metrology and Testing)
[22] STN EN 1015-10/A1 2007 *Methods of test for mortar for masonry. Part 10: Determination of dry bulk density of hardened mortar.* (Bratislava: Office of Standards, Metrology and Testing)
[23] STN EN 1015-11/A1 2007 *Methods of test for mortar for masonry*. Part 11: Determination of flexural and compressive strength of hardened mortar. (Bratislava: Office of Standards, Metrology and Testing)
[24] Sawsen C *et al* 2015 *Constr. Build. Mater.* **79** 229
[25] Chakraborty S *et al* 2013 *Constr. Build. Mater.* **38** 776
[26] Felekoglu B *et al* 2014 *Compos. Part B-Eng.* **60** 359
[27] Aitcin P C 1998 *High Performance Concrete* (London: CRC Press)
[28] Hospodarova V *et al* 2018 *Buildings* **8** 43
[29] Raut A N and Gomez C P 2016 *Constr. Build. Mater.* **126** 476
[30] Hosseinpourpia R *et al* 2012 *Constr. Build. Mater.* **31** 105
[31] Pereira-de-Oliveira L A *et al* 2012 *Constr. Build. Mater.* **27** 189
[32] Liuzzi S *et al* 2018 *Constr. Build. Mater.* **158** 24
[33] STN EN 998-1 2010 *Specification for mortar for masonry*. Part *1: Rendering and plastering mortar.* (Bratislava: Office of Standards, Metrology and Testing)
[34] Khedari J *et al* 2001 *Cem. Concr. Compos.* **23** 65
[35] Odera R S *et al* 2011 *J. Emerg. Trends Eng. Appl. Sci.* **2** 231
[36] Hosseinpourpia R *et al* 2012 *Constr. Build. Mater.* **31** 105
[37] Kumar M *et al* 2011 *Front. Chem. China* **6** 38
[38] Kumar M *et al* 2010 *Mater. Res.* **13** 177

IOP Conf. Series: Materials Science and Engineering **549** (2019) 012005 doi:10.1088/1757-899X/549/1/012005

Influence of thermal insulation from PUR foam to improve thermo-technical parameters of building envelope for broiler housing

T Szabóová[1] and M Balková[1]

[1] Department of Building Equipment and Technology Safety, Faculty of Engineering, Slovak University of Agriculture in Nitra, Tr. A. Hlinku 2, 949 76 Nitra, Slovakia

E-mail: timea.szaboova@uniag.sk

Abstract. The article deals with the thermo-technical assessment of a hall building for broiler breeding. There was a large heat losses in the examined object before the thermal insulation. For this reason, the cladding was from internal side insulated with sprayed PUR foam. In the building was assessed the hygiene criterion, thermo-technical properties of perimeter walls and the life of the structure before and after the insulation. It has been shown that the thermal insulation of the walls had a significant impact on improving the conditions in the hall. Due to the fact that from hygienic assessment and thermovision measurements have been proven heat losses and unsatisfactory condition of uninsulated parts of the hall envelope, it will be necessary to additionally insulate also the soffit part of the ceiling-roof construction and the plinth part.

1. Introduction

In the past, emphasis wasn´t placed to energy performance of buildings, which we feel currently mainly in old large area buildings with high energy costs [1]. Modern poultry housing is designed and constructed to reduce a heat loss and to improve energy efficiency [2]. Older hall buildings have large heat leaks, which we are currently trying to reduce through various measures. One of them is insulation of the envelope. It is necessary to solve the envelope of the hall building in relation to the fulfillment of the "hygienic criterion" in all details of building structures [3]. In the case of occurence of thermal bridges caused by failures during the operation of the building, it has proven successful the additional thermal insulation of the external cladding with the sprayed PUR foam, resistant to the aggressive environment [3]. The aim of this work is therefore to assess the hygiene criterion, the thermal properties of the walls and the service life of the hall building used for the fattening of broilers in the original state (before thermal insulation) and subsequently after thermal insulation with sprayed polyurethane foam from the inside.

2. Material and methods

The subject of the research was a hall building used for broiler breeding situated in the first temperature area with an altitude of 137 meters above sea level. The ground plan dimensions of the hall were 94 m x 15.2 m. The clear height of the hall was 3.1 m. The total area for chickens was 1428.8 m^2. The supporting structure of the hall object was made of a frame steel supporting structure. Longitudinal peripheral walls of the hall 100 mm thick were made with galvanized sheet on both sides with thermal insulation of mineral wool. The hall was measured by thermovision in the winter season (before its insulation). To reduce heat loss were subsequently walls of the building partially insulated from the inside by spraying thermal insulation foam and in the following year were made thermovision measurements after their thermal insulation. The roof structure wasn't additionally insulated. The surface temperatures of individual building structures (walls, ceiling and floor) were also measured by a thermal camera before and after thermal insulation of the hall. The heat loss from the hall was also significantly influenced by the rear steel gates, which were also insulated with sprayed thermal insulation foam. After their insulation, the leakage of heat from the hall through the openings was significantly reduced. The thermal bridges in the plinth part were caused by higher temperatures in the bedding and inappropriate wall solution. Since only the sheet-metal part of the wall construction was insulated in the lower part, there were still significant heat loss after the thermal insulation, which are evident from the thermovision pictures.

The hall building was considered in three ways. First of all, compliance with the hygiene criterion was assessed. Subsequently, the walls were assessed of thermo-technical point of view before and after insulation. In the end, the life of the hall building was determined before and after the insulation. The thermo-technical characteristics of the building materials of the wall structure before and after insulation are given in Tables 1 and 2.

Table 1. Thermo-technical characteristics of building materials for wall construction before insulation.

Layer	Material	t (m)	λ ($W \cdot m^{-1} \cdot K^{-1}$)	ρ_0 ($kg \cdot m^{-3}$)	c ($J \cdot kg^{-1} \cdot K^{-1}$)
1	Corrugated sheet	0.002	204	2700	880
2	Mineral wool	0.100	0.04	15	940
3	Corrugated sheet	0.002	204	2700	880

Table 2. Thermo-technical characteristics of building materials for wall construction after insulation.

Layer	Material	t (m)	λ ($W \cdot m^{-1} \cdot K^{-1}$)	ρ_0 ($kg \cdot m^{-3}$)	c ($J \cdot kg^{-1} \cdot K^{-1}$)
1	Polyurethane foam*	0.015	0.025	35	1500
2	Corrugated sheet	0.002	204	2700	880
3	Mineral wool	0.100	0.04	15	940
4	Corrugated sheet	0.002	204	2700	880

*Note: inner part of the walls

3. Results and discussion

3.1. Assessment of compliance with the hygiene criterion

The internal wall structure of the monitored hall was assessed from the hygienic point of view before and after her insulation. For the determination of the hygienic criterion of the wall structure of the hall building, it was necessary to proceed in accordance with [6].

3.1.1. Assessment of the wall structure in the place of the thermal bridge before insulation

Input parameters:

- internal air temperature at time of measurement $\theta_i = 17\ °C$
- external air temperature at time of measurement $\theta_e = -3.5\ °C$
- difference of internal and external temperatures $\Delta\theta_{ie} = 20.5\ °C$
- internal air humidity $\varphi_i = 50\%$
- external air humidity $\varphi_e = 38.5\%$

- critical surface temperature for mold formation [4] $\theta_{si,80} = 9.79$ ° C
- dew point temperature [4] $\theta_{dp} = 6.51$ °C
- safety surcharge for intermittent heating of stables with a drop in the internal air temperature θ_i up to 5 K [5]: $\Delta\theta_{dp} = 0.2$ K
- safety surcharge for intermittent heating with a drop in the internal air temperature θ_i up to 5 K [6]: $\Delta\theta_{si} = 0.5$ K
- surface temperature of uninsulated wall (Figure 1) $\theta_{si} = 6.5$ ° C

Requirement for critical surface temperature for mold formation $\theta_{si,80}$ [6]

$$\theta_{si} \geq \theta_{si,N} = \theta_{si,80} + \Delta\theta_{si} \tag{1}$$

$6.5°C < 10.29°C$ → The wall doesn't meet the standard requirement

Requirement for dew point temperature θ_{dp} [5]

$$\theta_{si} \geq \theta_{dp} + \Delta\theta_{dp} \tag{2}$$

$6.5°C < 6.71°C$ → The wall doesn't meet the standard requirement

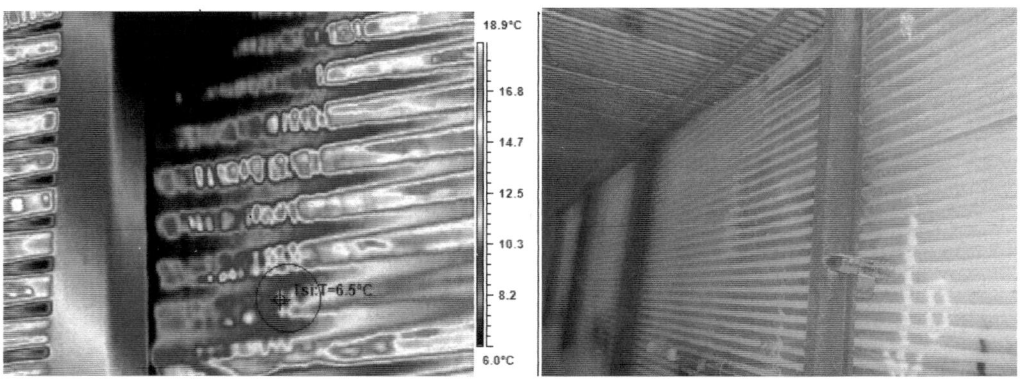

Figure 1. Thermogram of non-insulated internal wall.

3.1.2. Wall construction in the assessed place after thermal insulation

Input parameters were:

$\theta_i = 18$ °C, $\theta_e = -3.5$ °C, $\Delta\theta_{ie} = 21.5$ °C, $\varphi_i = 54$ %, $\varphi_e = 39$ %, $\theta_{si,80} = 12$ °C, $\theta_{dp} = 9$ °C, $\Delta\theta_{dp} = 0.2$ K, $\Delta\theta_{si} = 0.5$ K, $\theta_{si} = 17.4$ °C (the surface temperature of the insulated wall of Figure 2).

Requirement for critical surface temperature for mold formation $\theta_{si,80}$ [6]

$17.4°C \geq 12.5°C$ → The wall meets the standard requirement

Requirement for dew point temperature θ_{dp} [5]

$17.4°C \geq 9.2°C$ → The wall meets the standard requirement

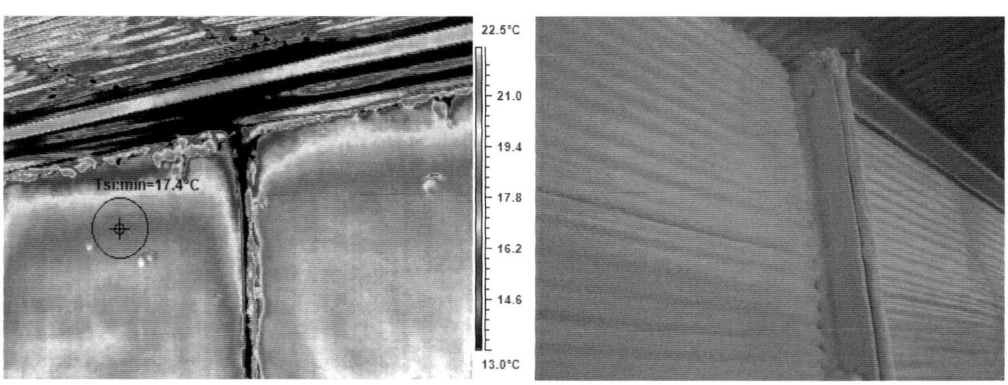

Figure 2. Thermogram of insulated wall from inside.

The surface temperature of the non-insulated peripheral wall of the hall building θ_{si} was only 6.5 ° C (Figure 1). Assessing the wall from the hygienic point of view, it was found that the monitored wall doesn't meet the standard requirements.

After insulation of the examined wall with thermal insulation foam the inside her surface temperature increased to 17.4 ° C (Figure 2), at nearly the same indoor air temperature. Subsequently, the wall was assessed from a hygienic point of view, and it was found that after wall insulation it meets the hygienic criterion as its surface temperature was higher than the critical surface temperature for mold formation $\theta_{si,80} = 12$ ° C and dew point temperature $\theta_{dp} = 9$ ° C.

3.1.3. Assessment of the plinth part of the wall

Input parameters:

$\theta_i = 17$ °C, $\theta_e = -3.5$ °C, $\Delta\theta_{ie} = 20.5$ °C, $\varphi_i = 50$ %, $\varphi_e = 38.5$ %, $\theta_{si,80} = 9.79$ °C, $\theta_{dp} = 6.51$ °C, $\theta_{dp} =$ 0.2 K (for thermal bridges $\Delta\theta_{dp} = 0$ K), $\Delta\theta_{si} = 0.5$ K, the surface temperature of the wall at the thermal bridge TM (Figure 3) $\theta_{si,TM} = 3$ °C, the surface temperature of the wall outside of the thermal bridge MTM (Figure 3) $\theta_{si,MTM} = 10.5$ °C.

Requirement for critical surface temperature for mold formation $\theta_{si,80}$ [6]

TM: $3°C < 9.79°C$ → The wall doesn't meet the standard requirement

MTM: $10.5°C \geq 10.29°C$ → The wall meets the standard requirement

Requirement for dew point temperature θ_{dp} [5]

TM: $3°C < 6.51°C$ → The wall doesn't meet the standard requirement

MTM: $10.5°C \geq 6.71°C$ → The wall meets the standard requirement

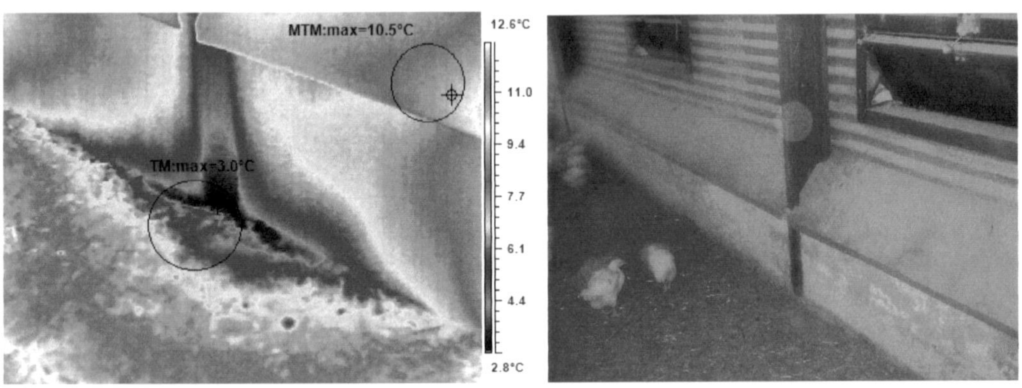

Figure 3. Detail of a thermal bridge of non-insulated plinth.
TM – measurement made at the point of thermal bridge, MTM – measurement taken outside the thermal bridge

A thermal bridge was detected by thermovision in the plinth part of the wall construction in contact wall with bedding, which was subsequently assessed from a hygienic point of view. The results show that the wall in question at the location of the thermal bridge doesn't meet the hygiene criterion. At this place there was a significant leakage of heat resulting in a low surface temperature $\theta_{si,TM} = 3$ °C, which was lower than the critical surface temperature for mold formation $\theta_{si,80} = 9.79$ °C and also as dew point temperature $\theta_{dp} = 6.51$ °C. Condensation of water vapor and the formation of molds is a prerequisite in the given location. Outside the thermal bridge, the wall met the hygiene criterion as its surface temperature was higher than the norm value of $\theta_{si,MTM} = 10.5$ °C.

3.1.4. Contact between insulated wall and uninsulated ceiling from corrugated sheet

Input parameters:

$\theta_i = 16.7$ °C, $\theta_e = -3.5$ °C, $\Delta\theta_{ie} = 20.2$ °C, $\varphi_i = 54$ %, $\varphi_e = 39$ %, $\theta_{si,80} = 10.7$ °C, $\theta_{dp} = 7.4$ °C, $\theta_{si} = 0.5$ K, $\Delta\theta_{dp} = 0.2$ K, surface temperature of the ceiling at the place of the thermal bridge TM (Figure 4)

$\theta_{si,TM}$ = 5.8 °C, surface temperature of insulated wall outside of the thermal bridge MTM (Figure 4), $\theta_{si,MTM}$ = 11.7 °C.

Requirement for critical surface temperature for mold formation $\theta_{si,\,80}$ [6]

TM: $5.8°C < 11.2°C$ → The wall doesn't meet the standard requirement

MTM: $11.7°C \geq 11.2°C$ → The wall meets the standard requirement

Requirement for dew point temperature θ_{dp} [5]

TM: $5.8°C < 7.4°C$ → The wall doesn't meet the standard requirement

MTM: $11.7°C \geq 7.6°C$ → The wall meets the standard requirement

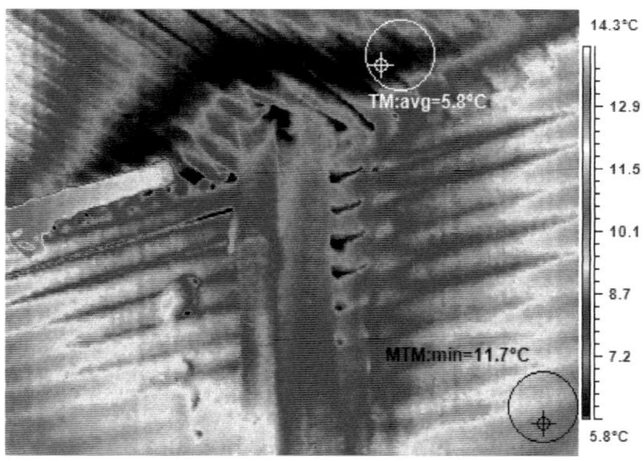

Figure 4. The contact of the uninsulated ceiling and insulated internal wall.
TM – measurement made at the point of thermal bridge, MTM – measurement taken outside the thermal bridge

Wall construction made from corrugated sheet was insulated with sprayed PUR foam, while the ceiling wasn't insulated. There was heat leakage at the point of contact between the ceiling and the wall (Figure 4). Place of the thermal bridge was considered hygienic, while from the point of view of mold formation it didn't meet the standard requirements, because the surface temperature at this critical point was only $\theta_{si,\,TM}$ = 5.8 °C. In terms of dew point temperature θ_{dp} = 7.4 °C and critical surface temperature for mold formation $\theta_{si,\,80}$ = 10.7 °C, the wall construction didn't fit. Insulated wall construction outside the thermal bridge meet hygienic criteria. Almost double value of the surface temperature of the thermally insulated part of wall $\theta_{si,\,MTM}$ = 11.7 °C above the surface temperature at the site of thermal bridge indicates a need for additional thermal insulation of the ceiling construction. For agricultural buildings with relative humidity over 75 % is recommended to double-shell roof construction system with ventilated air gap [7], to reduce the heat loss through the roof construction during the winter season, thereby reducing the cost of heating the building with gas [8].

3.2. Assessment of thermal performance of the hall building walls

The software „Teplo 2017 EDU" was used to assess the thermal performance of the hall building walls [9]. Due to the fact that the requirement to the temperature environment in the hall is different with respect to the age of the broilers, we used the calculation of the required insulation from temperature and humidity conditions for broilers in the 5th week of fattening. The desired ambient temperature in the hall for broiler at week 5th is θ_i = 20 ° C and the optimum advised value of relative air humidity for the broilers is 70 % [10]. In the outdoor environment we considered exterior temperatures θ_e = -3.5° C and air humidity 38.5 %. The marginal conditions for the Nitra region for winter are θ_e = -11 ° C. First of all, we assessed the wall of hall, which was not yet insulated from the inside. Due to the fact that there were significant heat leaks through the structure, the walls of the building were partially insulated from the inside with sprayed PUR foam of 15 mm thickness. Subsequently we assessed the wall after thermal insulation according to [11]. Although this standard is mainly used for residential buildings, we also used it for a hall building because we currently do not have an alternative for hall poultry buildings. Building insulation is important in terms of energy savings for heating in the winter.

At present, according to valid criteria, this thickness of insulation is not sufficient. After designing a larger insulation thickness of PUR foam to 50 mm, this thickness already suited. The recommended heat transfer coefficient of the external wall structure (U_{r1}) according to [11] is 0.22 W·m^{-2}·K^{-1}. This value is valid from 2016 to 2020. In order for the condition to be met, $U \leq U_{r1}$ must apply. In the event that this condition cannot be met, at least the condition that the maximum value of the heat transfer coefficient U_{max} = 0.46 W·m^{-2}·K^{-1} must be met. The recommended value of the thermal resistance of the external wall structure (R_{r1}) according to [11] is 4.4 m^2·K·W^{-1} (valid from 2016 to 2020) where the condition $R \geq R_{r1}$ must be met. The minimum thermal resistance value of R_{min} shall be at least 2.0 m^2·K·W^{-1}. The calculated and recommended heat transfer coefficient and thermal resistance values of the structure are shown in Table 3.

Table 3. Calculated and recommended heat transfer coefficient and thermal resistance values of the structure.

	U^a (W·m^{-2}·K^{-1})	U_{r1} (W·m^{-2}·K^{-1})	R^b (m^2·K·W^{-1})	R_{r1} (m^2·K·W^{-1})
Wall before insulation	0.375	0.22	2.50	4.4
Wall after insulation	0.306	0.22	3.10	4.4
Wall after insulation, design	0.214	0.22	4.5	4.4

[a] U – Heat transfer coefficient of the structure, W·m^{-2}·K^{-1}
[b] R – Thermal resistance of the structure, m^2·K·W^{-1}

Figures 5 – 7 show the temperature course in a typical structure site under steady-state design conditions before thermal insulation, after thermal insulation and after thermal insulation design with greater thickness.

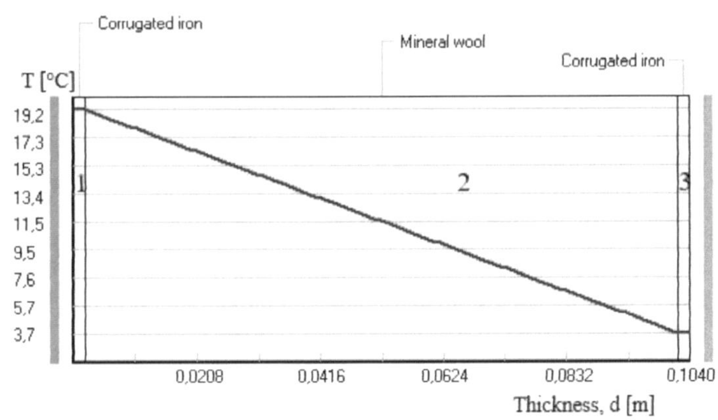

Figure 5. Temperatures in a typical structure site under steady-state design conditions before thermal insulation.

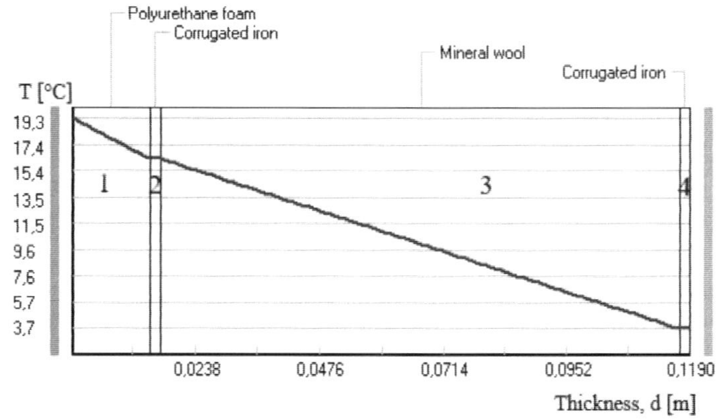

Figure 6. Temperatures in a typical structure site under steady-state design conditions after thermal insulation.

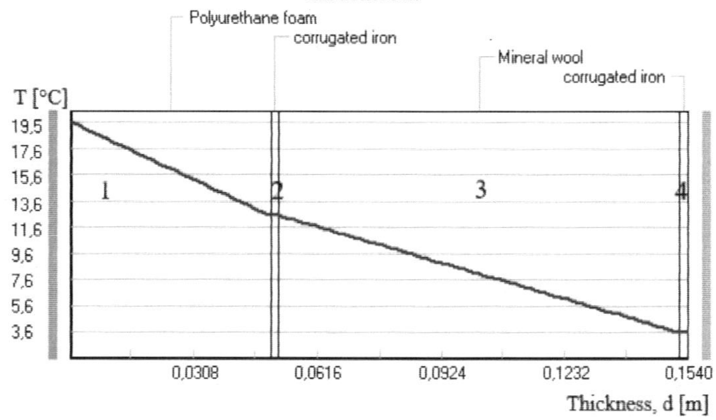

Figure 7. Temperatures in a typical structure site under steady-state design conditions after thermal insulation with greater thickness.

3.3. Determining the lifetime of the hall building before and after thermal insulation

We also assessed the investigated hall building in terms of the service lifetime of the building. The original object had several shortcomings. Some shortcomings have been eliminated by the thermal insulation of the perimeter walls. Due to the fact that the building was partially insulated, the lifetime of the building also increased. The wear of the building is determined by the Bradáč cubic method. The input data for determining the service life of the building are as follows for the building before thermal insulation:

Year of valuation of building : 2013, Year of final inspection: 1990.
The age of the building (V) is 23 years.
Basic life of the building (ZZ) [12]: 100 years.
Basic time of another duration of the building (TT) [12]: 78 years.
State of the components of the long-term lifetime (Q) is 73 %.
Time of another duration of the building: T = TT x Q / 100.
T = 78 x 73/100 = 57 years
Total estimated lifetime Z = V + T
Z = 23 + 57 = 80 years
The estimated lifetime of the original building is 80 years.

The input data for determining the service life of the building are as follows for the building after thermal insulation:

Year of valuation of building : 2014, Year of final inspection: 1990.
The age of the building (V) is 24 years.
Basic life of the building (ZZ) [12]: 100 years.
Basic time of another duration of the building (TT) [12]: 77 years.
State of the components of the long-term lifetime (Q) is 90 %.
Time of another duration of the building: T = 77 x 90/100 = 69 years.
Total estimated lifetime Z = 24 + 69,3 = 93 years.
The estimated lifetime of the building was increased to 93 years after thermal insulation.

4. Conclusion

From thermovision measurements and subsequent assessment of the uninsulated wall construction of a hall building used for poultry breeding from the hygienic point of view, we found out that in the wall construction there was heat leakage through thermal bridges, while the hygienic criterion wasn't met in these places. After spraying the wall construction with foam thermal insulation from the inside of the cladding, thermal bridges were eliminated. This improved the thermo-technical properties of the wall construction and the wall meets all requirements. Since from the point of view of the financial intensity of insulation was insulated only the wall construction of the hall, it is necessary to additionally insulate also the soffit part of the ceiling-roof construction and plinth. From the hygienic assessment and from the thermovision measurements, were proved heat losses and unsatisfactory condition of the uninsulated parts of the envelope. The hall building was also assessed in terms of the life of the building. The estimated lifetime of the original building was 80 years. After thermal insulation of the perimeter walls, the lifetime of the building was increased to 93 years. In case that the other construction deficiencies would be removed, it means that would be also insulated the ceiling-roof construction and the plinth part of the construction, the life of the building would be even higher.

References

[1] Fuliotto R, Cambuli F, Mandas N, Bacchin N, Manara G, Chen Q 2010 *Experimental and numerical analysis of heat transfer and airflow on an interactive building facade* (Energy and Buildings) **42** pp 23-28
[2] Knížatová M, Mihina Š, Brouček J, Karandušovská I, Mačuhová J 2010 *The influence of litter age, litter tempremature and ventilation rate on ammonia emissions from broiler rearing facility.* (Czech: Czech Journal of Animal Science) **55** pp 337-345
[3] Reichstädterová T 2013 *Energy audit hall facilities* (Nitra: Slovak University of Agricultural in Nitra Press) p 127
[4] STN 73 0540-3: 2012 *Thermal protection of buildings. Thermal performance of buildings and components. Part 3: Properties of environments and building products*
[5] STN 73 0565/Z2: 1998 *Thermo-technical properties of engineering structures and buildings. Stables (canceled without refund)*
[6] STN 73 0540-2: 2012 *Thermal protection of buildings. Thermal performance of buildings and Components. Part 2: Functional requirements* (in Slovak)
[7] Oláh J and Černík P 1986 *Designing and repairing flat roofs* (Bratislava: ALFA Press) p 232
[8] Karandušovská I, Pogran Š, Knížatová M 2009 *Assessment of pollutant production in dairy and poultry farms: scientific monograph* (Nitra: Slovak University of Agricultural in Nitra Press) p 170
[9] Teplo 2017 EDU, simulation software (in Czech)
[10] Pogran Š, Bieda W, Gálik R, Lendelová J, Švenková J 2011 *Quality of the indoor environment of stable buildings* (Nitra: Slovak University of Agricultural in Nitra Press) p 242
[11] STN 73 0540-2/Z1: 2016 *Thermal protection of buildings. Thermal performance of buildings and Components. Part 2: Functional requirements*
[12] Bradáč A 2001 *Valuation of buildings in a cost manner. Determining the starting price, lifetime and wear of buildings* (Bratislava: STU in Bratislava Press) pp 22-29

IOP Conf. Series: Materials Science and Engineering **549** (2019) 012006 doi:10.1088/1757-899X/549/1/012006

Possibilities of use of glass recyclate from photovoltaic panels for concrete masonry units

M Stehlík[1], J Knapová[1] and V Kostka[1]

[1] Brno University of Technology, Faculty of Civil Engineering, Veveří 331/95, 602 00, Brno, Czech Republic

E-mail: stehlik.m@fce.vutbr.cz

Abstract. The paper deals with the possibilities of use of glass recyclate from photovoltaic panels for concrete masonry units. It compares different recipes and its physical and mechanical properties with the focus on the compressive strength, density. It then compares the values of these recipes with the values of commonly used composite materials for masonry units without recyclates. Recycling of materials from photovoltaic panels is a highly discussed topic nowadays. The paper presents possibility for secondary use of glass from these panels in building industry, namely the substitution of aggregate in concrete with glass recyclate.

1. Introduction

The Czech Republic is one of the best in Europe to sort waste and recycle. And it is great that this trend does not tend to cease [1]. However, regarding the recycling of industrial waste and its possible subsequent use, here we do have considerable space for improvement. This paper deals with the recycling of glass from photovoltaic panels and studies the possibilities of use of this glass for concrete masonry units.

Photovoltaic panels installed in Europe, including the Czech Republic, will begin to reach the limits of their lifetime in few years. Therefore, the topic of recycling of photovoltaic panels is more than actual and the possibilities as well as the limits of recycling are increasingly being discussed. Waste material from photovoltaic panels consists mostly of glass, for which we can find a wide range of secondary use. Here we focus of the usage in the building industry, specifically as a substitute for aggregates in concrete. The paper examines the mechanical properties of concrete samples made from glass recyclate, that has supplemented or completely replaced the aggregate in the concrete recipe.

2. Recycling of photovoltaic panels

Recently, the recycling of solar panels has been a major topic and it brings several serious questions and unsolved issues.

EU Directive on waste electrical and electronic equipment aims to contribute to sustainable production and consumption by the prevention of electrical waste and by the re-use, recycling and other forms of recovery of such wastes to reduce the disposal of waste and to contribute to the efficient use of resources. Since 2018, this EU Directive dealing with the disposal of solar panels requires recycling of at least 80 % of the panel and minimum of 85 % of waste recovery. Recycling means secondary material use; the waste recovery (use of waste) means both material and energy use. Table 1 shows the evolution of the quota over the time.

Table 1. EU quota for recycling of photovoltaic panels.

	2012–2015	2015–2018	Since 2018
Waste recovery of PV panels	75 %	80 %	85 %
Recycling of PV panels	65 %	70 %	80 %

source: https://www.resolar.cz/cs/co-nabizime

There are many types of photovoltaic panels – Monocrystalline (Mono-SI), Polycrystalline (Poly-SI), Thin-film (TFSC, CdTe, CIS, CIGS, a-Si), or Concentrated (CVP, HCVP). Various types from different manufacturers are installed in the Czech Republic. Classic solar panels of first and second generation are largely made up of glass (typically 70 %, up to 95 % in Thin-film PV panels) [3]. In order to meet the quotas, it is therefore necessary to ensure primarily the recycling of glass.

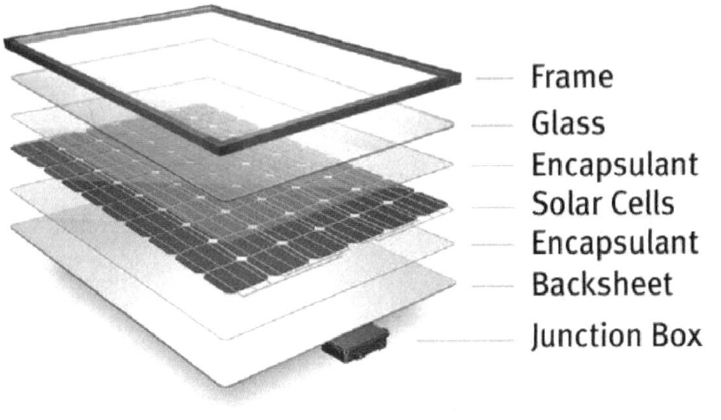

Figure 1. Structure of a photovoltaic panel.
source: http://www.dupont.com/

For the presented research, crushed glass from photovoltaic panels was used for production of glass recycled concrete samples. The glass we get from the PV panels is not a 100 % SiO_2. Table 2 shows the chemical composition of glass from the two types of panels commonly used in the Czech Republic.

Table 2. Chemical composition of PV panels glass.

	Loss on drying (105°C)	Loss by ignition 1100°C	SiO_2	Al_2O_3	Fe_2O_3	TiO_2	MnO	CaO
LDK SOLAR	0.23	1.08	69.5	1.230	0.172	0.023	0.005	10.20
QS SOLAR	0.22	0.50	71.0	0.499	0.110	0.023	0.006	8.45
	MgO	K_2O	Na_2O	Li_2O	Cr_2O_3	BaO	ZrO_2	SrO
LDK SOLAR	1.44	0.032	12.9	<0.002	0.005	0.005	0.008	0.008
QS SOLAR	4.04	0.171	12.4	<0.002	0.005	0.009	0.009	0.005

Figure 2. PV panels – producer QS Solar and LDK Solar Co.

3. Concrete recipes

Glass recyclate from photovoltaic panels was obtained from BAMBAS Elektroodpady s.r.o., in two fractions, 1-2 mm and 4-8 mm.

By the sieving method (ČSN EN 933-1) we confirmed that in the case of fraction 1-2 mm 86 % of the recyclate falls into the indicated fraction, in the case of 4-8 mm 84 % falls into the indicated fraction [4].

Table 3. Results of sieving method for glass recyclate of fraction 1-2 mm.

Sieve Designation [mm]	Weight of the Partial Residue R_i [g]	Partial Residue on the Sieve [%]	Total Residue on the Sieve [%]	Total Sieve Drop [%]
16	0.00	0.00	0.00	100.00
8	0.00	0.00	0.00	100.00
4	0.00	0.00	0.00	100.00
2	1.09	0.05	0.05	99.95
1	1717.43	85.90	85.95	14.05
0.5	281.14	14.10	100.05	0.00
0.25	0.97	0.05	100.10	0.00
0.125	0.17	0.00	100.10	0.00
0.063	0.02	0.00	100.10	0.00
0	0.17	0.00	100.10	0.00

Table 4. Results of sieving method for glass recyclate of fraction 4-8 mm.

Sieve Designation [mm]	Weight of the Partial Residue R_i [g]	Partial Residue on the Sieve [%]	Total Residue on the Sieve [%]	Total Sieve Drop [%]
16	0.00	0.00	0.00	100.00
8	9.59	0.50	0.50	99.50
4	1685.24	84.30	84.80	15.20
2	269.32	13.50	98.30	1.70
1	31.26	1.60	99.90	0.10
0.5	3.83	0.20	100.10	0.00
0.25	0.23	0.00	100.10	0.00
0.125	0.38	0.00	100.10	0.00
0.063	0.00	0.00	100.10	0.00
0	0.57	0.00	100.10	0.00

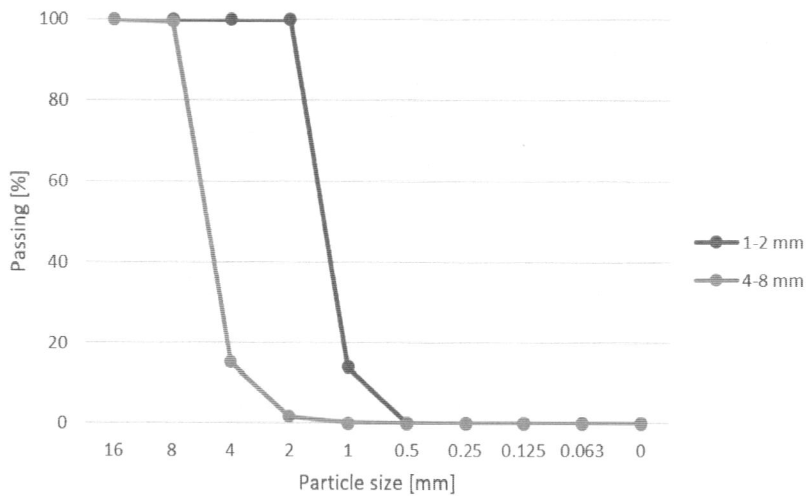

Figure 3. Grain size distribution curves of the glass recyclate fractions.

Two recipes were used to produce concrete samples with glass recyclate, that we used for further research. The first recipe (A) contained glass recyclate of both of mentioned fractions and the aggregate of fraction of 4-8 mm. The second one (B) contained only the glass recyclate of both fractions, without the addition of aggregate. A detailed description of both recipes is given in the Table 5.

Table 5. Recipes of concrete specimens with glass recyclate.

Recipe designation	Cement CEM II 32,5 R [kg.m^{-3}]	Glass		Aggregate	Water [kg.m^{-3}]
		1-2 mm [kg.m^{-3}]	4-8 mm [kg.m^{-3}]	4-8 mm [kg.m^{-3}]	
A	350	700	300	1000	150
B	350	950	1150	-	150

4. Performed tests

Specimens made from concrete with glass recyclate (both recipes A and B) and the recyclate itself were subjected to the following test:

- Tests for mechanical and physical properties of aggregates – Determination of loose bulk density and voids (ČSN EN 1097) [5]
- Determination of bulk density of fresh concrete – 48 hours old (ČSN EN 206+A1, ČSN EN 12350) [6],[7]
- Determination of bulk density of concrete – after 28 days (ČSN EN 206+A1, ČSN EN 12390) [6],[8]
- Determination of compressive strength of concrete (ČSN EN 206+A1, ČSN EN 12390-3) [6],[9]

5. Results and discussion

According to the standard ČSN EN 1097, we determined the loose bulk density and voids of glass recyclate from PV panels [5]. Where the void content indicates the ration of grain gap volume to total volume of mass. The measured results of bulk density and void content is shown in the Table 6.

Table 6. Measured results of loose bulk density and voids of glass recyclate.

Bulk density	fraction 1-2 mm [kg.m^{-3}]	fraction 4-8 mm [kg.m^{-3}]
Loose recyclate	1086	1265
Compacted recyclate	1261	1413
Void content	fraction 1-2 mm [%]	fraction 4-8 mm [%]
Loose recyclate	60	53
Compacted recyclate	53	48

Figure 4. Determination of bulk density of glass recyclate of fractions of 4-8 mm and 1-2 mm.

The values of *fresh concrete density*, i.e. samples at the age of 48 hours, reached an average of 2160 kg.m^{-3} for the recipe A. For the recipe B, that contained only the glass recyclate, the average value was 1880 kg.m^{-3}.

Further test results are presented in the following Table 7. There are presented values for each specimen, as well as average values for *recipe A* and *B*. The compressive strength was determined after 28 days.

Table 7. Tests results and measured values.

Specimen	Weight [g]	Density [kg.m^{-3}]	Specimen Dimensions [mm]	Compressive Strength [MPa]
A1	7213.2	2137.2	150x150x150	20.9
A2	7227.0	-	150x150x150	22.6
A3	7243.6	-	150x150x150	22.0
A – average	7227.9	**2140**	150x150x150	**21.8**
B1	6220.4	1843.1	150x150x150	11.5
B2	6286.4	1862.6	150x150x150	11.9
B3	6198.6	1836.6	150x150x150	11.9
B4	6229.8	1845.9	150x150x150	11.3
B – average	6233.8	**1850**	150x150x150	**11.7**

The results of compressive strength, that we obtained from the measurements, show satisfactory values. For recipe A, with both glass recyclate and one fraction of aggregate, the average measured strength is 21.8 MPa. This value corresponds to the mean values of ordinary normal weight concrete. Recipe B, where all the aggregate is replaced by the recyclate from PV panels, reached the average value of 11.7 MPa. However, it also has a lower density and is therefore lighter. Both recipes, even the mix B with lower value of compressive strength, are suitable for load-bearing masonry units.

For illustrative comparison, we can take a short look at compressive strength and density of the widely used lightweight aerated concrete YTONG. See the data in the Table 8. We can see that the compressive strengths of different types of YTONG reach just half (or in some types quarter) of the value measured for the concrete with glass recyclate from PV panels. (This comparison only aims to illustrate differences in compressive strength and density, it does not seek to compare mentioned materials overall. More properties such as thermal conductivity and frost resistance need to be studied and compared.)

Table 8. Data and values for lightweight aerated concrete YTONG for peripheral and load-bearing walls.

Type	Density [kg.m^{-3}]	Compressive Strength [MPa]
Standard P2-400	400	2.7
Univerzal P3-450	450	3.5
Statik P4-550	550	5.0
Statik Plus P6-650	650	6.5

data source: data sheet of the producer, www.ytong.cz

6. Conclusion

The results obtained from measurement proved that strengths of concrete units with glass recyclate from photovoltaic panels are satisfactory for use in building foundations as well as for load-bearing structures of low-rise buildings.

The admixture of polymer fibres could bring further improvements in mechanical properties of the material. Combination of coarse concrete recyclate with fine glass recyclate comes into consideration when preparing new possible recipe.

One limitation will always be the smoothness of glass surface in connection with cement and possible reaction glass – calcium hydroxide. Several scientific studies have already addressed this issue [10], [11], [12], [13].

Acknowledgements

The paper was created within the project No. LO1408 "AdMaS UP - Advanced Building Materials, Structures and Technologies", supported by Ministry of Education, Youth and Sports under the „National Sustainability Program I".

References

[1] Deník.cz. 2017. *Češi jsou hned po Belgičanech nejlepší ve třídění odpadu.* May 11 2017. [Online]. Available from: https://www.denik.cz/ekonomika/cesi-jsou-hned-po-belgicanech-nejlepsi-ve-trideni-odpadu-20170511.html.

[2] The European Parliament and the Council. 2012. *Directive 2012/19/EU of the European Parliament and of the Council of 4 July 2012 on waste electrical and electronic equipment (WEEE).* Text with EEA relevance. [Online]. Available from: http://data.europa.eu/eli/dir/2012/19/oj.

[3] IQ Energy. 2019. *Technologický vývoj fotovoltaických panelů.* [Online]. Available from: http://www.iqenergy.cz/produkty-a-reseni/fotovoltaika/technologicky-vyvoj/#.XHLXV-hKhPY.

[4] ČSN EN 933-1. Tests for geometrical properties of aggregates – Part 1: Determination of particle size distribution – Sieving method.

[5] ČSN EN 1097. Tests for mechanical and physical properties of aggregates.

[6] ČSN EN 206+A1. Concrete – Specification, performance, production and conformity.

[7] ČSN EN 12350. Testing fresh concrete.

[8] ČSN EN 12390. Testing hardened concrete.

[9] ČSN EN 12390-3. Testing hardened concrete – Part 3: Compressive strength of test specimens.

[10] Afshinnia K, Rangaraju P R. 2016. Impact of combined use of ground glass powder and crushed glass aggregate on selected properties of Portland cement concrete. *Construction and Building Materials* **117**, pp 263-272. https://doi.org/10.1016/j.conbuildmat.2016.04.072.

[11] Corinaldesi V, Nardinocchi A, Donnini J. 2016. Reuse of recycled glass in mortar manufacturing.
European Journal of Environmental and Civil Engineering. **20**, pp. 140-151. https://doi.org/10.1080/19648189.2016.1246695.

[12] Ismail Z Z., Al-Hashmi E A. 2009. Recycling of waste glass as a partial replacement for fine aggregate in concrete. *Waste Management.* **29**(2), pp. 655-659. https://doi.org/10.1016/j.wasman.2008.08.012.

[13] De Castro S., De Brito J. 2013. Evaluation of the durability of concrete made with crushed glass aggregates. *Journal of Cleaner Production.* **41**, pp. 7-14. https://doi.org/10.1016/j.jclepro.2012.09.021.

IOP Conf. Series: Materials Science and Engineering **549** (2019) 012007 doi:10.1088/1757-899X/549/1/012007

Construction-technical inspection of cemetery church of St. Peter and Paul in Želiv

J Daňková[1], P Mec[1], T Majstríková[1] and M Škopcová[1]

[1] Department of Building Materials and Diagnostics of buildings, VSB – Technical University of Ostrava, 17. listopadu 2172/15, 708 00 Ostrava – Poruba, Czech Republic

E-mail: jana.dankova@vsb.cz

Abstract. This article deals with the construction-technical inspection of Cemetery Church of St. Peter and Paul in Želiv. The church is a cultural heritage and its history dates back almost for 900 years. The construction-technical inspection was performed with the goals to survey construction structures and define their major defects and failures before the commencement of constructional renovation. The survey also included measuring humidity of masonry and wooden parts of the structure.

1. Introduction

The Želiv Monastery is situated at the border of Bohemia and Moravia, at the junction of the Želivka and Trnávka Rivers, eleven kilometres westwards of Pelhřimov. The Cemetery Church of St. Peter and Paul is situated at the edge of the eastern part of the monastery premises and continues partly preserved remains of a historical fortification of the monastery (Figure 1). It is an oriented church of western, Latin rite.

The history of the Church of St. Peter and a Paul in Želiv has not been scientifically studies yet. However, it can be expected that it is closely related to the beginnings of history of the Premonstratensian Monastery in Želiv, or even to the beginning of the settlement in that location [1,2]. Although the Cemetery Church of St. Peter and Paul is often presented as “Roman”, it is very likely that it is a Baroque structure from the period of a large monastery refurbishment performed approximately in the 2nd half of the 17the and at the beginning of the 18th century (Figure 2). The last larger constructional modifications of the Church of St. Peter and Paul were implemented in the 20th century. The “Roman” designation may refer to an older, Roman foundation of the church. The goal of the construction-technical inspection was to survey the existing constructional structures and identify major defects and failures of structures.

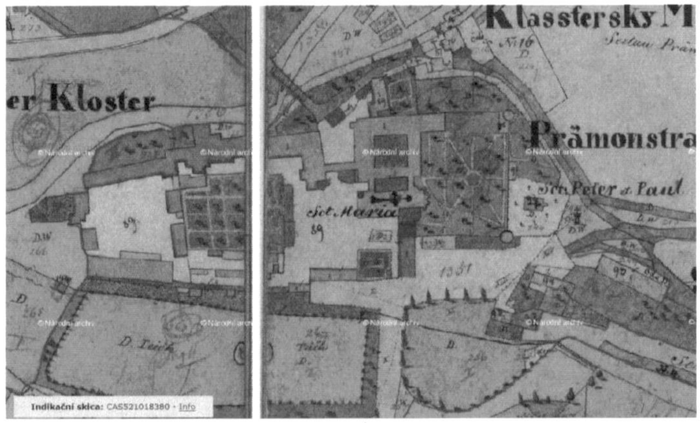

Figure 1. The Church of St. Peter and Paul in Želiv: historical map of 1838 [3].

Figure 2. A picture of the church in the historical veduta of 1736 [4].
(note: the church is circled in the figure)

2. Materials and methods

2.1. Constructional history of the church

Historians currently tend to believe that the monastery in Želiv was founded in 1139 and its founder was the Premyslid prince Soběslav I, however, the year of 1144 is also mentioned as a year of monastery foundation [2]. As admitted by some interpretations of historical sources [2], the first monastery in Želiv had already been finished in the stage of a wooden temporary structure in 1139. The first written mention of the monastery is from 1226 [2]. The constructional history of the church is bound to the history of settlement or the history of the monastery. The monastery was destroyed by wars and fires many times to be subsequently renewed as to the construction and spirituality. The Roman architecture of the monastery disappeared even before the period of Hussite wars as in 1278 the monastery was pillaged in relation to the battle on the Moravian Field and in 1375 the monastery was burnt, including the monastery Roman basilica [5].

At the beginning of the 14th century during Hussite wars the monastery was highly destroyed and in 1467 – 1468 the assets of the monastery was pledged to the noble man Burian Trčka of Lípa [6]. Burian Trčka of Lípa belonged among leading noble men in the country. He recognised the church reformation and he did not belong in his religious belief to the Catholic church as presented at that times by the Pope. This fact may relate to the Cemetery Church of St. Peter and Paul as the symbol of the Holy Trinity, situated in the west face wall of the church gable can indicate that the church was Protestant in the era of the Trček family reign.

At the end of the 16th century the monastery was gradually renewed. In 1644 the monastery and the whole region were attacked and pillaged by the Swedes and Premonstratians returned to the monastery only in 1653. The war events left the monastery highly dilapidated but in about 1655 a large Baroque renovation of the monastery was commenced, more or less preserved until the present days. In 1712 the monastery convent as well as the church burnt down. In April 1714 the monastery was visited by the recognised architect Jan Blažej Santini who designed a refurbishment of the monastery church in the spirit of Baroque Gothic. Repairs of the church were finished in 1720, but the works on monastery buildings and the interior of the church continued. There is currently a hypothesis that Santini also designed and implemented constructional modifications to the Church of St. Peter and Paul [7].

The third and the last fire in the monastery broken down on 12 August 1907. The Trčka's castle, the office building and other buildings were damaged by the fire. The convent and the church were only preserved from fire. One of the larger front towers of the church was damaged by fire. During the World War I a temporary military hospital was situated in the monastery. In 1939 – 1945 the monastery buildings were occupied by German Nazis. After 1950 a forced work concentration cam was established in the Želiv monastery. Since 1954 the monastery buildings were used for the operation of a mental hospital.

The monastery premises did not see any major constructional modifications in the period of 1933 – 1991. The technical condition of buildings was bad after the return of Premonstratians to the monastery in 1991. The famous Baroque monastery gardens disappeared completely, the valuable Baroque inventory, the library and interior of historical buildings especially suffered from irreplaceable losses. On the other hand, this situation of a neglected constructional maintenance and renovation caused that the artistic and material authenticity of constructions and materials from the Baroque monastery refurbishment have been preserved to a large extent.

2.2. Description of the construction

Geographically, Želiv is situated at the border of Bohemia and Moravia, at the junction of the Želivka and Trnávka Rivers, at the latitude of 406 m. Geologically, the territory of the monastery premises and its surrounding is situated in the system of the Bohemian massif. Crystalinicum and prevaris paleozoicum prevail, capping masses and postvaris magmatites partly occur in the location [8].

The Cemetery Church of St. Peter and Paul in Želiv is a single aisle, oriented construction with an offset presbytery. A younger polygonal apse continues the presbytery in its end. A narthex is built to the western face wall. The building has a double sloping roof, the roofs of presbytery and apse end with a hip in the eastern side. Above the presbytery there is a spirelet. The aisle ground plan has dimensions of 8.3 m x 11.49 m. The presbytery ground plan including apse has dimensions of 5.6 m x 7.2 m (Figure 3). No archaeological surveys have been performed in the Church of St. Peter and Paul that would identify hidden structure under the terrain level. The load-bearing circumferential walls of the church aisle and the presbytery consist of rocky masonry with the treatment of gaps evoking Roman masonry (Figure 4, 5). The secondary use of the building material from an older constructional stage of the church cannot be excluded (Figure 5). On the periphery of the masonry on its outer side there is an open ditch as a measure against higher masonry humidity. The thickness of the rocky masonry is 1 meter and the masonry is ended with a string course. The headway of the triumph arch is 7.060 m above the church aisle floor. The masonry of later structures (polygonal apse) is made of bricks or of mixed bricks with stones, the thickness of this masonry is 0.700 – 0.950 m. The church aisle has a flat wooden cassette ceiling and it originates from the latest refurbishment in the 1930s. The narthex, presbytery and apse are arched. The ceiling in the presbytery consists of an arched cross vault with stripped vaults, sacristy situated in the apse is arched with a concha with stripped vaults (Figure 6). The building is not heated, the ventilation is ensured through the doors and windows and venting holes in the rocky masonry (Figure 4, 5). During the survey the temperature and average air humidity in the interior was 8°C, 60-65 %. The daily temperature and average humidity in the outdoor environment was 20 – 26°C, 55% at the time of measurement.

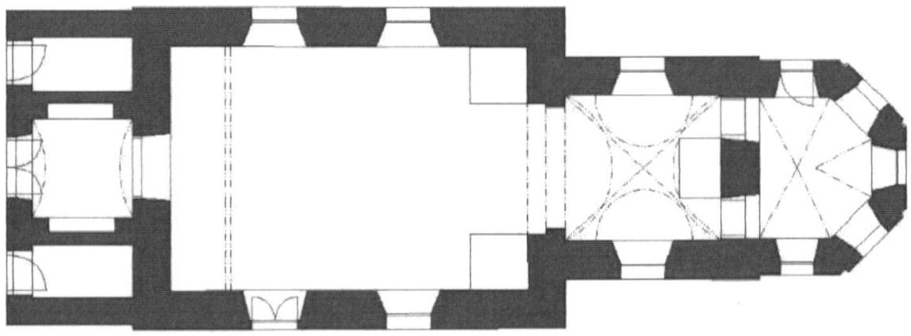

Figure 3. The ground plan of the church.

Figure 4. An eastern view of the church.

Figure 5. A detail of the rocky masonry with a venting hole.

Figure 6. The vault in the apse.

2.3. Defects and failures of the construction

The most significant failures are the ones in the roof cladding. During the inspection leaks in the roofing in the whole area of the roof cladding have been identified, and defects in the details of the gutter, corner and roof valley. Failures which resulted from the leakage of the roof cladding can be seen in constructional structures and materials in the interior and exterior. They especially include the degradation of plasters and cracks in the load-bearing structures, especially vaults (Figure 7).

A major problem is dampness of masonry, plasters as well as other materials. The dampness is manifested in dampness maps, complete degradation can be seen locally as well as blooming of salts in stones and bricks. Plaster, masonry and stone moister by weight has been determined for the purposes of sanitation proposal. Results of plaster moisture determination are given in Table 1. Moisture plaster and masonry was determined in the laboratory by the gravimetric test.

The wood moisture measurement was carried out in structural elements in the interior. Non-destructive measurement with resistance moisture meters was used to determine the moisture content. The resulting wood moisture ranged in 13.9 – 31.7 % by weight. A large range of wood moisture values is caused by various wood surface treatments (e.g. polychromy), potential occurrence of metallic joining elements in the wood etc.

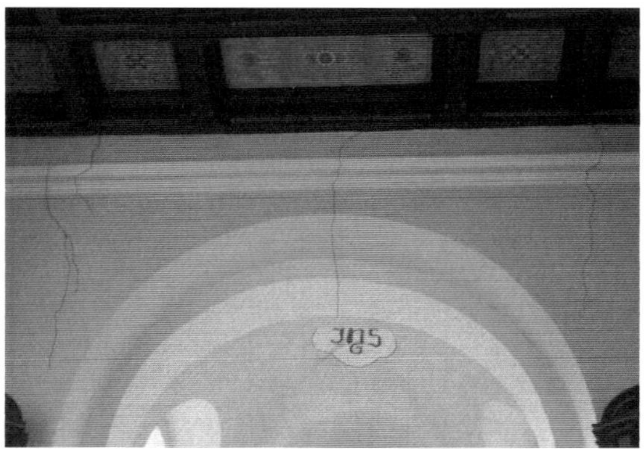

Figure 7. Cracks in the vault of the triumphal arch.

Table 1. Plaster moisture determination.

Sample	Humidity (% wt.)	Degree of moisture according to ČSN P 73 0610
S1	3.18	low
S2	3.34	low
S3	3.19	low
S4	3.85	low
S5	7.15	increased
S6	0.94	very low
S7	2.52	very low
S8	2.91	very low
S9	1.11	very low
S10	1.90	very low

3. Conclusion

The Church of St. Peter and Paul in Želiv is a cultural heritage with extensively reserved authentic form and material composition from the Baroque refurbishment in the 17th – 18th centuries. The defects and failures identified by the construction-technical inspection are not serious and do not present acute risks to the use of the structure. The reconstruction of the roof structures should be carried out simultaneously with securing the stability of vaults. The occurrence of manifestations of higher moisture in masonry, plasters and wooden structures can be efficiently managed by an adjustment of the interior climate to ensure a suitable temperature and indoor air humidity. In this case it is not recommended to control moisture in masonry and plasters by grouting or other radical methods, e.g. by insertion of waterproofing layers. Grouting would damage authentic historical constructional structures and grouting of the rocky and mixed masonry with a thickness of 1 m is difficult and provides little reliable effect.

Acknowledgements

The work was supported from financial funds of the Conceptual Development of Science, Research and Innovations for 2019, allocated to VSB – TU Ostrava by the Ministry of Education, Youth and Sport in the Czech Republic

References

[1] Hejhal P 2010 *Zaměřeno na středověk* (Praha: Nakladatelství Lidové noviny) pp 581–585
[2] Hejhal P and Šrámek J 2014 *Sborník k poctě J. Kalfersta* (Hradec Králové: Muzeum východních Čech v Hradci Králové) pp. 32–54
[3] *Mapa kláštera v Želivi* (online, cit. 2019-04-16) available: http://archivnimapy.cuzk.cz
[4] *Veduta* (online, cit. 2019-04-11) available: http://www.wikiwand.com
[5] Petrů V 1898 *Klášter Želiv* (Praha: Cyrillo-Methodějská knihtiskárna a nakladatelstvi) p. 32
[6] Charouz O Praem and Jindřich Z 1995 *Oživené dědictví – Premonstrátský Želiv včera a dnes* (Želiv: Opatství Želiv) p. 30
[7] HORYNA M 1998 *JB Santini-Aichel–Život a dílo* (Praha: Karolinum)
[8] *Geologická mapa* (online, cit. 2019-04-07) available: http://www.geologicke-mapy.cz

IOP Conf. Series: Materials Science and Engineering **549** (2019) 012008 doi:10.1088/1757-899X/549/1/012008

Effect of industrial by-products incorporated in composites on strength and leaching parameters due to microbiologically induced corrosion

M Smolakova[1] A Estokova[1] A Luptakova[2] and J Junak[1]

[1] Technical University of Kosice, Faculty of Civil Engineering, Institute of Environmental Engineering, Kosice, Slovakia
[2] Institute of Geotechnics, Slovak Academy of Science, Kosice, Slovakia

E-mail: michaela.smolakova@tuke.sk

Abstract. The durability of cement-based materials is an increasingly important characteristic for civil engineering structures due to the growing demand for their long service life and minimal maintenance. Mortars and concrete with high acid resistance are required for the construction and the repair of sewer structures, in which biogenic sulphuric acid attack is the major degradation mechanism. Important progress has been achieved in this regard with hybrid binders and materials known as supplementary cementitious materials (blast furnace slag, silica fume, fly ash etc.). These mainly industrial wastes and by-products increase many important properties of concrete structures such as acid resistance, permeability, strength and many more. This study investigates the effect of industrial by-products such as blast furnace slag and fly ash on strength and leaching parameters of composites exposed to bacterial influence. The medium with bacteria A. thiooxidans was used for microbiologically induced corrosion and the experiments were conducted during the period of 9 months. The sample with 95 wt. % of slag was found more resistible in bacterial environment compared to the sample with hybrid binder.

1. Introduction

It is a well-known fact that concrete structures made with ordinary Portland cement (OPC) deteriorate when exposed to various acidic environments. Cement-based materials can come into contact with different aggressive acidic media, which can be harmful to cementitious materials and these includes biogenic mineral and organic acids [1,2].

Sewer networks and wastewater systems worldwide suffer with significant corrosion and deterioration due to sulphuric acid attack, resulting in large economic losses [1,3]. Sulphuric acid in sewer systems is generated by sulphur-oxidising bacteria and it has been identified as the corroding acid [1,4,5]. This acid can also be present in groundwater or produced from the oxidation of sulphur bearing compounds in back fill, such as pyrite [1,6]. The dissolution of hydrogen sulphide can also contribute to formation of sulphuric acid with a low pH on the concrete walls of geothermal wells [1,7]. Therefore, biogenic sulphuric acid is a major cause of biodegradation and microbiologically induced corrosion (MIC) of concrete and cement-based materials. MIC often leads to a reduction in strength, serviceability and even loss of the structure [1].

Increased resistance to microbiologically induced corrosion can be achieved by, among other things, the addition of supplementary cementitious materials (SCM) or hybrid binders [8,9,10]. Both

hybrid binders and supplementary cementitious materials (SCMs) have been added to concrete and mortars as a partial replacement of cement to lower cement consumption and improve fresh and as well as hardened properties [11]. Hybrid binders are based on the blend of alkali-activated aluminosilicates with less than 30% of Portland cement clinker. Typical mixtures of hybrid binders contain fly ash and clinker, activated with Na_2SO_4 and Na_2CO_3 [12,13] or blast furnace slag with metakaolin and clinker, activated with Na_2CO_3 and K_2CO_3 [12,14]. From economic point of view are industrial by-products such as fly ash or slag interesting materials for binder mass production, even though the composition and quality of fly ash and blast furnace slag can be changing [12].
Granulated blast-furnace slag is nowadays commonly used pozzolanic material, which is often used in combination with Portland cement for many different applications. Cement based materials made with blast furnace slag have many advantages, including improved durability, workability and better sulphate resistance, particularly at higher levels of replacement [15,16,17].

The current experiment was conducted to verify the difference in microbiologically induced corrosion between samples made with high share of different waste materials. Two mortar mixtures one with hybrid binder made with slag and fly ash and the other with 95 wt. % of blast furnace slag and 5 wt. % of Portland cement were subjected to biogenic sulphuric acid for 270 days. Leaching trends of Ca^{2+} ions along with degradation depth, compressive strength and degree of hydration of samples were evaluated.

2. Material and methods

2.1. Materials and exposure conditions
Three different types of cementitious materials, namely: ordinary Portland cement of strength grade 52.5 R (OPC), fly ash (FA) and blast furnace slag (BFS), were used. The samples were prepared according to standard STN EN 196-1 [18]. Two mixtures were designed for the experiment, in the first mixture (H-B) the Portland cement was replaced by hybrid binder by 100 wt. % and in the second mixture (C-BFS) the OPC was replaced by blast furnace slag by 95 wt. % The test specimens have been designed in cooperation with cement factory plant and the dimensions of samples were approximately 160 mm × 40 mm × 40 mm.

Table 1. Chemical composition of used materials (% by mass).

Components	Sample H-B			Sample C-BFS	
	Fly ash	Blast furnace slag	Clinker	Blast furnace slag	Portland cement
CaO	3.36	39.24	66.13	39.55	57.15
SiO_2	51.42	40.13	21.69	38.95	19.01
Al_2O_3	26.39	7.19	5.31	8.33	4.02
Fe_2O_3	7.27	0.27	2.83	0.54	2.69
MgO	2.10	10.04	1.44	10.11	1.37
K_2O	3.28	0.52	1.00	0.48	1.12
SO_3	0.87	1.50	0.53	0.57	1.49
Other	5.31	1.11	1.07	1.47	13.15

Used hybrid binder is produced according to patent application on the base of industrial by-products and wastes [12,19]. The content of clinker is under 30%. Mineralogical composition of fly ash in the hybrid binder is quartz SiO_2, mullite A_3S_2, hematite Fe_2O_3, magnetite Fe_3O_4, granulated blast furnace slag is fully glassy and clinker is composed typically by C_3S 69 % wt., C_2S 11 % wt., C_3A 9 % wt. and C_4AF 8 % wt. Highly alkaline waste water from red mud pond is used for activation of fly ash and granulated blast furnace slag [12]. Blast furnace slag used for the mixture C-BFS is resistant to mold and is defined as a fungistatic building material according to CSN 72 4310 [20]. BFS is suitable as an antifungal admixture for cements, concretes, mortars, dry plaster mixtures, building binders, cement-containing mixtures and building materials. Blends with BFS are also characterized by increased chemical resistance, especially to sulphate and chloride ingress. Chemical composition of main components of hybrid binder, blast furnace slag and Portland cement is given in table 1.

In the experiment, the bacterial culture of sulfur-oxidizing bacteria *Acidithiobacillus thiooxidans*, originated from the local sources (the shaft Pech, the locality Smolnik, Eastern Slovakia) was used. The selective nutrient medium by Waksman and Joffe (pH 4.0) was used for the preparation of the active bacterial culture as well as for the isolation and cultivation of *A. thiooxidans* in the presence of the samples. The prepared samples were exposed to the medium of activated bacteria for 270 days. The volume ratio of solid samples to the liquid phase was set to 1:10. Experiment was carried out in covered tanks under aerobic atmosphere at laboratory temperature. pH value of the liquid phase was kept on an optimal level of 4.0. The dissolved concentrations of calcium ions in leachates were regularly measured through experiment [21].

2.2. Experimental techniques for investigation

Several test methods were used in this study to investigate the effect of aggressive bacterial medium and its consequences on durability properties of tested samples.

X-ray fluorescence analysis (XRF) was used to analyse the chemical composition of the leachates. SPECTRO iQ II equipment (Ametek, Germany) has been fitted with silicon drift detector (resolution of 145 eV at 10 000 pulses) and the samples were measured using standard procedure at two voltages and corresponding currents. The concentrations of the measured elements were calculated from the calibration methods for liquids [21].

The compressive strength tests were carried out according to the methodology described in the standard EN 196-1 [18]. Compressive strength of first set of specimens was determined after the 28 days long curing period and the compressive strength of second set of samples was determined after 270 days of bacterial exposure by using the instrument ADR 2000 (ELE International, England).

The samples before and after the bio-corrosion experiment were researched also by thermal analysis. Simultaneous DSC/TG analysis was performed by using the STA 449 F3 Jupiter (Netsch, Germany). The samples of 9 mg were heated in Al_2O_3 crucibles in the temperature range 26-800 °C, at heating rate 30 K/min under nitrogen atmosphere.

Based on the thermal analysis results, the degree of hydration α of the samples was calculated using Bhatty´s method [19] following the equations (1) and (2).

$$W_B = Rh + Rx + 0.41\ (Rc) \quad (\%) \tag{1}$$

$$\alpha = \frac{W_B}{0.24} \quad (\%) \tag{2}$$

Equation (1) shows the calculation of the chemically bound water (W_B) from relative mass losses in the particular regions on the TG curve Rh (calcium-silicate hydrates), Rx (portlandite) and Rc (carbonates). The value 0.41 is conversion factor to calculate the chemically bound water derived from the carbonated porlandite. In equation (2), the value of 0.24 represents the maximum chemically bound water required to entirely hydrate a cement particle. This value can vary between 0.23 and 0.25 according to different authors [22,23,24]. For Bhatty, the constant value of 0.24 is used in the equation. Regarding the final calculation presented by such a method, the degree of hydration is directly proportional to chemically bound water as only a conversion factor equal to 0.24 is considered [24].

The degradation depths of samples, after the biogenic sulphuric acid exposure, were evaluated using equation (3), by dividing the total amount of leached calcium by the theoretical initial total calcium content of specimens [21,25].

$$\text{Degradation depth} = \frac{c(Ca)_L\ (mg/cm^2)}{c(Ca)_I\ (mg/cm^3)}\ (cm) \quad (3)$$

Where $c(Ca)_L$ is the amount of leached-out Ca^{2+} ions per unit area (mg/cm^2) of the sample and $c(Ca)_I$ represents an initial calcium content in the sample per 1 cm^3 of sample.

The calculation of the degradation depth (DD) of samples assumes that all portlandite (CH) was dissolved from the surface layers of samples after the 270-day leaching and that the Ca^{2+} ions were fully leached out in a degradation depth [21,25]. Considering the leached-out concentrations of Ca^{2+} ions, assuming that they origin from the decomposition of portlandite, DD of tested specimens with different share of pozzolanic materials were calculated, according to Roziere et al. [25]. The theoretical DD could actually estimate the degradation depth of samples and based on this assumption degradation depths could predict the deterioration stage of material.

3. Results and discussion

The more intensive leaching trend of Ca^{2+} ions was observed for the sample C-BFS with 95 wt. % of blast furnace slag compared with the sample H-B with hybrid binder (figure 1).

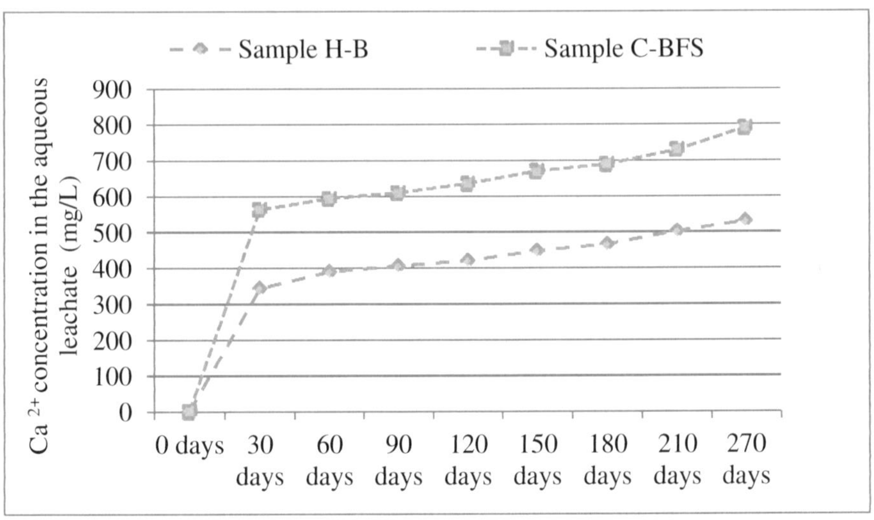

Figure 1. Dissolving trends of calcium during bio-corrosion simulation.

The increasing leaching of Ca^{2+} ions was observed continuously for both samples placed in the bacterial environment until the end of the experiment at 270 days. For both samples, the H-B and the C-BFS, the highest value of concentration of Ca^{2+} ions was observed at the end of the exposure (531.2 and 790.1 mg/L, respectively).

The leaching trends were similar, even though the C-BFS sample had higher leaching performance of calcium ions throughout the experiment compared with the H-B sample. It is important to note, that the specimen C-BFS had 2.3 higher initial calcium content that the specimen H-B (table 2), so it was expected that the values of total leached out Ca^{2+} ions would be higher. However, the dissolved amount of calcium from the C-BFS sample was only 1.5 times higher. This points to the higher stability of the C-BFS sample in bacterial environment even though the absolute mass of the calcium dissolved was higher when compared to the H-B sample. This is in accordance with the degradation depths as reported in table 2.

Table 2. Calculated degradation depths of samples.

Composite sample	Leached out Ca^{2+} (mg/cm^2)	Initial amount of Ca^{2+} (mg/cm^3)	Degradation depth (mm)
H-B	4.611	87.707	0.525
C-BFS	6.858	199.422	0.344

Although the most intensive Ca^{2+} leaching was observed for sample C-BFS exposed to bacterial medium for 270 days, sample with hybrid binder H-B was observed to have higher degradation depth (0.525 mm). The value of degradation depth of specimen H-B had increased for 35% compared to the C-BFS specimen. Sample made with blast furnace slag was more durable against the influence of biogenic sulphuric acid, from the degradation depth point of view.
Compressive strengths of investigated samples made with pozzolanic materials after 28 and 270 days are shown in figure 2. An initial value (28-day) of compressive strength was higher for the sample H-B made with hybrid binder. In comparison the compressive strength after 28 days of curing in water for sample C-BFS was only 12.50 MPa.

Results for both samples after 270 days of exposure to the biogenic sulphuric acid produced by *A. thiooxidans* were visibly different. The 270 day compressive strength for specimen with 95 wt.% had increased by 72% from 12.50 to 44.76 MPa. For the sample H-B the decrease by 24% in compressive strength was observed. The decrease in compressive strength of H-B sample could manifest a bio-deterioration process due to bacterial exposure. Increase in compressive strengths of C-BFS specimen with slag addition, after 270 days of exposure to the biotic environment, can likely resulted in an additional hydration process, which was probably more intensive compared to the biological corrosive process. Corrosion as well as hydration processes takes place at the same time, yet after 270 days long biological exposure the hydration process exceeds the corrosion one.

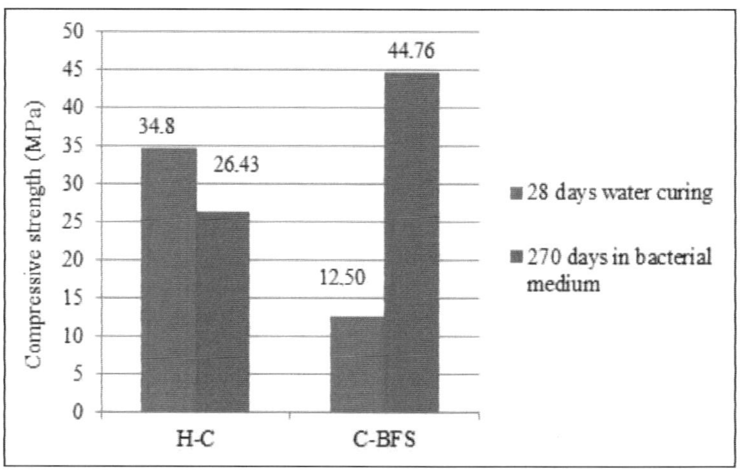

Figure 2. Compressive strengths of the samples.

Regarding processes that take place in the presence of mineral admixtures, as complexity increases, use of additional techniques is required in understanding their development in a better way. The use of differential thermal and thermogravimetric analysis (DSC/TG) is proposed as one of the most effective tests for the evaluation of hydrated compounds and hydration degree [24]. Hydration degree is an important parameter and also it is linked with the evaluation of the compressive strength of cement-based materials. The various regions of dehydration (Rh), dehydroxylation (Rx) and decarbonation

(Rc) that correspond to the hydrated compounds are defined according to the limit temperatures selected (table 3).

Table 3. Temperature regions of the decomposition of main hydration products.

Region	Temperature (°C)
Rh	105-400
Rx	400-500
Rc	500-800

TG curves of studied samples with the Rh, Rx and Rc decomposition regions are shown in the figure 3.

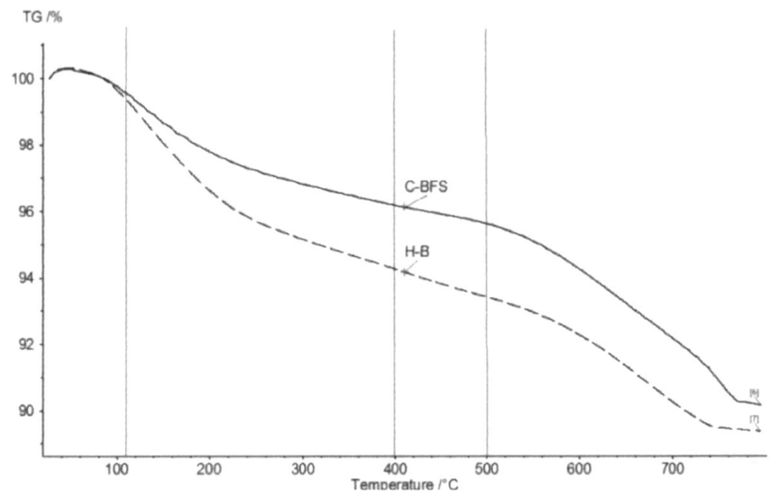

Figure 3. TG curves of the samples after the experiments.

The relative values of each decomposition region from TG curves of samples, the chemically bound water and the degree of hydration calculated according to Bhatty [22] are shown in table 4.

Table 4. Chemically bound water and degree of hydration of studied samples.

Sample	**Exposure time (days)**	**Weight losses in regions (%)**			**Chemically bound water and degree of hydration according to Bhatty (%)**	
		Rh	Rx	Rc	W_B	α
H-B	0	1.18	0.50	2.58	2.74	11.41
	270	3.82	3.18	2.82	8.16	33.98
C-BFS	0	1.74	0.31	0.96	2.44	10.18
	270	5.74	3.18	1.71	9.62	40.09

The degree of hydration calculated according Bhatty was higher for the H-B sample after 28 day curing in water, which is consistent with the compressive strength measurements. After the bacterial influence, the degree of hydration as well as the chemically bound water increased for the both mixtures, and the C-BFS sample showed higher degree of hydration (40.09 %), which is again consistent with the strength development. Sample C-BFS had higher compressive strength after 270 days of bacterial exposure, which relates to higher amount of hydration products.

4. Conclusion

The influence of biogenic sulphuric acid on cement composites with incorporated industrial by-products on leaching of calcium ions, degradation depth, strength development and degree of hydration were investigated in this paper. The following conclusion can be drawn based on the experimental analyses and results:

- Chemical composition of leachates confirmed the corrosive process was taking place, because of the significant leaching of Ca^{2+} ions.
- Evaluation of damaged depth according to leaching of calcium ions showed better resistance of sample with the addition of BFS compared to the sample made with hybrid binder in the aggressive biotic environment.
- Investigation of compressive strength of samples showed the increase in compressive strength for sample with 95 wt. % of blast furnace slag after 270 days of bacterial exposure and decrease in compressive strength for the sample with hybrid binder. Assumption that hydration was a dominant process compared to deterioration, due to increasing trend in compressive strength for sample with slag can be made.
- Results from thermal analysis completed the compressive strength results and confirmed that the degree of hydration is linked with the development of strength of studied samples.
- Evaluation of durability of samples made with different industrial by-product showed better resistance of sample made with 95 wt. % BFS in the terms of all studied parameters.

Acknowledgments

This research was funded by VEGA - Scientific Grant Agency of the Ministry of Education, Science, Research and Sport of the Slovak Republic and the Slovak Academy of Sciences, grants number 2/0142/19 and 1/0648/17.

References

[1] Aiken T A, Kwasny J, Sha W and Soutsos M N 2018 *Cem. Concr. Res.* **111** 23-40
[2] Miyamoto S, Minagawa H and Hisada M 2014 *Constr. Build. Mater.* **67** 47-54
[3] O'Connell M, McNally C and Richardson M G 2010 *Cement Concrete Comp.* **32** 479-485
[4] De Belie N, Monteny J, Beeldens A, Vincke E, Van Gemert D and Verstraete W 2004 *Cem. Concr. Res.* **34** 2223-36
[5] Huber B, Hilbig H, Mago M M, Drewes J E and Müller E 2016 *Cem. Concr. Res.* **87** 14-21
[6] Bassuoni M T and Nehdi M L 2007 *Cem. Concr. Res.* **37** 1070-84
[7] Pyatina T and Sugama T 2016 *Adv. Cem. Res.* **28** 433-457
[8] Moffatt E G and Thomas M D A 2018 *Cem. Concr. Res.* **113** 65-73
[9] Vaclavik V, Dvorsky T, Simicek V, Ondova M, Valicek J, Kusnerova M and Gola L 2016 *Sol. St. Phen.* **244** 77-87
[10] Vaclavvk V, Kusnerova M, Dvorsky T, Simicek V, Valicek J, Gola L and Harnicarova M 2018 *Improved Performance of Materials* **72** 107-118
[11] Chu S H and Kwan A K H 2019 *Constr. Build. Mater.* **197** 716-724
[12] Strigac J and Martauz P 2013 *Proc. Int. Workshop on Performance-based Specification and Control of Concrete Durability RILEM (Zagreb)* (France: RILEM publications) p 25
[13] Fernández-Jimenénez A, Sobrados I, Sanz I and Palomo A 2013 *Proc. Int. Congress on the Chemistry of Cement (Madrid)* (Spain: Instituto de Ciencias de la Construcción) 1-7
[14] Bernal S A, Herfort D and Skibsted J 2013 *Proc. Int. Congress on the Chemistry of Cement (Madrid)* (Spain: Instituto de Ciencias de la Construcción) 1-7
[15] Barnett S J, Soutsos M N, Millard S G and Bungey J H 2006 *Cem. Concr. Res.* **36** 434-440
[16] Whittaker M, Zajac M, Haha M B and Black L 2016 *Constr. Build. Mater.* **119** 356-369
[17] Vaclavik V, Ondrasikova I, Dvorský T and Cernochova K 2016 *Int. J. Env. Res. Pub. He.* **13** 1-12

[18] STN EN 196-1: 2016 *Methods of testing cement - Part 1: Determination of strength*
[19] Strigáč J and Martauz P 2012 Patent application SK PP 50024-2012 *Combined binder on the base of wastes*
[20] CSN 72 4310: 1977 *The testing of mould proofness of building products and materials*
[21] Estokova A, Smolakova M and Luptakova A 2018 *Buildings* **8** 1-10
[22] Bhatty J I 1986 *Thermochim. Acta* **106** 93-103
[23] Taylor H F 1997 *Cement chemistry* (United Kingdom: Thomas Telford)
[24] Monteagudo S M, Moragues A, Gálvez J C, Casati M J and Reyes E 2014 *Thermochim. Acta* **592** 37-51
[25] Rozière E, Loukili A, El Hachem R and Grondin F 2009 *Cem. Concr. Res.* **39** 1188-98

Experimental determination of the influence of additives on shrinkage in self-compacting concrete

M Alexa[1], B Kucharczyková[1], D Kocáb[1] and P Daněk[1]

[1] Brno University of Technology, Faculty of Civil Engineering, Veveří 331/95, Brno, Czech Republic

E-mail: martin.alexa@vutbr.cz

Abstract. The paper deals with the influence of additives on the progress of shrinkage and mass losses in self-compacting concrete. Limestone and fly ash at varying amounts were used as an additive in the production of four types of self-compacting concrete. Volume changes were observed from an early stage of setting (the measurement was commenced within 1 hour of mixing) using shrinkage drains with a moveable wall, followed by a long-term manual measurement using an external strain gauge. For the first 48 hours the volume-change measurement was continuous, and for the next few months it was performed manually. Mass losses were determined simultaneously with shrinkage – during the first 48 hours by weighing on a special weighing table capable of continuous recording, and manually afterwards.

1. Introduction

The fresh-state properties of self-compacting concrete (SCC) make it a very popular building material which is finding very broad use. It is generally agreed that the first SCC was developed at the University of Tokyo in Japan in the late 1980s [1]. Given the typical aspects of SCC mixtures, especially the higher content of small particles and a lower water-cement ratio, these concretes are more vulnerable to shrinkage than ordinary ones [2,3]. Shrinkage is usually defined as a reduction in volume in unloaded concrete as it ages [4]. This is mainly due to changes in the water content of the concrete structure. Shrinkage is generally divided into these categories: plastic shrinkage, chemical shrinkage, autogenous shrinkage, carbonation shrinkage, and desiccation shrinkage [4]. More pronounced shrinkage can produce stress inside the concrete which may exceed its tensile strength in some places and cause microcracks and even cracks to form. These microstructure defects can lead to an overall drop in the material's load-bearing capacity and, more importantly, reduce its durability [5].

Concrete shrinkage as such cannot be eliminated, but it is possible to minimise its negative effects. Curing the concrete as it sets and hardens is one of the effective means of containing its shrinkage. Another option is to adjust the formula – for instance it is possible to reduce the generation of hydration heat or to use a shrinkage-reducing agent (SRA). SRA generally operate by reducing surface tension in the pore structure and thus containing shrinkage and minimising the risk of cracking. However, the way SRA affect pore structure delays hydration and is thus detrimental to durability [6]. Because they are easily water-soluble, they can be washed away from the material during rain or by other running water [7,8]. Using SRA can also increase the total costs of the concrete, which is why they need to be used with caution [9].

IOP Conf. Series: Materials Science and Engineering **549** (2019) 012009 doi:10.1088/1757-899X/549/1/012009

Additives, such as fly ash or limestone replace Portland cement in order to improve the concrete's fresh-state and hardened-state properties. These fine particles can improve the compaction of the cement matrix, which reduces the concrete's permeability and improves its durability [10]. Fly ash, commonly added as a substitute of 25 – 30 % of cement volume, holds several advantages for concrete durability, which can emerge during setting and hardening. These are mainly better workability, smaller water requirement, smaller degree of segregation, less hydration heat, higher long-term strength, substantial reduction in permeability (and thus better resistance to chlorides), less expansion due to the reaction of alkaline silicon dioxide, or increased sulphate resistance [11, 12].

In concretes where the replacement is 40–60 %, fly ash is usually combined with finely-milled limestone with an average particle size of 1 micrometre (5–15 % of total cement volume). The combination of limestone and fly ash produces a mixture with better workability and improves the material's durability as well [13, 14, 15].

2. Experiment

The main goal of the experiment was to observe shrinkage in self-compacting concretes and determine the influence of additives on its measured value. Over the first 48 hours the specimens were kept in moulds with the top surface exposed at a laboratory temperature of (21±2) °C and relative humidity of (55±10) %. Once demoulded, the specimens remained on racks under the same conditions and were left to dry freely through all surfaces.

2.1. Material

The experiment was performed with four mixtures of self-compacting concrete. For better orientation during result evaluation they were identified as SCC1 through SCC4. All four concretes were made with the same coarse aggregate with the fraction of 4/8 mm and 8/16 mm and the same sand of 0/4 mm. The plasticiser was likewise the same and all concretes contained it at an amount of 1.5 % of cement mass. Seeing as commercial products were used, their names and composition will not be revealed in this paper. Table 1 therefore shows only the percentages of all the components (cement and additives); the figures represent ratios of the mass of each component to the total mass of all components per 1 m^3 of fresh concrete.

Table 1. Average content of cement and additives in each concrete (mass %).

	Relative content of components in the concrete formula			
Component per 1 m^3	**SCC1**	**SCC2**	**SCC3**	**SCC4**
Cement (%)	18.5	17.5	18.5	18.5
Fly ash (%)	4.5	0	0	3
Ground limestone (%)	0	3	4.5	1.7
w/c ratio (-)	0.38	0.33	0.31	0.32

2.2. Testing

For the first 48 hours shrinkage was measured using shrinkage drains with a moveable wall produced by Schleinbinger Geräte [16], which made it possible to commence measurement of relative length changes very soon after the concrete had been cast. The shift of the walls was recorded using induction transducers and a Quantum data logger. A similar test is described in [17]. The inside of the drains was lined with a Teflon sheet to minimise friction between the mould and the specimen, allowing the concrete to shrink freely (see figure 1). The spaces between the drain and the moveable wall were sealed with Vaseline to stop the concrete from leaking underneath and thus preventing the wall from moving.

Measuring dollies were embedded in the top surface of the specimens (reaching approx. 25 mm deep) to be used for shrinkage measurements after demoulding. While the specimens were being cast, the dollies were held in place by an auxiliary frame to prevent them from moving out of place. There were three dollies positioned along the length axis, making two measuring bases of 200 mm (see figure 1). These bases were used for measuring relative length changes using an external Hollan-type strain gauge with a digital dial; see figure 2. There was no need to vibrate the specimens, since they were made from SCC. During the experiment the specimens were not cured in any way.

During the first 48 hours, mass losses were measured using a special weighing table [17] and the readings were being recorded using the same data logger as deformation. Once the specimens were demoulded, the measurements were performed manually at planned intervals using a set of digital scales with an accuracy of 0.1 g.

Figure 1. Shrinkage drains prior to casting with measurement dollies in position (left) and after casting mounted on a weighing table (right).

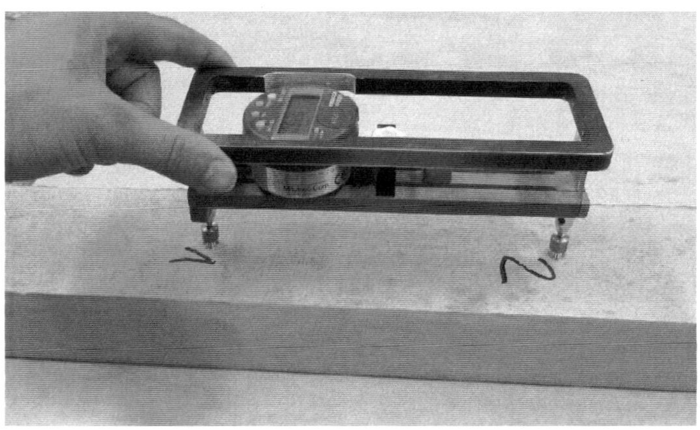

Figure 2. Manual shrinkage measurement using an external strain gauge and measuring dollies.

3. Results

The values of mass loss and shrinkage were measured for approximately 200 days. Figure 3 shows the final shrinkage values and mass losses over the first 48 hours of ageing, figure 4 shows average values of shrinkage and mass losses over the entire testing period (i.e. until the age of 200 days), and figure 5 plots the temperature developing inside the specimens during the first 48 hours.

The values show that SCCs shrink rapidly during the first 12 hours, after which the shrinkage slows down and increases only gradually. Relative strain at the age of 48-hours in SCC1 reached approx. 400 μm/m, in SCC2 through SCC4 it was around 900 μm/m and never exceeded 1100 μm/m. After 200 days of measurement the value of relative strain in SCC1 was approx. 1200 μm/m, while in SCC2 through SCC4 it ranged within 1400 – 1700 μm/m. SCC1 was rather different from the others in terms of its w/c ratio – in SCC1 the w/c value was 0.38, while in the other three it was 0.31 to 0.33. After the SCC1 specimens had been cast, the concrete bled a little, which most likely affected its shrinkage as well.

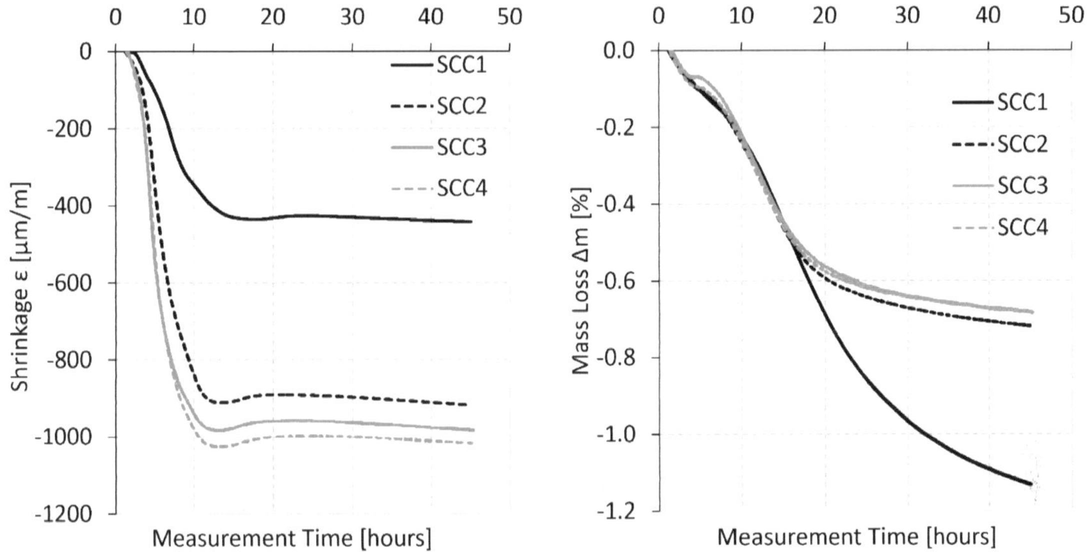

Figure 3. Development of shrinkage and mass losses during the first 48 hours.

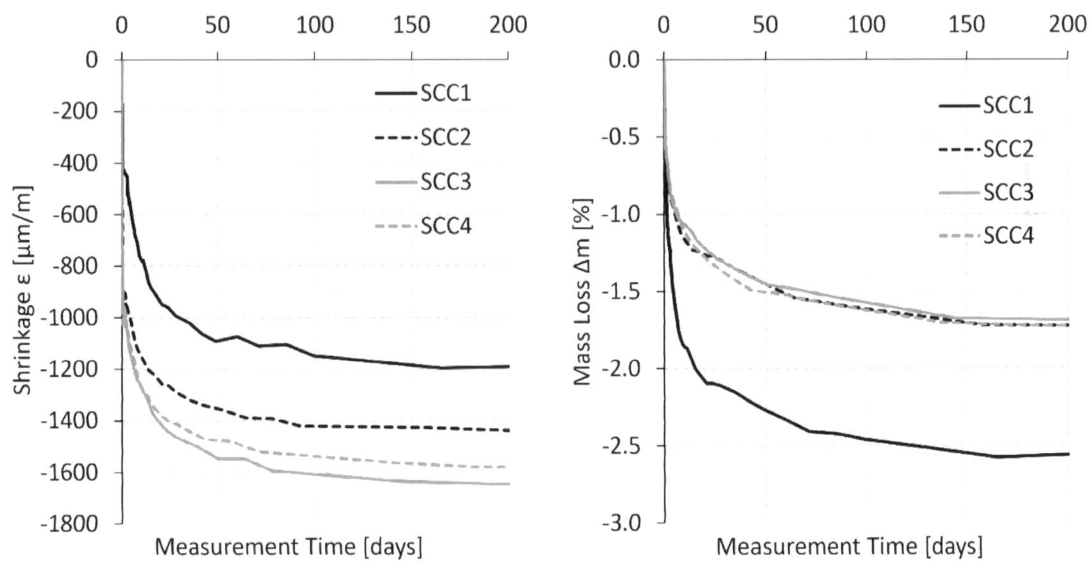

Figure 4. Development of shrinkage and mass losses during the first 200 days.

Mass losses show a similar tendency as shrinkage. In the first 48 hours the mass loss in all the concretes was almost 50 % of the total steady value measured at the age of 200 days. The 48-hour mass losses in SCC1 were approx. 1.1 % while in concretes SCC2 through SCC4 the loss was about 0.7 %. Such a marked difference was likely caused by desiccation of the top surface of SCC1 specimens. At the age of 200 days the mass loss in the first concrete was approx. 2.5 %, in the others it was 1.7 %.

The shrinkage and mass loss data measured on all three specimens of each concrete at the age 48 hours and 200 days were verified for normality on a level of significance of 0.05. Next, a two-sample t-test was performed to compare the average values of shrinkage and mass losses at the above-stated stages of maturation on a significance level of 0.05. The statistical analysis showed that SCC2 and SCC3 did not differ with statistical significance neither in shrinkage, nor in mass loss value regardless of age. However, SCC1 does bear a statistically significant difference from the others at the age of 48 hours and both in shrinkage values and in mass losses. At the age of 200 days it differs with a statistical significance from SCC3 and SCC4, and in mass losses it differs from all the other concretes.

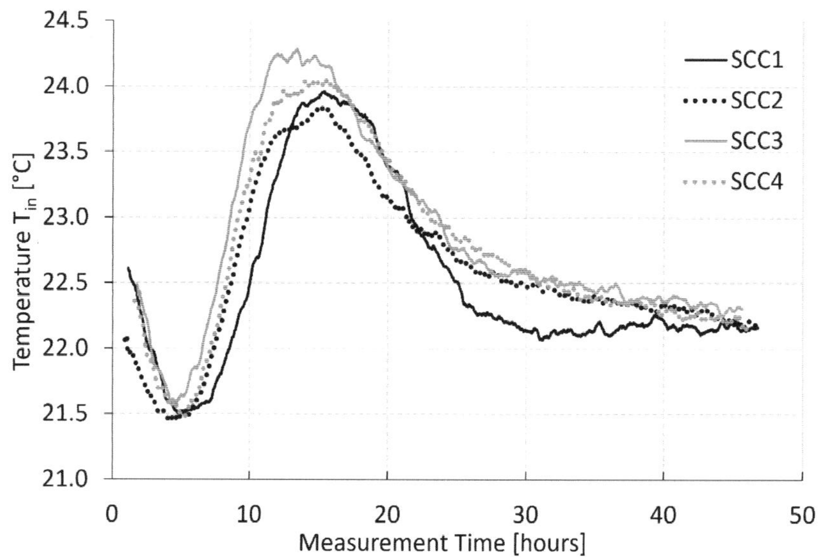

Figure 5. Development of internal temperature during the first 48 hours.

4. Conclusion

In conclusion, an influence of the additives on the short-term and long-term values of shrinkage cannot be clearly proved because the shrinkage did not exhibit any significant variation. Only the first set of specimens (SCC1) differed from the others, which may likely have been caused by the higher w/c ratio. SCC1 also manifested bleeding after it was cast in the shrinkage drains. The water, which thus gathered on the surface of the specimens, temporarily acted as curing water, which probably reduced the initial shrinkage value and thus it swayed the overall long-term progress of shrinkage. The bleeding in SCC1 also had a clear effect on the progress and final value of mass losses.

A two-sample t-test performed at a significance level of 0.05 showed that the values of shrinkage and mass losses determined from SCC1 specimens at the age of 48 hours and 200 days bear a statistical difference from those measured on the specimens of the other concretes (with the only exception being SCC2 at the age of 200 days). The measurement results show that the type and amount of additives do not affect the process of shrinkage at the selected ages of the concretes tested herein. It appears that the w/c ratio and the correct SCC mixture design play a more important role in this regard. Bleeding is considered undesirable in SCC, since it indicates that the concrete has a tendency to segregate. In terms of shrinkage measurement, it also distorts the overall progress as well as the absolute values of shrinkage and mass losses.

Acknowledgement
This paper has been written as a part of project No. 17-14302S, supported by the GAČR - Czech Science Foundation and project No. FAST-J-19-5995, supported by Brno University of Technology.

References

[1] Okamura H and Ouchi M 2003 Self-Compacting Concrete *Journal of Advanced Concrete Technology* **1**(1) pp 5-15

[2] Goodier C I 2003 Development of self-compacting concrete *Proceedings of the Institution of Civil Engineers - Structures and Buildings* **156**(4) pp 405-414

[3] Dang Y, Qian J, Qu Y, Zhang L, Wang Z, Qiao D and Jia X 2013 Curing cement concrete by using shrinkage reducing admixture and curing compound *Constr. Build. Mater.* **48** pp 992-997

[4] Abdalhmid J M, Ashour A F and Sheehan T 2019 Long-term drying shrinkage of self-compacting concrete: Experimental and analytical investigations *Constr. Build. Mater.* **202** pp 825-837

[5] Dang Y, Qian J, Qu Y, Zhang L, Wang Z, Qiao D and Jia X 2013 Curing cement concrete by using shrinkage reducing admixture and curing compound *Constr. Build. Mater.* **48** pp 992-997

[6] Demir İ, Sevim Ö and Tekin E 2018 The effects of shrinkage-reducing admixtures used in self-compacting concrete on its strength and durability *Constr. Build. Mater.* **172** pp 153-165

[7] Mora-Ruacho J, Gettu R and Aguado A 2009 Influence of shrinkage-reducing admixtures on the reduction of plastic shrinkage cracking in concrete *Cem. Concr. Res.* **39** pp 141-146

[8] Maltese C, Pistolesi C, Lolli A, Bravo A, Cerulli T and Salvioni D 2005 Combined effect of expansive and shrinkage reducing admixtures to obtain stable and durable mortars *Cem. Concr. Res.* **35** pp 2244-2251

[9] Ferreira S 2014 ABC's of Crackless Bridge Decks with applications in accelerated bridge construction *Bridge Contractors / Caltrans Liaison Committee Meeting* PE California Department of Transportation March 21

[10] Altoubat S, Talha Junaid M, Leblouba M and Badran D 2017 Effectiveness of fly ash on the restrained shrinkage cracking resistance of self-compacting concrete *Cem. Concr. Comp.* **79** pp 9-20

[11] Hwang S D and Khayat K H 2008 Effect of mixture composition on restrained shrinkage cracking of self-consolidating concrete used in repair *ACI Materials journal* **105**(5) pp 499

[12] Yasumoto A, Edamatsu Y, Mizukoshi M and Nagaoka S 1998 A study on the shrinkage crack resistance of self-compacting concrete *Special Publication* **179** pp 651-670

[13] Altoubat S A and Lange D A 2002 Grip-specimen interaction in uniaxial restrained test *ACI SPECIAL PUBLICATIONS* **206** pp 189-204

[14] Paillere A, Buil M and Serrano J J 1989 Effect of fiber addition on the autogenous shrinkage of silica fume *Materials Journal* **86**(2) pp 139-144

[15] Esping O 2008 Effect of limestone filler BET(H2O)-area on the fresh and hardened properties of self-compacting concrete *Cem. Concr. Res.* **38** pp 938-944

[16] Kucharczyková B, Vymazal T, Daněk P, Misák P amd Pospíchal O 2009 SOP 01/09: Standard operating procedure for determination of shrinkage and swelling of concrete, BUT (in Czech)

[17] Vymazal T, Daněk P, Misák P and Kucharczyková B 2015 Způsob kontinuálního měření hmotnostních úbytků cementových kompozitů v raném stadiu tuhnutí a tvrdnutí a zařízení k provádění tohoto způsobu, Czech patent: 304898, CZ, BUT (in Czech)

IOP Conf. Series: Materials Science and Engineering **549** (2019) 012010 doi:10.1088/1757-899X/549/1/012010

Non-destructive diagnostics of steel-reinforced concrete structures: detecting and locating reinforcement

P Cikrle[1], S Hüblová[1], D Kocáb[1] and O Karel[1]

[1] Brno University of Technology, Faculty of Civil Engineering, Veveří 331/95, Brno, Czech Republic

E-mail: petr.cikrle@vutbr.cz

Abstract. The paper describes the possibilities of non-destructive determination of the location and depth of reinforcement in concrete structures using electromagnetic cover meters. It is currently one of the most popular methods for determining the location, cover depth and diameter of reinforcement. However, there are not many sources that would give a detailed description of specific measurement results. This paper describes the use of the Profometer PM-630 cover meter for in-situ measurement of a steel-reinforced-concrete road bridge deck. The paper details the results of measurements of cover depth, bends in reinforcement, and discusses the strong and weak points of the method.

1. Introduction

Determining the precise location, cover depth, and diameter of concrete reinforcement in existing structures can be rather difficult. Especially in densely reinforced elements, where there are several layers of reinforcement, clear results may not be achieved even by a destructive technique. Still, non-destructive methods assume a strong position in building diagnostics. Certain limitations notwithstanding, non-destructive techniques can determine (or, in some cases, approximate) the cover depth and diameter of reinforcement bars as well as their type and quality [1,2].

One of the most common non-destructive methods for locating reinforcement is the use of electromagnetic cover meters. It has been used for many decades; however, until recently, the imaging capability of available instruments had been rather limited. This is why most measurement outputs consisted of photographs of the tested surface with lines drawn to indicate the rebar locations. This may be the reason for the near absence of any detailed measurement results, including a thorough discussion of the method's strong points and shortcomings, or suggestions as to the suitability of its use across various cases. A significant advance in the instruments was made in the form of the Ferroscan cover meter introduced by the company Hilti, which was capable of performing area scans [3]. More progress was made by Proceq, who have marketed the Profometer PM-630 (or Profometer PM-650). A more detailed description of the development and history of measuring instruments is presented [4].

Nowadays, when radiography is only rarely used to inspect and locate reinforcement, it also possible to use e.g. the Radar PS 1000 produced by Hilti. Its use is described in [5,6]. It is a good choice to use the instruments in combination, as described in [7], or comparing their results [8].

2. Using a cover meter to locate reinforcement

2.1. Method and instrument

The measurement examples described here testify to how widespread the use of cover meters is and what innovations have taken place in recent years. The paper describes a practical example of the use of the Profometer PM-630 – specifically detecting and locating reinforcement in a steel-reinforced-concrete bridge deck.

Detecting concrete reinforcement by a cover meter works on the principle of registering eddy currents that occur in conductors placed in an electromagnetic field. It used to be common that instruments exploited the magnetic properties of the reinforcement. The Profometer PM-630, which was used for measurement in this paper, however, uses pulse-induction technology – it is described e.g. in the manual to the instrument [9]. The probe of the Profometer PM-630 holds two coils without a magnetic core, which is beneficial, as it is not susceptible to interference from other magnetic fields.

Modern magnetic cover meters have their strong and weak points. Advantages certainly include the measurement accuracy and speed, immediate visual output, and the option to correct for the proximity of adjacent reinforcement when measuring the cover depth of individual rebars (applies to the Profometer PM-630). A drawback is the inability to detect rebars if they are too close (regardless of whether they are behind or next to each other), shorter range compared to e.g. radiography (most probes cannot scan deeper than 100 mm, deep probes reach a maximum of 200 mm), or the near impossibility of measurement in the vicinity of other electromagnetic fields [10].

Prior to the actual measurement, it is recommended to test the instrument's sensitivity on a model element, especially in cases where the rebars are close. It is also good to make sure there is no magnetic aggregate present, as it could significantly distort the results; see [11]. The correct use of the Profometer PM-630, including the principle of operation and the method, are described in the user manual [9]. It should be noted that in order for this result to deliver accurate results, the operator has to be trained and experienced in its use. Even though the Profometer PM-630 can locate concrete reinforcement very accurately, a novice user may still arrive at very wrong results.

2.2. Detecting and locating reinforcement

This paper discusses detecting and locating reinforcement bars using the example of finding the beginning of bends in straight rebars in a reinforced-concrete slab near the abutment. Tests were performed on a load-bearing steel-reinforced concrete bridge spanning a stream near the town of Nosislav near Brno. It consists of a slab with the thickness of 280 mm resting on bitumen sheets on solid abutments. The inner distance between the abutments is 3.0 m and the bridge is 8.0 m wide. The deck consists of two 4.0 m slabs separated by an expansion joint.

The reinforcement was measured using a Profometer PM-630. Besides locating bends, the goal was to ascertain the number of profiles, their spacing and cover depth. Learning the reinforcement type and diameter was easy, as a part of the deck near the ledge was missing a piece of cover, revealing several reinforcement bars made of type 10216 (E) smooth circular reinforcement with the diameter of 16 mm.

The methodology of locating the bends was first tested in a laboratory. When reinforcement is being located, the instrument is pushed along the surface of the slab waiting for the signal to peak. This occurs when the instrument, moving in a perpendicular direction to the reinforcement, is at the shortest distance from a bar. The position of the reinforcement is always drawn onto the concrete surface, having been determined using several line scans.

During cover depth measurement it is possible to locate inward-bending bars with an accuracy of $\pm$ 10 mm, as their cover depth is increasing. A 1.2 m long line scan was performed in the middle of the bridge. It identified 12 profiles with the spacing of 80 - 90 mm. The average cover depth was only approx. 12 mm. The next step involved making several line scans at a distance of 0.12 m, 0.22 m, 0.26 m, 0.29 m, 0.32 m and 0.40 m from the abutment. Figure 1 shows where on the slab the line scans

IOP Conf. Series: Materials Science and Engineering **549** (2019) 012010 doi:10.1088/1757-899X/549/1/012010

were performed. In addition, one of the bends was examined (specifically bar no. 8 – this is an internal designation of the rebars starting from the ledge inward) by chiselling its cover away; see figure 2.

Figure 1. A view of the deck with line scans shown in blue.

Figure 2. Bend in rebar no. 8 – it begins precisely in the position of the black marker line at a distance of 0.32 m from the abutment.

3. Results and discussion

Figure 3 shows the key line scans (made at a distance of 0.12 m, 0.22 m, 0.29 m and 0.32 m from the abutment) exported from the cover meter. The vertical axis is the cover depth and the horizontal axis is the length of the scan. It is clear that bent bars are not recorded on the top scan (closest to the abutment) while on the bottom scan they are already at the level of the other bars. Figure 4 shows the results of cover depth measurement for rebars no. 1 through 12, out of which rebars no. 2, 8 and 11 bend at a distance of 0.32 from the abutment and rebar no. 5 at 0.29 from the abutment.

It is, of course, possible to measure bent reinforcement only up to a certain depth, as stated above. By default, if reinforcement is being located by the electromagnetic method, rebars placed deeper inside the concrete will appear overshadowed by parallel rebars nearer to the surface. During the destructive measurement of the bend in rebar no. 8, the beginning of the bend was measured with an accuracy of ± 10 mm at a distance of 0.32 m from the abutment. The electromagnetic cover meter appeared to be very accurate in identifying bends in reinforcement in the present case, and can be used in all cases where the individual rebars are easily distinguishable.

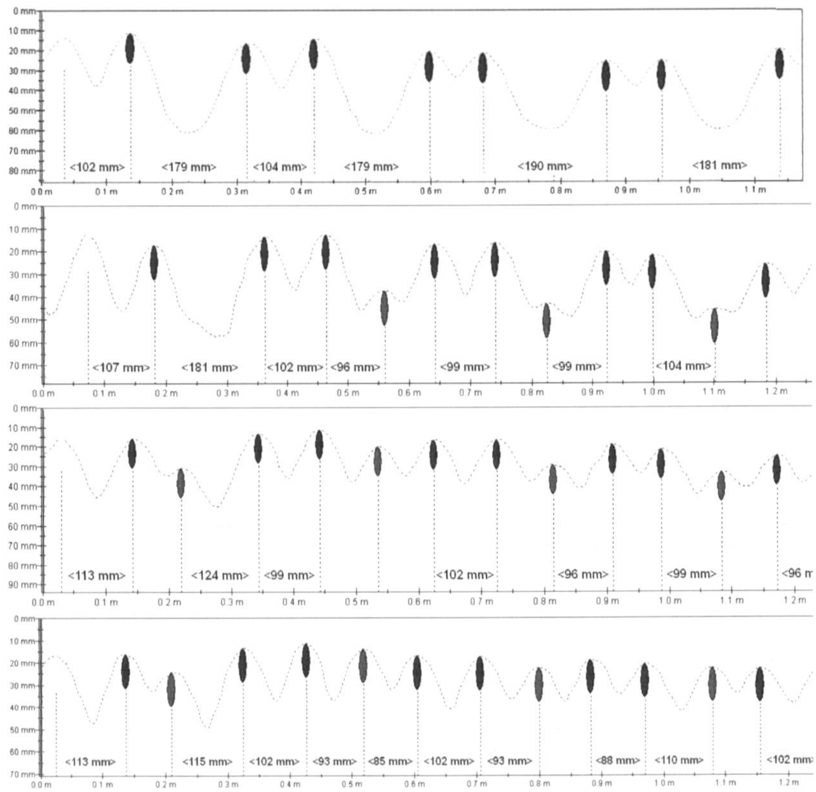

Figure 3. A series of line scans (top to bottom) at a distance of 0.12, 0.22, 0.29 and 0.32 m from the abutment – the bent shear reinforcement (red) is slowly rising to the surface to the same level as continuous reinforcement.

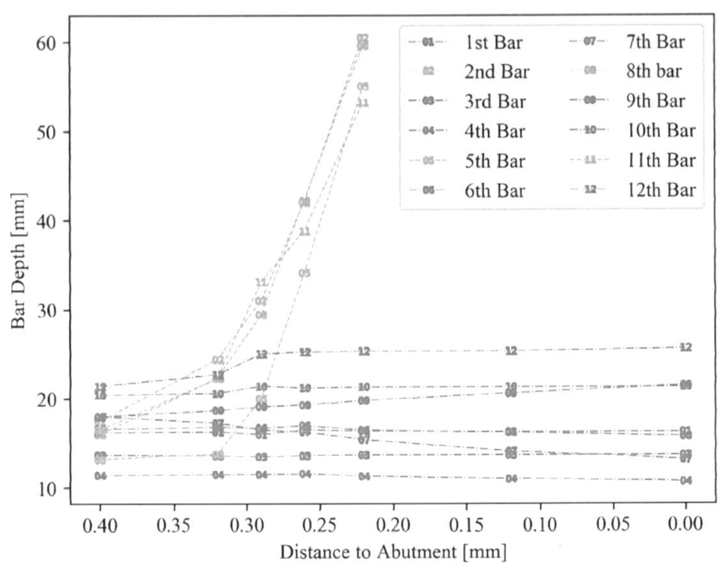

Figure 4. A plot of the cover depth for rebars no. 1 through 12 – rebars no. 2, 8 and 11 bend at a distance of 0.32 m, rebar no. 5 at 0.29 m from the abutment.

4. Conclusion

In conclusion, this particular example of detecting bends in the reinforcement of a bridge deck showed that the method and instrument are an ideal non-destructive choice. The distance of the bend from the abutment was measured with an accuracy of ± 10 mm, and was confirmed by a destructive probe. The method is very accurate for this purpose and can be effectively used in all cases where it is possible to clearly discern between the individual rebars in the slab. The Profometer PM-630 thus appears ideal not only for locating reinforcement, but also for ascertaining the location of bends. In terms of quality, it performs far better than previous models of cover meters.

Acknowledgement

This paper has been worked out under the specific research project No. FAST-S-19-6002 at Brno University of Technology.

References

[1] Malhotra V and Carino N 2004 Handbook on nondestructive testing of concrete (Boca Raton, Fla.: CRC Press)
[2] Bungey J, Millard S and Grantham M 2006 Testing of concrete in structures (New York: CRC Press)
[3] HILTI PS 200: PS 200 Ferroscan TechRentals, www.hilti.cz/c/CLS_MEA_TOOL_INSERT_7127/CLS_CONCRETE_SCANNERS_7127/r6436646
[4] Cikrle P and Anton O 2015 Development of nondestructive methods for testing of concrete since 1990 (original: Vývoj nedestruktivních metod pro zkoušení betonu od roku 1990) *Beton TKS* **3**(2015) pp 3-7 (in Czech)
[5] Gaydecki P, Silva I, Fernandes B and Yu Z 2000 A portable inductive scanning system for imaging steel-reinforcing bars embedded within concrete *Sensors and Actuators A: Physical* **84** pp 25-32
[6] Balayssac J and Garnier V 2018 Non-destructive testing and evaluation of civil engineering structures (Kidlington, Oxford, UK: Elsevier)
[7] Zhou F, Chen Z, Liu H, Cui J, Spencer B and Fang G 2018 Simultaneous Estimation of Rebar Diameter and Cover Thickness by a GPR-EMI Dual Sensor *Sensors* **18**(9): 2969
[8] Gregory J and Riley P 2003 Rebar locators *Structure magazine* **2003** pp 28-30
[9] Proceq: Profometer PM-600/PM-630 Operating instructions, https://www.proceq.com/compare/rebar-detection-and-cover-measurement
[10] Cikrle P 2017 Importance of non-destructive diagnostic methods for surveying steel- reinforced concrete structures – habilitation thesis (Brno: BUT) (in Czech)
[11] ČSN 73 2011:2012 Non-destructive testing of concrete structures (Prague: UNMZ) (in Czech)

Experimental measurement of the diameter and cover depth of steel reinforcement using an electromagnetic concrete cover meter

S Hüblová[1], P Cikrle[1], O Karel[1] and D Kocáb[1]

[1] Brno University of Technology, Faculty of Civil Engineering, Veveří 331/95, Brno, Czech Republic

E-mail: sabina.hublova@vutbr.cz

Abstract. In spite of the fact that concrete cover meters are some of the most commonly used non-destructive means of measuring reinforcement diameter, position, and cover, there are barely any specific measurement results available in published literature. This paper focuses on the use of a Profometer PM-630 concrete cover meter for in-situ measurements of such things as the cover depth of stirrups in the column of a high-rise building or determining the diameter of main reinforcement in hollow-core ceiling slabs.

1. Introduction

Unless reinforcement is completely exposed, it is difficult to assess its quality and type in an existing concrete structure. Where no technical documentation is available and destructive tests cannot be performed, non-destructive tests assume a prominent role in building diagnostics [1]. One of the most widespread non-destructive methods for locating steel reinforcement is the use of electromagnetic cover meters (rebar locators). However, literary sources provide almost no results or examples of measurement. The main reason for this may be the rather primitive imaging capability of existing cover meters. Therefore, measurement outcomes have thus far been presented mainly as photographs or lines drawn directly onto the surface of the structure being tested, or into a technical drawing of the building.

Progress in reinforcement visualization was made by the introduction of the Ferroscan by Hilti, which enabled the area scanning of reinforcement [2]. The Profometer PM-630 (or PM-650) from the company Proceq expanded the possibilities of the method's application including visual outputs of the measurement. It also greatly improved the accuracy of estimating the reinforcement diameter because it allows for correction in the distance of adjacent rebars, thus increasing the quality of the measurement outputs [3]. In the case of this instrument, the quality and accuracy of results is proportional to the qualification and experience of the operator. The history and development of these instruments is described e.g. in [1]. Besides electromagnetic cover meters, other non-destructive instruments that can locate the reinforcement in concrete is e.g. the PS 1000 radar manufactured by Hilti, which is described in more detail in [4,5]. In an ideal case, both instruments would be used at the same time (e.g. there was an experiment performed using a GPR-EMI prototype dual sensor [6]) and their results compared, as discussed in [7].

The examples discussed in this paper document innovation in the applicability of electromagnetic cover meters. The paper is focused on practical examples of using a Profometer PM-630 cover meter for determining the cover depth and diameter of steel reinforcement.

2. Method description and examples of measurement

Cover meters use the principle of eddy currents that occur in conductors placed in an electromagnetic field. Originally, the method exploited the magnetic properties of the material being detected. The Profometer PM-630 operates on the pulse induction principle, as described in [3]. The probe houses coils without a magnetic core, which means it is unaffected by magnetic fields in the surrounding area. Advanced signal processing allows locating a rebar as well as measuring of cover and rebar diameter.

Some of the strong points of the instrument include immediate visual output, high accuracy, the option to make corrections for the distance of adjacent reinforcement, and the considerable speed of measurement. Measurements can suffer e.g. by mis-detecting rebars that are too close, whether next to or behind one another. Another limiting factor can be the instrument's range (in normal probes about 60 – 100 mm, in depth probes approximately 150 mm to 220 mm). The measurements can also be impaired by strong electromagnetic fields near the structure being tested [8].

Prior to the actual measurement it is important to become well acquainted with the instrument and follow the manufacturer's instructions. It is recommended to test the instrument's sensitivity on a model element with a known position of reinforcement, especially if the rebars are close to each other. Where the concrete is not familiar to the user, it is also necessary to ascertain whether there is any magnetic aggregate present, as it could sway the results [9]. The principle and procedure of correct measurement are described by the manufacturer in [3]. This paper presents practical examples of using a Profometer PM-630. Specifically, it describes measurements of the coverage of stirrups in the column of a high-rise building in Bratislava and measurements of hollow-core ceiling slabs of a high-rise building in Brno.

2.1. Determining cover depth

The building in Bratislava was being surveyed for structural stability, which involved measuring the cover depth of stirrups in an 800 × 500 mm column. Figure 1 shows the column and the line scans of all of its four sides performed from the floor upwards.

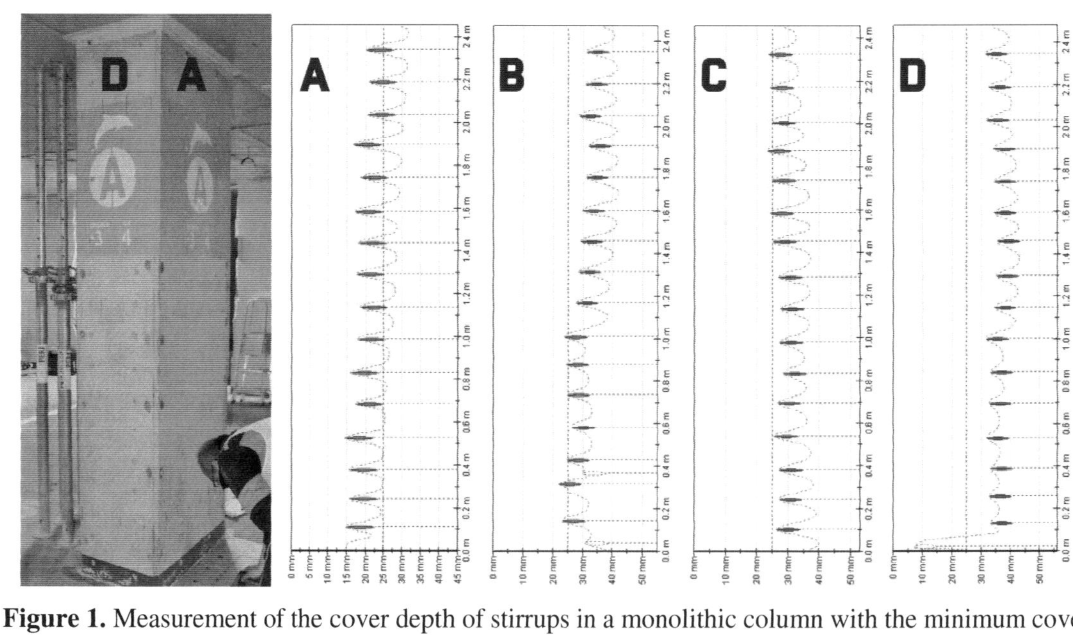

Figure 1. Measurement of the cover depth of stirrups in a monolithic column with the minimum cover depth set at cp = 25 mm – stirrups which are nearer to the surface are in red.

Another example discussed herein is the measurement of the main reinforcement in a column with a 500 × 800 mm cross-section, which is basically identical with that shown in figure 1. This measurement was made about 1.50 m above the floor. When measuring cover depth it is important to remember whether only one rebar is being measured or whether there are others nearby. If there is only one rebar, the results are clear. The operator only needs to enter the diameter and the instrument shows the cover depth. If there is interference from adjacent reinforcement (i.e. if the minimum distance between the main reinforcement bars is not kept), the instrument returns a depth value which is smaller than the real one. This is because more rebars at a greater depth behave the same as one rebar at a shallower depth. Therefore, when measuring a densely reinforced element, it is important to make calibration measurements on a model (followed by a correlation), which has the same reinforcement the building being diagnosed. However, the diameter is not always known and the building plans are not always available. If the diameter is incorrectly estimated, the cover value will be incorrect [9].

2.2. Determining reinforcement diameter

The determination of reinforcement diameter is affected by multiple parameters, some of which are more easily foreseeable (e.g. longitudinal or transverse rebar spacing), some less so (e.g. the effect of welded meshes, overlapping reinforcement, etc.). The diameter value may also appear higher due to the action of electromagnetic fields (crane rails, interconnected reinforcement in multi-storey buildings). This is why it is better to combine the method with other techniques, e.g. a destructive probe.

Modern instruments have rebar diameter probes, which make it easy to determine the diameter of the reinforcement, however, it is necessary for the individual bars to be far enough apart [9]. An example of determining reinforcement diameter is a measurement made on hollow-core ceiling slabs in a high-rise building at Šumavská street in Brno. According to available information, these were precast elements with 6 hollows, made in 1975, spanning 6.0 m in length and with a width of 1.2 m (see figure 2).

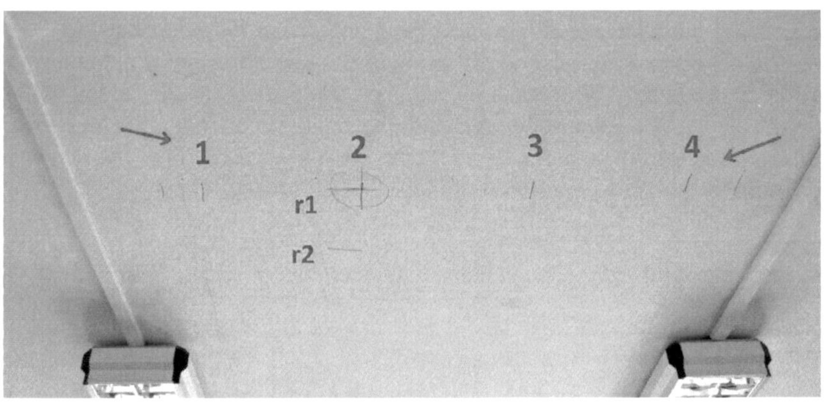

Figure 2. View of the bottom surface of a 1,20 m wide ceiling slab showing the position of the main reinforcement (1, 2, 3, 4) as well as distribution reinforcement (r1, r2) – the red arrows show the joint between the slabs, the unmarked reinforcement belongs to the adjacent slabs.

3. Results and discussion

3.1. Determining cover depth

Every line scan of the column made from the floor upwards (see section 2.1) shows 16 stirrups with an average spacing of 150 mm. The dotted lines show cover depth according to the building design documentation; cp = 25 mm. All stirrups which are closer than 25 mm from the surface are in red, stirrups located deeper are in blue (figure 1). The measurements show that while the cover depth of side A is clearly smaller than prescribed, the situation is reversed on the opposite side C. Values of deviation from the prescribed cover depth are plotted in Figure 3. It shows that the column's rebar cage was off-centre and closer to the side A.

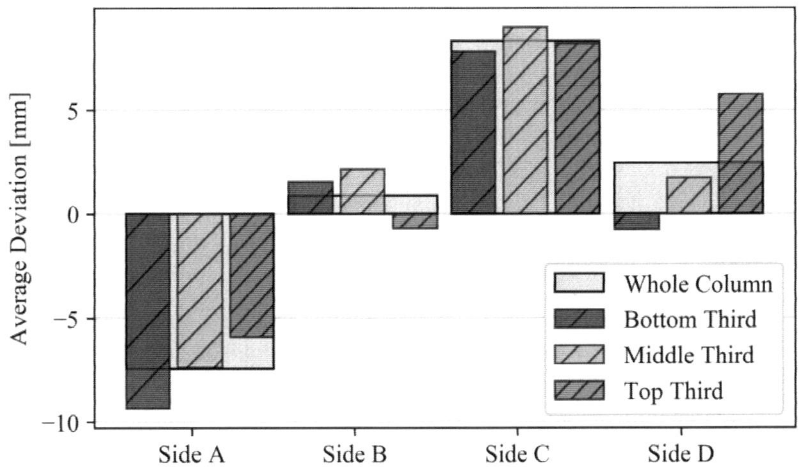

Figure 3. Plot of deviations from the prescribed stirrup cover of cp = 25 mm – the cover depth was not kept on side A (stirrups in the bottom part of the column were closest to the surface), the cover depth on the opposite side C was highest.

Figure 5 shows the results of main reinforcement measurements performed on all four sides of the column at a height of 1.5 m above the floor. An advantage of the Profometer PM-630 is its ability to record an uninterrupted linear scan of up to 15 m. Figure 4 therefore shows a continuous scan of the column's entire circumference. The scan shows that the main rebars nearer to the left edge of every face always had deeper cover as opposed to those on the right. The measured values were then used to create a cross-section through the column, see Figure 5. It clearly show how the rebar cage is rotated out of alignment with the column's intended outline.

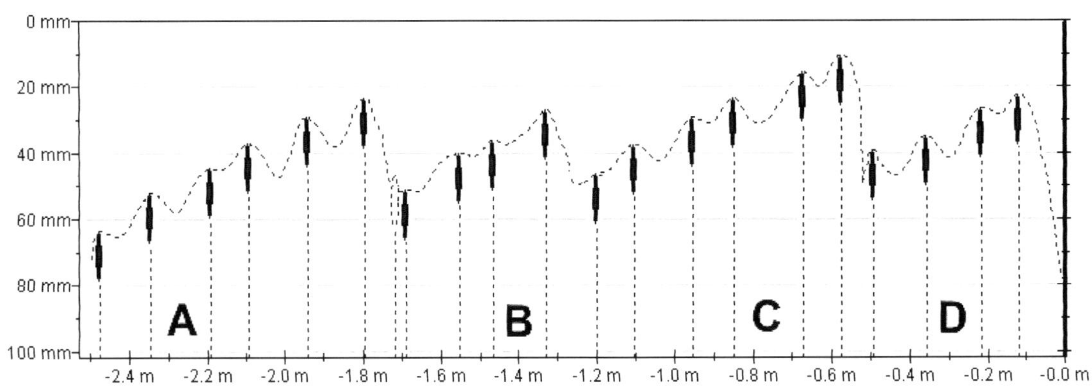

Figure 4. Continuous line scan of the main reinforcement around all four sides of the column – the cage's deformation is regularly repeated and ranges between 10 mm and 60 mm.

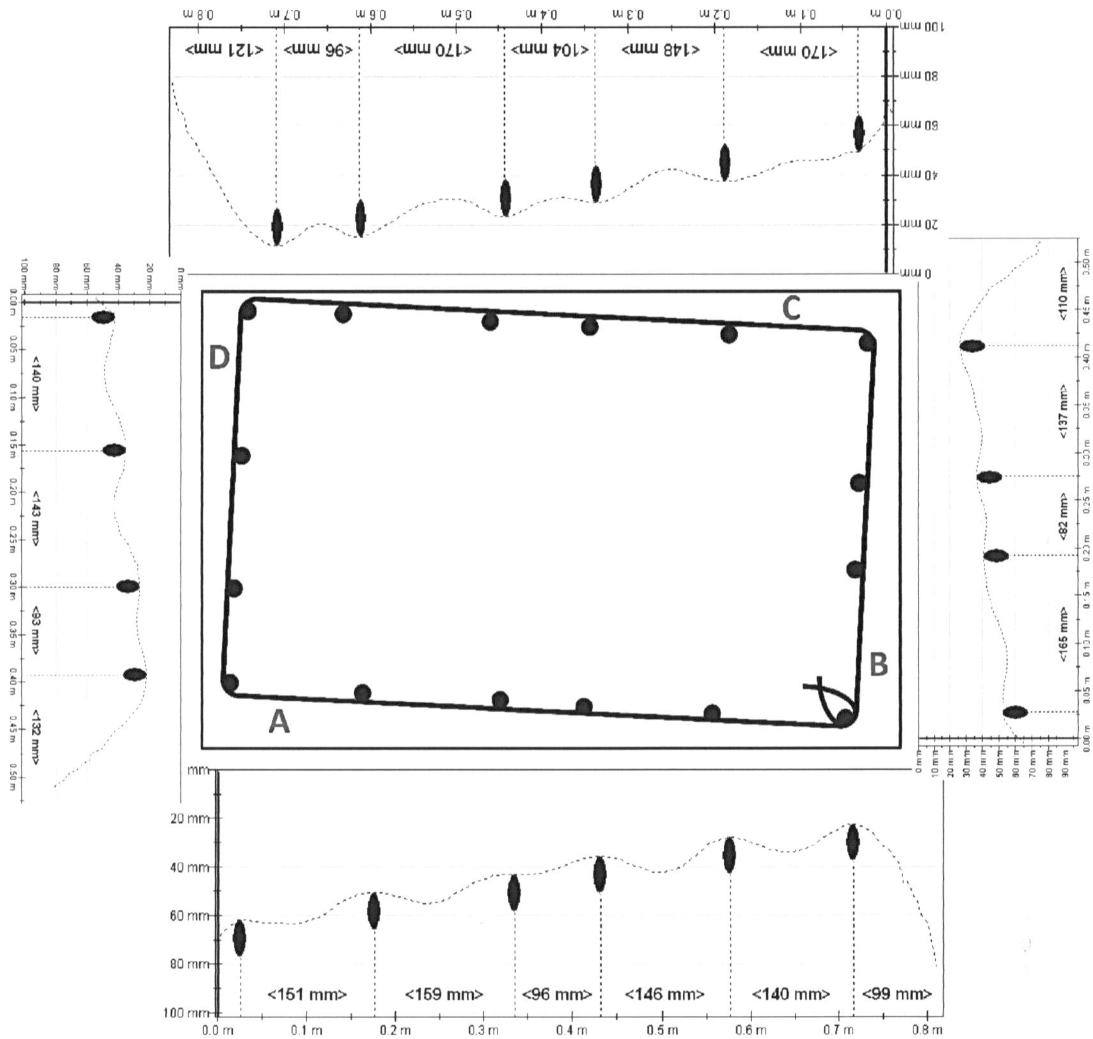

Figure 5. A series of line scans to determine the position of main reinforcement in the column of 500 × 800 mm performed in sequence on sides A, B, C, and D.

This example is only one of many measurements made on dozens of the columns in the building in Bratislava. It was intentionally picked to demonstrate the use of the Profometer PM-630 for measuring reinforcement cover depth. It is clear that this cover meter is very fast and accurate in determining whether the prescribed cover depth has been kept and allows for immediate visualisation and logging of the results.

3.2. Determining reinforcement diameter
Measurements of the hollow-core slab (see section 2.2) showed that there is only a minimum of distribution reinforcement – only two stirrups in the middle for the entire span, see figure 6, and then several stirrups near the ends where the slab is seated against the girder. Because the stirrups are far enough apart (approx. 200 mm), as are the bars of the main reinforcement (approx. 380 mm), the stirrup measurement could be done without correction for the proximity of adjacent reinforcement. The result is the diameter of 7 mm (see figure 6). The true diameter may vary only to a small degree; between 6 and 8 mm. Next, the profiles of the main reinforcement were measured, however, only on

two of the inner rebars, because the rebars near the edge were too close to those in adjacent slabs. The diameter of the inner rebars was measured at 18 to 19 mm; see figure 7.

Once the diameter is measured with a Profometer PM-630, it is then verified by a destructive probe where a small portion of the cover is chiselled away to reveal the actual reinforcement. In this case, however, the documentation contained all the necessary information – it was a ribbed reinforcement of the type 10425 (V), 4 × Ø 18 mm. The destructive probe was therefore not necessary, as there was no doubt as to the veracity of the measured data, not even in the case of stirrups. It should be noted, however, that in older buildings some of the planned reinforcement may not have been available in sufficient amounts and could have been replaced by other, whether in in terms of the number of rebars, or their type and diameter.

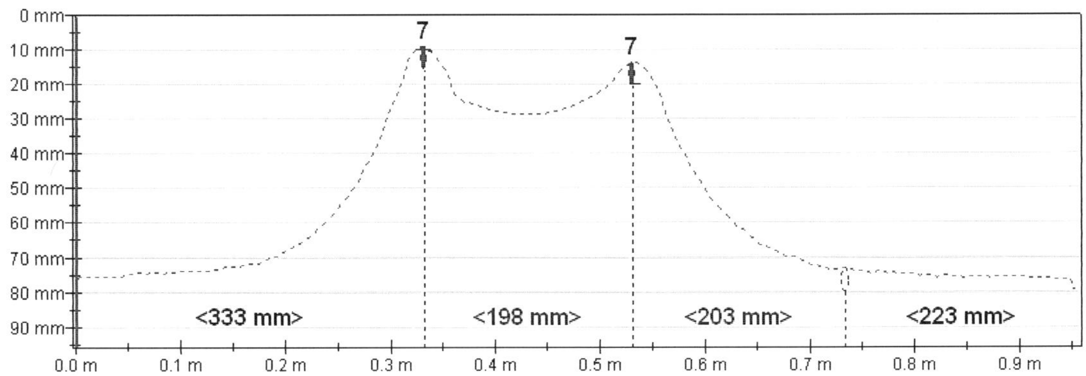

Figure 6. Measurements of the position, cover depth, and diameter of stirrups in the ceiling slab – diameter was determined at 7 mm, cover depth at 9 mm to 13 mm.

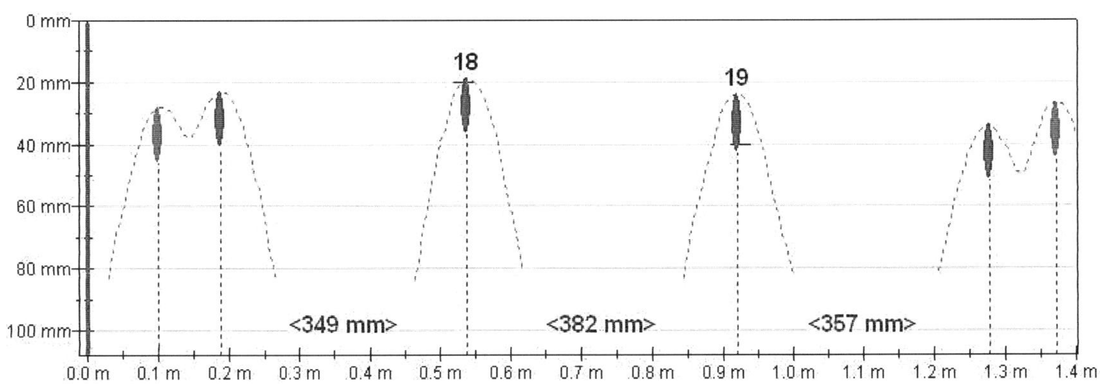

Figure 7. Measurements of the position, cover depth, and diameter of main reinforcement in the ceiling slab – the red interrupted line shows the edges, the diameter was measured at 18 mm and 19 mm.

4. Conclusion

The determination of the reinforcement diameter and depth is objectively a rather complicated task. However, nowadays there are highly sophisticated electromagnetic cover meters available, which can determine the diameter and cover depth of reinforcement non-destructively and with good accuracy. Still, it should be noted that the more advanced the instrument, the more qualified the operator, who therefore needs to have the appropriate training and experience.

Based on the experiments described here, it can be concluded that electromagnetic cover meters are well suited for determining the reinforcement diameter in hollow-core ceiling slabs. This is because such slabs meet basically all the requirements for making the measurements correctly; most notably because they possess the sufficient spacing in the main and distribution reinforcement. Because the building's project documentation was available and was therefore able to provide enough information about the reinforcement being measured, it was not necessary to verify the diameter destructively. The example of measuring reinforcement in concrete columns showed that the instrument is fast and accurate in revealing errors in the positioning of reinforcement cages inside formwork.

If one were to choose the most important task for the cover meter, it would likely be measuring cover depth. This is because of its accuracy, because most professional instruments can (as long as the conditions for accurate measurement are kept) determine cover depth of up to 50 mm with a ± 1 mm accuracy. The Profometer PM-630 is thus an ideal instrument for monitoring cover depth, but also for verifying reinforcement diameter providing all rebars can be satisfactorily identified, making it far better than any of the older types of cover meters.

Acknowledgements
This paper has been worked out under the specific research project No. FAST-S-19-6002 at Brno University of Technology.

References

[1] Cikrle P and Anton O 2015 Development of nondestructive methods for testing of concrete since 1990 *Beton TKS* **3**(2018) pp 3-7 (in Czech)
[2] HILTI PS 200: PS 200 Ferroscan TechRentals, www.hilti.cz/c/CLS_MEA_TOOL_INSERT_7127/CLS_CONCRETE_SCANNERS_7127/r6436646
[3] Proceq: Profometer PM-600/PM-630 Operating instructions, https://www.proceq.com/compare/rebar-detection-and-cover-measurement
[4] Balayssac J and Garnier V 2018 Non-destructive testing and evaluation of civil engineering structures (Kidlington, Oxford, UK: Elsevier)
[5] Gaydecki P, Silva I, Fernandes B and Yu Z 2000 A portable inductive scanning system for imaging steel-reinforcing bars embedded within concrete *Sensors and Actuators A: Physical* **84** pp 25-32
[6] Zhou F, Chen Z, Liu H, Cui J, Spencer B and Fang G 2018 Simultaneous Estimation of Rebar Diameter and Cover Thickness by a GPR-EMI Dual Sensor *Sensors* **18**(9): 2969
[7] Gregory J and Riley P 2003 Rebar locators *Structure magazine* **2003** pp 28-30
[8] Cikrle P 2017 Importance of non-destructive diagnostic methods for surveying steel- reinforced concrete structures – habilitation thesis (Brno: BUT) (in Czech)
[9] ČSN 73 2011:2012 Non-destructive testing of concrete structures (Prague: UNMZ) (in Czech)

Testing and mechanical properties of high strength concrete

O Sucharda[1], V Bilek[1] and P Mateckova[2]

[1] Department of Building Materials and Diagnostics of Structures, Faculty of Civil Engineering, VSB Technical University of Ostrava, Ludvíka Podéště 1875/17, 708 33, Ostrava-Poruba, Czech Republic
[2] Department of Structures, Faculty of Civil Engineering, VSB Technical University of Ostrava, Ludvíka Podéště 1875/17, 708 33, Ostrava-Poruba, Czech Republic

E-mail: oldrich.sucharda@vsb.cz

Abstract. The paper deals with mechanical properties of high-strength concrete, which are illustrated in various laboratory tests. The aim of the paper is to compare the possibilities of concrete compressive and tensile strength assessment and modulus of elasticity assessment. Results of nine type laboratory tests are presented in summary. Tested material properties are complemented with testing of small structural element, the beam without shear reinforcement. The behaviour of structural element is analysed also with a non-linear numerical model that respects the quasi-brittle properties of the concrete, where fracture-plastic model for concrete is used.

1. Introduction

The aim of current material research in the field of concrete structures is to develop concrete composition that is environmentally friendly using recycled materials or have improved performance properties [1], [2]. One of the most common improved mechanical property is compressive strength, whereby concrete with strengths above 50 MPa is referred as high strength concrete. In the case of improvements in other material properties such as durability, lifetime, workability, HPC concrete can be discussed. The HPC parameters for processing [3], mechanical properties [4], [5] or aspects of lifetime [6] cannot be used as for conventional concrete. Improved mechanical properties of concrete can also be utilised in sophisticated design methods of building construction, which, however, require detailed knowledge of mechanical parameters and quasi-brittle behaviour [7], [8]. Common testing of mechanical properties, where basic compressive strength tests and the modulus of elasticity tests are performed, are insufficient. There are also different values of results depending on the type of test. This knowledge about material properties is of limited value and insufficient for nonlinear computer simulations of structural stress-strain state. Essential mechanical properties for nonlinear analysis is especially tensile strength, softening functions are taken usually according to the recommendation [9]. Testing of mentioned material characteristics is technically difficult and there is a large dispersion of measured values. Fracture mechanics provides a detailed description of crack damage and propagation in quasi-brittle composites. The research is often focused on fracture mechanics area [10], [11]. The use of fracture mechanics high strength and high performance concrete is limited with the availability of suitable fracture mechanical characteristics of materials. Fracture mechanical parameters are influenced by number of factors, such as concrete recipe, manufacturing and testing method. Certain characteristics are available in databases and recommendations for common concrete, however often has a number of limitations (scope of research, size or geometry of samples) and are insufficient for non-linear analysis of structural elements.

In current research great attention is paid to the testing of mechanical properties, where a number of recommendations and approaches can be found in the Model Code 2010 [12]. It is also important to note that concrete is often combined with fibres. In this case the determination of mechanical properties becomes a complex task. Mechanical properties should be determined using a comprehensive set of tests and inverse analysis is also used.

Shear failure can be considered as specific problem of concrete structures. The scope of the issue is very large [13], [14], [15], where the mechanism of collapse is affected by concrete, the shape of the cross-section, the position of the load and the reinforcement. The development and testing of the mechanical properties of new types of concrete is only one part of the innovative structural design and complex nonlinear task. The most important recommendations for advanced design include the Model Code 2010 [12], which outlines the basic principles of using advanced numerical models and nonlinear analysis. The material models respect real behaviour and the possibility of disruption in pressure and tension. Nonlinear numerical analysis is used especially in specific problems [16] and in design optimization [17], [18].

2. Experimental details and materials

The objective of the presented research is detailed description of mechanical properties of particular HPC. The composition of analysed HPC is listed in Table 1.

Table 1. Composition of HPC [kg/m^3].

CEM I 42,5 R (Mokrá)	575
Metakaoline (Metaver I)	80
PCE superplasticizer (ACE Glenium 300)	20
Water	155
Sand 0/4 (Lípa)	680
Crushed aggregates (4/8 Litice)	200
Crushed aggregates (8/16 Litice)	810

Figure 1. HPC concrete structural element – RC beam without shear reinforcement.

The experimental program included testing of common laboratory samples and testing of a structural element. Structural element is shown in Figure 1. Laboratory tests includ a comprehensive set of experiments to determine compressive strength, tensile strength, and modulus of elasticity. Tests are carried out in a variant mode (cube/cylinder/beam). The tests are schematically shown in Figure 2 and Figure 3. Compressive strength was assessed in a variant for cubes, beams and cylinders. Both static and dynamic modulus of elasticity was tested. The tensile strength is then determined on the basis of the split tensile test and three and four-point bending test. Testing was based on the Model Code 2010 recommendations.

For the split-tensile test specimens of cube or cylinder shape can be used. With regard to laboratory testing capacity in the presented research split tensile strength was performed on cubes. All bending tests were performed on beams with cross-section 150x150 mm and length of 700 mm. The three-point bending test has a span of 600 mm. In the four-point bending test the distance between the supports and the load is 200 mm.

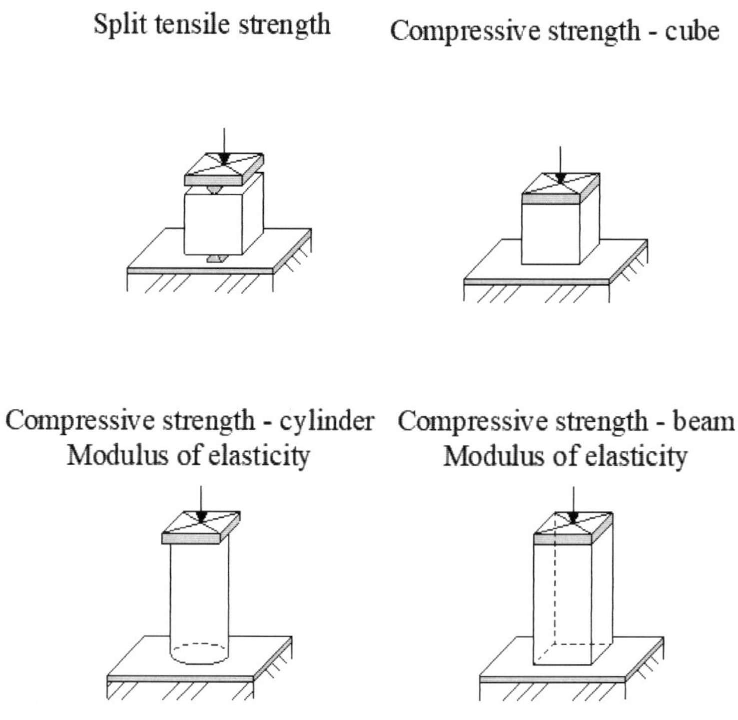

Figure 2. Laboratory set of tests – strength and modulus of elasticity.

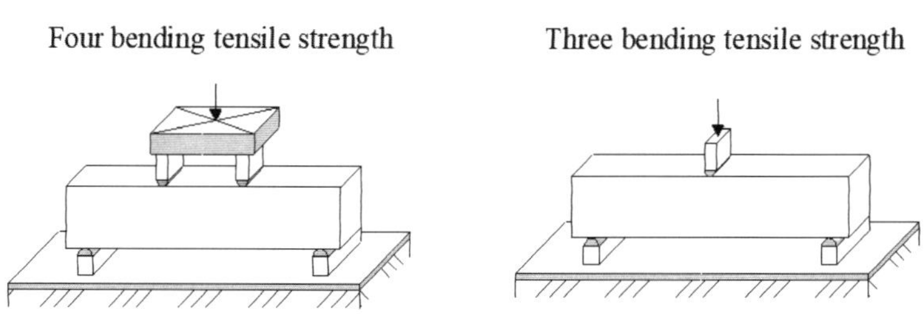

Figure 3. Laboratory set of tests – bending tests.

3. Results and material properties of HPC

Laboratory test results are summarized in Table 2. The information in Table 2 is supplemented with graph in Figure 4. The resulting ratio of compressive strength for cube and cylinder specimen is 0.76. The compressive strength difference for cylinder and beam specimen is 15.1 MPa. The modulus of elasticity is similar to value given in Model Code 2010 for particular compressive strength. Dynamic modulus of elasticity is larger than static according to the assumption.

Table 2. Mechanical properties of HPC.

Compressive strength - cube	102.0	MPa
Split tension strength	5.8	MPa
Compressive strength - cylinder	77.6	MPa
Modulus of elasticity - static - cylinder	42.1	GPa
Modulus of elasticity - dynamic - cylinder	52.1	GPa
Compressive strength – beam	92.7	GPa
Modulus of elasticity - static - beam	39.5	GPa
Modulus of elasticity - dynamic - beam	55.0	GPa
Bending tensile strength – 3-point	7.25	MPa
Bending tensile strength – 4-point	7.04	MPa

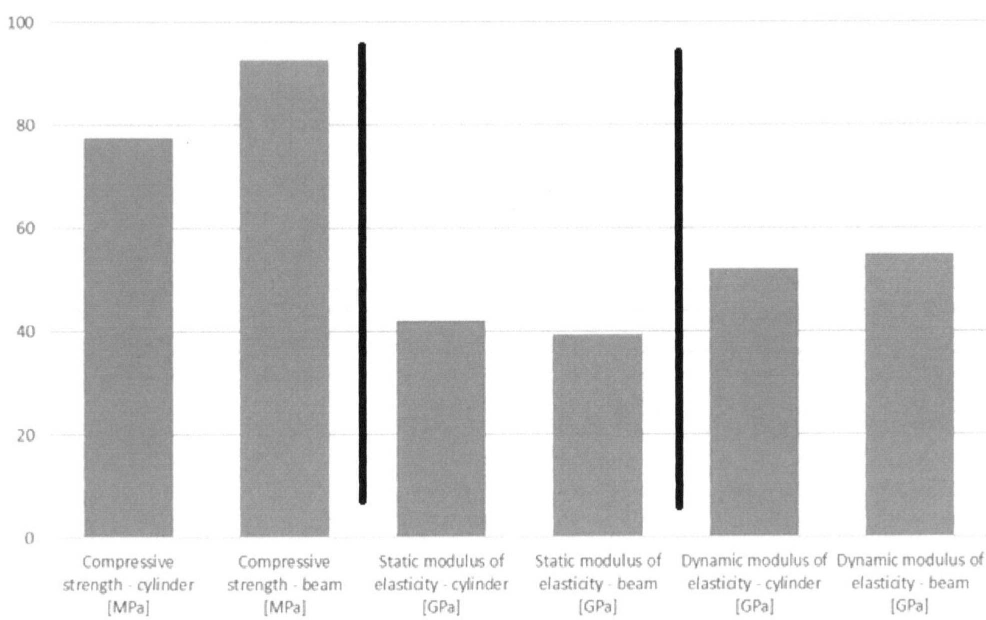

Figure 4. Comparison of results of laboratory tests of modulus of elasticity and strength for cylinders/beams.

4. Non-Linear Analysis and Beam without Shear Reinforcement

The laboratory program also included testing of a small beam without shear reinforcement. The beam is designed to verify mechanical properties based on laboratory using non-linear analysis. The beam is of rectangular cross-section with dimensions 100 x 190 mm, where the span between supports is 900 mm. The three point bending test scheme is shown in Figure 5. The beam was reinforced with three B500B bars of 10 mm diameter on the bottom surface, the reinforcement cover was 20 mm. The collapse of the RC beam itself was caused by the formation of shear cracks. The maximum load capacity was 67 kN, the shear crack is shown in Figure 6. Photo of texture concrete shown in Figure 7.

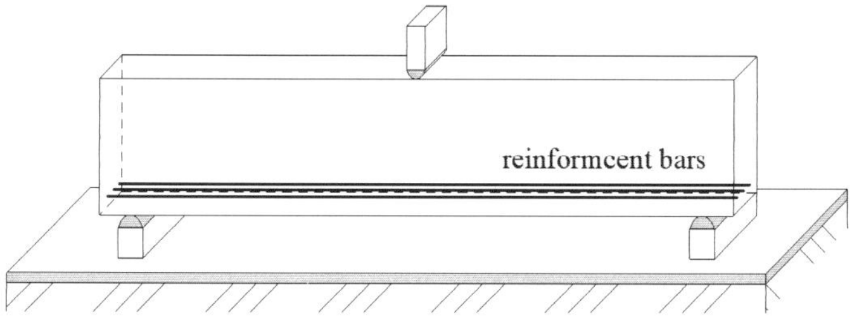

Figure 5. RC beam without shear reinforcement – span 900 mm.

For numerical analysis of the structural element 2D nonlinear model was used with concrete – fracture - plastic material model in ATENA [19]. The load was applied with deformation. Fracture energy was determined according to the ATENA and Model Code 2010 recommendations. The split tensile strength value for calculation fracture energy is used, which has been modified by however the coefficient 0.7. The calculation respects also further the recommendations [18] for nonlinear analysis. Comparison of calculation and experiment proved that the collapse mechanism is similar in both cases. The shear crack in the computational model is shown in Figure 8. The maximum load achieved in the calculation is 67 kN. The experiment maximum load capacity was also 67 kN, so the difference of results for the numerical model and experiment is to 1 %, only tenths of kN.

Figure 6. Shear crack of beam.

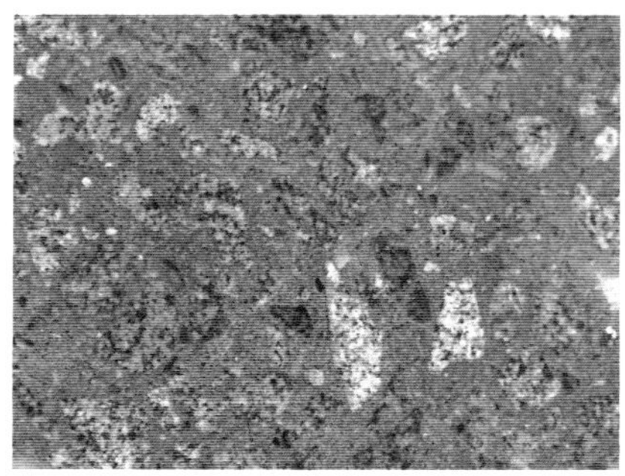

Figure 7. Photo of texture concrete.

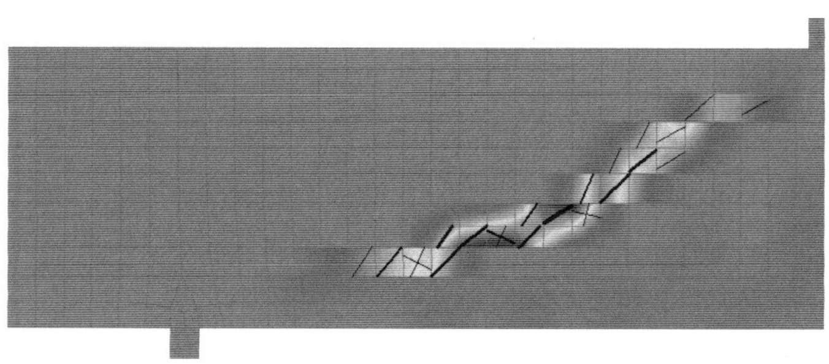

Figure 8. 2D Computer model with crack.

5. Conclusion

The paper deals with mechanical properties of particular HPC, which are assessed on a number of laboratory tests and verified on a small structural element. The results of compressive strength tests for cubes and cylinders have a 0.76 ratio. This is smaller ratio factor than the common concrete value. The modulus of elasticity is also sensitive area of mechanical properties. The resulting value is similar to a value given in Model Code 2010. The ratio of static/dynamic modulus of elasticity ratio is 0.8. The number of samples does not allow wider statistic assessment. Nonlinear analysis results with input data based on laboratory tests well captured the type of collapse and formation of shear crack and maximum load capacity in three point testing of small structural element. However, can be advisable to use for more complex beams and structures 3D model and special determination of fracture-mechanical parameters. The authors will focus on this area in further research.

Acknowledgment

The work was supported by the means of conceptual development of science, research and innovation assigned to VŠB-TUO by the Ministry of Education, Youth and Sports of the Czech Republic.

References

[1] Aïtcin P C and S Mindess 2011 *Sustainability of Concrete* (New York: Spon Press)
[2] Aïtcin P C and R J Flatt 2015 *Science and Technology of Concrete Admixtures* (Sawston, Cambridge : Woodhead Publishing)
[3] Piasta W, Budzyński W and J Góra 2015 The effect of selected aggregates on the properties of high performance concrete [Wpływ wybranych kruszyw na właściwości betonu wysokowartościowego] *Cement, Wapno, Beton*, 2015 (**3**), pp 171-178.
[4] Sucharda O, Pajak M, Ponikiewski T and P Konecny 2017 Identification of mechanical and fracture properties of self-compacting concrete beams with different types of steel fibres using inverse analysis *Construction and Building Materials* vol. **138**, pp. 263-275.
[5] Nitesh, K J N S , Rao, S.V., and P R Kumar 2019 An experimental investigation on torsional behaviour of recycled aggregate based steel fiber reinforced self compacting concrete *Journal of Building Engineering* vol **22** pp 242-251. DOI: 10.1016/j.jobe.2018.12.011
[6] Lehner P, Ghosh P and P Konečný 2018 Statistical analysis of time dependent variation of diffusion coefficient for various binary and ternary based concrete mixtures *Construction and Building Materials*, vol **183**, pp 75-87. DOI: 10.1016/j.conbuildmat.2018.06.048
[7] Bažant ZP and J Planas 1998 *Fracture and size effect in concrete and other quasibrittle materials* (Boca Raton: CRC Press)
[8] Bažant ZP and E Becq-Giraudon 2002 Statistical prediction of fracture parameters of concrete and implications for choice of testing standard *Cem Concr Res* vol **32** (4) pp 529–556
[9] Hoover C.G. and ZP Bažant 2014 Cohesive crack, size effect, crack band and work-of-fracture models compared to comprehensive concrete fracture tests *International Journal of Fracture* vol **187** (1) pp 133-143
[10] Lehky D, Kersner Z and D Novak 2013 FraMePID-3PB software for material parameter dentification using fracture tests and inverse analysis *Adv. Eng. Software* vol **72**. pp 147–154
[11] Strauss A, Zimmermann T, Lehký D, Novák D and Z Keršner 2014 Stochastic fracture mechanical parameters for the performance-based design of concrete structures *Structural Concrete* vol **15** (3) pp 380-394.
[12] Model Code 2010 (2012) - Final draft, fib, Bulletin no. 65 and 66, 1-2
[13] Special Issue on Shear, *Structural Concrete*, 1/2018, vol **19** iss. 1, pp. 1-328, february 2018, doi.org/10.1002/suco.201870014
[14] Augustín T, Fillo L, Halvonik J and M Marčiš 2018 Punching resistance of flat slabs with openings -experimental investigation *Solid State Phenomena* vol **272** pp 41-46. DOI: 10.4028/www.scientific.net/SSP.272.41

[15] Vida R and J Halvonik 2017 Shear assessment of concrete bridge deck slabs, *Key Engineering Materials* vol **738**, pp 110-119 DOI: 10.4028/www.scientific.net/KEM.738.110
[16] Králik J 2016 Nonlinear analysis of NPP safety against the aircraft attack *AIP Conference Proceedings*, **1738**, art no 480079, DOI: 10.1063/1.4952315
[17] Wieczorek B 2017 Numerical Analysis of the Inner Slab-Column Connection of RC Structures Loaded Eccentrically Procedia Engineering vol **190**, pp 668-675. DOI: 10.1016/j.proeng.2017.05.395
[18] Sucharda O and P Konecny 2018 Recommendation for the modelling of 3D non-linear analysis of RC beam tests *Computers and Concrete. (Techno Press)* vol **21** (1), pp 11-20. 2018. DOI: 10.12989/cac.2018.21.1.011
[19] Červenka V, Červenka J and R Pukl 2002 ATENA - A tool for engineering analysis of fracture in concrete *Sadhana - Academy Proceedings in Engineering Sciences*, **27**, pp 485-492

IOP Conf. Series: Materials Science and Engineering **549** (2019) 012013 doi:10.1088/1757-899X/549/1/012013

Measurement of acoustic emission at fibre reinforcement concrete slab with subsoil

O Sucharda[1], L Pazdera[2], Z Marcalikova[3], L Topolar[2], M Kozielova[3] and V Bilek[1]

[1] Department of Building Materials and Diagnostics of Structures, VSB-Technical University of Ostrava, Ludvíka Podéště 1875/17, 708 33 Ostrava-Poruba, Czech Republic
[2] Institute of Physics, Faculty of Civil Engineering, Technical University of Brno, Veveří 331/95, 602 00 Brno, Czech Republic
[3] Department of Structures, Faculty of Civil Engineering, VSB-Technical University of Ostrava, Ludvíka Podéště 1875/17, 708 33 Ostrava-Poruba, Czech Republic

E-mail: oldrich.sucharda@vsb.cz

Abstract. The paper deals with the possibility of using the acoustic emission (AE) measurement for the identification of cracks and failure in fibre concrete slabs in interaction with subsoil. In this specific task it is not possible to check visually cracking on the slab bottom surface. The research task included an experimental load test of a slab and supplemented laboratory testing to determine the mechanical properties of fibre concrete. The concrete used has a dosage of 25 kg/m^3 fibres. Laboratory tests included cube and cylindrical compressive strength, flexural strength and split tensile strength. Homogeneity of the concrete mixture, i.e. uniform distribution of the fibres, was proven with testing of the split tensile strength both evenly and perpendicularly to the direction of filling the concrete mixture into the form. Experimental load test of slab on subsoil was performed on fibre concrete slab with plan dimensions 2000 x 2000 mm and thickness 150 mm. Concrete slab was loaded in the middle with force load step 75 kN. AE measurements were provided using track sensors.

1. Introduction

Diagnostics of buildings and identification of failure are important areas of research, especially in case of quasi-brittle materials [1, 2], where the diagnostic methods can be combined with advanced structural analysis. Using new materials and advanced structural design structures are more subtle and the intensity of internal forces is closer to material failure. New mixtures of concrete and composites as quasi-brittle materials may have different behavior or character of failure in the structure then common concrete [3, 4]. The group of quasi-brittle composites includes also fibre concrete [5-7]. Fibre concrete exists in a number of variations with a variety of mechanical properties [8-12].

Great attention is paid to concrete structures [13-16] and structural health monitoring using non-destructive methods. These include ultrasound, computed tomography or acoustic emission (AE) measurements since the 1970s [17, 18]. The advantages of AE method are that, depending on the measurement configuration, it enables identification of the occurrence of the disorder, its quantification and its continuous monitoring [19-21]. A number of recommendations and standards can be found [22-29]. It should be noted, however, that in view of the universality of the measurement method and the very principle of acoustic emission, it is always necessary to find an appropriate method of evaluation that takes into account specific boundary conditions of the solved task [30-34].

In the paper slab in interaction with the subsoil [35-36] is analyzed. With this structural type it is not possible visually identify the crack occurrence on the bottom surface. The crack formation itself does not necessarily endanger the safety of the construction, but it will affect the structural service life. Especially in the case of complicated geological conditions (groundwater level, aggressiveness of the environment), accelerated degradation of materials can occur at the point of crack formation.

The very possibility of identifying cracking in concrete and fiber reinforcement concrete in a solved task is illustrated on a set of experiments [37]. Taking into account the measurements so far and the number of tests, the present research has focused on further verifying possibility used acoustic emission and the possibility / form of presentation of measurement results.

2. Acoustic emission

Acoustic emission is a non-destructive testing method (NDT). It belongs to passive methods and allows not only identification and localization of crack formation. The acoustic emission itself can be interpreted by means of a number of parameters. They are AE counts, duration, amplitude, rise time, and the measured area under the corrected signal envelope that is also called relative energy. Basic parameters of acoustic emission are the change of ring down counting which can be recorded:

$$\frac{dn}{dt} = \alpha \times \Delta K_{\mathrm{I}}^{\mathrm{m}} \tag{1}$$

where ΔK_{I} is the change of the stress intensity, α and m constitute material constants. The AE measurements made within load test of slab were provided with acoustics sensors type MTPA-15 with an integrated 35 dB pre-amplifier (type) attached to the surface of the sample with beeswax. DAKEL-XEDO© universal measuring and diagnostic system developed by ZD Rpety-Dakel was used for the processing and recording of AE signals and the XEDO-AE unit was used for the assessment of acoustic emission parameters.

3. Laboratory tests

Knowledge of the mechanical properties of the fibre concrete is necessary for a wider evaluation and analysis of the concrete slab testing. Testing of mechanical properties was conducted in the laboratories of the Faculty of Civil Engineering, VŠB-TU Ostrava. The used concrete had a dosage of 25 kg/m^3 of Dramix® 3D 65/60 BG fibre. Basic properties of fibres are shown in Table 1 and the fibres shape is in Figure 1.

Table 1. Basic characteristics of the Dramix® 3D 65/60 BG fibre.

Fibre shape	curved ends
Length [mm]	60
Diameter [mm]	0.9
Aspect ratio	67
Tensile strength [N/mm^2]	1160
Effect on consistency [s]	8
Effect on concrete strength [kg/m^3]	15
Elastic modulus [GPa]	200

Figure 1. Fibres Dramix® 3D 65/60 BG.

The concrete recipe is presented in Table 2. Concrete compressive strength was tested on 6 cubes and 6 cylinders. Another sample series of cubes 150 × 150 × 150 mm was tested for split tensile strength. There were 12 cube samples in total, 6 samples were used to determine the split tensile strength perpendicular to the filling direction and 6 samples to determine the split tensile strength parallel to the filling direction. This way also homogeneity of the fibres in the concrete mixture was verified. Another series of samples were beams with dimension 150 × 150 × 600 mm, for determining the flexural strength. Two types of tests were performed, i.e. a three-point bending tensile test with a force in the middle of the beam and with a span between the supports 500 mm and a four-point bending test with forces 175 mm from support, where the span between supports was 500 mm. Test results are shown in Table 3.

Table 2. Material composition of concrete.

Description	Type/Quantity
Consistency	S3
Cement	CEM I 42.5 R
Minimal cement content	300 kg
Water cement ratio	0.6
Aggregates 0/4 (mined)	870 kg
Aggregates 4/8 (mined)	150 kg
Aggregates 8/16 (mined)	820 kg
Water	189 l
Plasticizer	2.9 l (Stacheplast)

Table 3. Fibre concrete strengths determined from laboratory tests.

Type of test	Number of samples	Average [MPa]	Standard deviation [MPa]
Compressive strength (Cube 150 × 150 × 150 mm)	6	26.49	1.292
Compressive strength (Cylindrical 150 × 300 mm)	6	21.88	0.692
Split tensile strength - perpendicular to filling direction	6	2.26	0.158
Split tensile strength - parallel to the filling direction	6	1.73	0.237
Three-point bending tensile test	2	4.26	-
Four-point bending tensile test	1	3.99	-

4. Experimental measurement on slab on subsoil

A special testing device for testing concrete slabs in interaction with the subsoil (Figure 2), is located in the campus of the Faculty of Civil Engineering.

IOP Conf. Series: Materials Science and Engineering **549** (2019) 012013 doi:10.1088/1757-899X/549/1/012013

Figure 2. Photo of FRC slab with sensors.

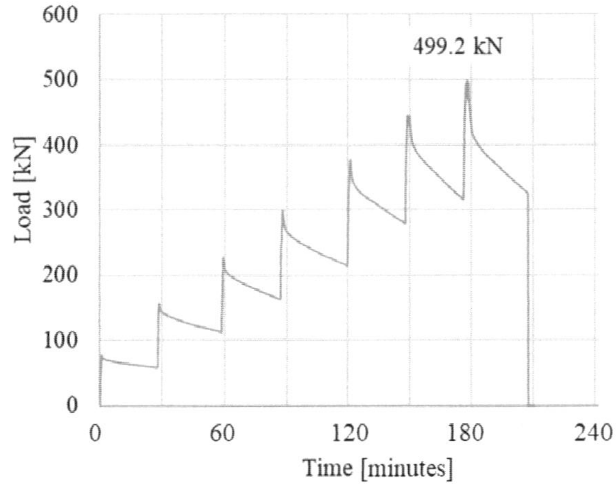

Figure 3. Load - time diagram.

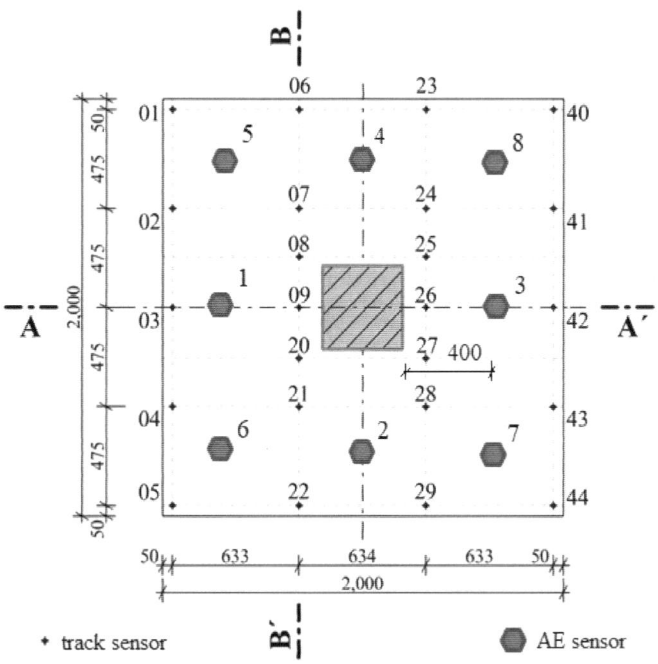

Figure 4. Schematic diagram of the track and sensors position.

Tested concrete slab was with plan dimensions 2000 × 2000 mm with thickness 150 mm. Load was applied in steps of 75 kN until it was completely damaged, i.e. until slab bearing capacity was reached. Time step of each load step was 30 min.

Figure 3 shows the load - time diagram of the concrete slab test. The maximum achieved load value was 499.2 kN, when the slab was damaged. Measurement was terminated after significant pressure drop in the hydraulic press and the slab was pressed into the subsoil. The deformations were measured by the net of 24 track sensors - potentiometers. 8 AE sensors were installed on the slab. The scheme diagram of the track sensors and AE sensors position is in the Figure 4. The maximum deformation 26.72 mm was measured in section A-A´, corresponding to middle part of slab pushed into subsoil. The maximum negative deformation -13.06 mm was measured at lifted up corners. Selected deformations are shown in Figure 5. Description the *x*-axis is the coordinates of plane in the FRC slab.

The $x = 0$ coordinate is the centre of the slab. Coordinates y-axis are the vertical deflection of the slab. Based on polynomial approximation maximum slab deformation 29.65 mm was calculated.

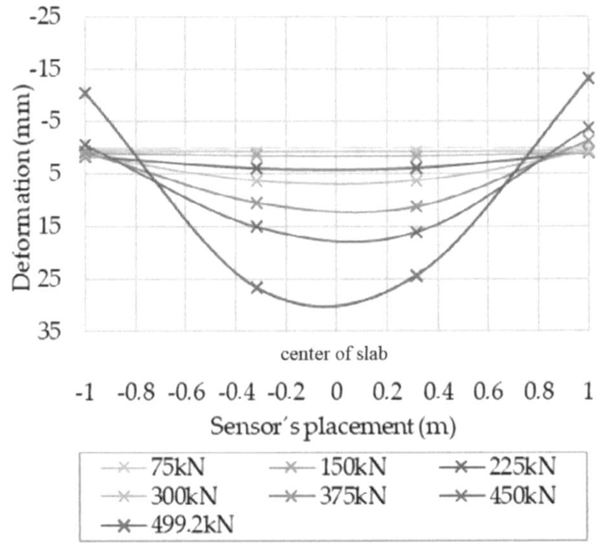

Figure 5. Deformation corresponding to track sensors – section A-A´.

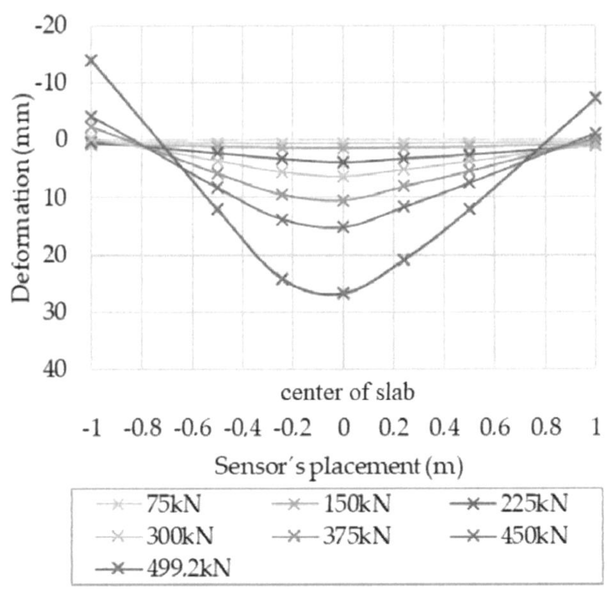

Figure 6. Deformation corresponding to track sensors – section B-B´.

For section B-B´ the maximum deformation 26.72 mm was measured, corresponding to middle part of slab pushed into subsoil and the maximum negative deformation -13.93 mm was measured, corresponding to slab border part with lifted up corners. The deformation sections are shown in Figure 6. The course of the test and the deformation evaluation shows that deformations up to the load 225 kN are very small. The size of acoustic emission parameters measured is also very low. For increasing load, the deformation increases more significantly and the size of the acoustic emission parameters also increases. At a load of 450 kN, the deformation in the middle of the slab is around 15 mm. The slab is damaged with cracks but is still entire. With the next load step, which ends with 499.2 kN, the deformation grows very fast and cracks are very distinct. This is evident also in the measurement of acoustic emission. Failure of the fibre concrete slab is shown in the Figure 7.

Figure 7. Failure of the fibre concrete slab with sensors.

The details of the AE measurements are in the Figure 8. The first higher AE values are visible for the load value 150 kN. For the load of 75 kN the AE count can be considered as a measurement noise. The highest AE intensity is for the load 375 kN and 450 kN for sensor 3, 4, 6, 8. This is consistent with deformation measurements where rapid growth occurs at the same load step. As for the sensor 7, the AE values are very high during the whole experiment and the values from this sensor are not taken into consideration.

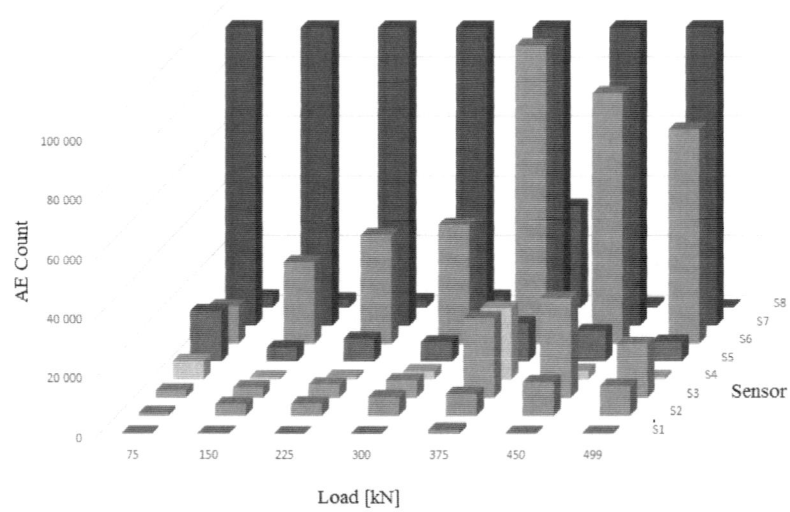

Figure 8. Acoustic emission measurement - AE count / Sensors / Load.

5. Conclusion

The paper was focused on the possibility of measuring acoustic emission in slab interacting with the subsoil. Acoustic emission measurement proved that it is possible to identify the crack occurrence in fiber concrete slab. Acoustic emission measurements have also confirmed the results of previous research in the project of focus on the possibility of identification crack in concrete [37]. The first cracks were formed on the bottom surface of the slab, which is in contact with the subsoil. Final break of tested slab and the exhausting of the load capacity is also very obvious in the AE graphs. Size of acoustic emission and the growth of the deformations is similar. Acoustic emission measurements contributed to a better understanding of damage mechanism and slab structure collapse and can be used to identify cracking. Wider range of conclusion requires more types of tested slabs and wider material variations. This is the aim of further research together with using AE measurement for crack localization.

Acknowledgments

The work was supported by the means of conceptual development of science, research and innovation assigned to VŠB-TUO by the Ministry of Education, Youth and Sports of the Czech Republic.

References

[1] Aroch R, Sokol M. and Venglar M 2016 Structural Health Monitoring of Major Danube Bridges in Bratislava *Procedia Engineering* vol 156 pp 24-31 DOI: 10.1016/j.proeng.2016.08.263

[2] Erdelyi J, Kopacik A, Kyrinovic P, Sokol M and Venglar M 2017 Long-term monitoring of the slovak national uprising bridge *International Multidisciplinary Scientific GeoConference Surveying Geology and Mining Ecology Management* SGEM vol 17 iss 22 pp 389-396 DOI: 10.5593/sgem2017/22/S09.049

[3] RILEM 2018 About Rilem [Online] url: http://www.rilem.net [Accessed on 4 May 2018]

[4] Karihaloo B 1995 *Fracture Mechanics of Concrete* Longman Scientific & Technical (New York)

[5] Brandt A M 2008 Fiber reinforced cement-based (FRC) composites after over 40 years of development in building and civil engineering *Composite Structures* vol 86 pp 3-9 DOI: 10.1016/j.compstruct.2008.03.006

[6] Koniki S and Ravi P D 2018 A study on mechanical properties and stress-strain response of high strength concrete reinforced with polypropylene–polyester hybrid fibres *Cement, Wapno, Beton* iss 1 pp 67-77

[7] Sucharda O, Bilek V, Smirakova M, Kubosek J and Cajka R 2017 Comparative Evaluation of Mechanical Properties of Fibre Reinforced Concrete and Approach to Modelling of Bearing Capacity Ground Slab *Periodica Polytechnica Civil Engineering* vol 61 iss 4 DOI: 10.3311/PPci.10688

[8] Zaborski A 2016 Constitutive model for restricted compression of fiber concrete *Cement, Wapno, Beton* iss 1 pp 46-52

[9] Sucharda O and Bilek V Aspects of Testing and Material Properties of Fiber Concrete *25th Concrete Days 2018* pp 1-6

[10] Kormanikova E and Kotrasova K 2011 Elastic mechanical properties of fiber reinforced composite materials *Chemicke Listy* vol 105 iss 17

[11] Sucharda O, Lehner P, Konecny P and Ponikiewski T 2018 Investigation of Fracture Properties by Inverse Analysis on Selected SCC Concrete Beams with Different Amount of Fibres In *Ecf22 - Loading And Environmental Effects On Structural Integrity, Procedia Structural Integrity* vol 13 pp 1533-1538 DOI: 10.1016/j.prostr.2018.12.313

[12] Sucharda O, Konecny P, Kubosek. Done P. J. 2015 Finite element modelling and identification of the material properties of fibre concrete In *Conference: 23rd Conference of Italian Group of Fracture* Location: Sicily *Procedia Engineering*, vol 109, pp 234-239.

[13] Augustin T, Fillo L, Halvonik J and Marcis M 2018 Punching resistance of flat slabs with openings - experimental investigation, *Solid State Phenomena* vol 272 pp 41-46 DOI: 10.4028/www.scientific.net/SSP.272.41

[14] Vida R and Halvonik J 2017 Shear assessment of concrete bridge deck slabs, *Key Engineering Materials* vol 738 pp 110-119 DOI: 10.4028/www.scientific.net/KEM.738.110

[15] 2018 Special Issue on Shear *Structural Concrete* vol 19 iss 1 pp 1-328

DOI:10.1002/suco.201870014

[16] Model Code 2010 - Final Draft, fib, Bulletin No 65 and 66 1-2. 2012

[17] Gholizadeh S, Lemana Z and Baharudin B T H T A 2015 Review of the application of acoustic emission technique in engineering *Structural Engineering and Mechanics* vol 54 no 6 pp 1075-1095 DOI: http://dx.doi.org/10.12989/sem.2015.54.6.1075 1075

[18] Unnþórsson R 2013 *Hit Detection and Determination in AE Bursts* Ed. Sikorski, W. Acoustic Emission - Research and Applications

[19] Topolář L, Rypák P, Timčaková K, Pazdera L and Rovnaník P 2016 Non-traditional Non-destructive Testing of the Alkali-activated Slag Mortar during the Hardening *Materiali in tehnologije* vol 50 no 1 pp 7-10

[20] Pazdera L, Topolář L, Smutný J and Timčaková K 2015 Nondestructive Testing of Advanced Concrete Structure during Lifetime *Advances in Materials Science and Engineering*, vol 2015, no 7 pp 1-5

[21] Topolar L, Pazdera L, Kucharczykova B, Smutny J and Mikulasek K 2017 Using Acoustic Emission Methods to Monitor Cement Composites during Setting and Hardening *Appl. Sci.* vol 7 iss 5 pp 451 DOI:10.3390/app7050451

[22] NDIS 2421 2000 Recommended Practice for In-Situ Monitoring of Concrete Structures by Acoustic Emission, Japanese Society for Non-Destructive Inspection

[23] JCMS-III B5706 2003 Monitoring method for active cracks in concrete by acoustic emission, Federation of Construction Materials Industries (Japan)

[24] Recommendation of RILEM TC 212-ACD: Acoustic emission and related NDE techniques for crack detection and damage evaluation in concrete. Test method for classification of active cracks in concrete structures by acoustic emission 2010

[25] European Norms 2009 EN1330-9 nondestructive testing-terminology-part 9: terms used in AE testing

[26] EN 13554 Nondestructive testing-acoustic emission-general principles

[27] ASTM E1316, Standard definitions of terms relating to AE

[28] ASTM E650, Standard guide for mounting piezoelectric AE sensors

[29] ASTM E750, Standard practice for characterizing AE instrumentation

[30] Lamonaca F, Carrozzini A, Grimaldi D and Olivito R S 2015 Improved monitoring of acoustic emissions in concrete structures by multi-triggering and adaptive acquisition time interval *Measurement: Journal of the International Measurement Confederation* vol 59 pp 227-236 DOI:10.1016/j.measurement.2014.09.053

[31] Carpinteri A, Xu J, Lacidogna G and Manuello A 2012 Reliable onset time determination and source location of acoustic emissions in concrete structures *Cement and Concrete Composites* vol 34 iss 4 pp 529-537 DOI:10.1016/j.cemconcomp.2011.11.013

[32] Ono K 2016 Calibration Methods of Acoustic Emission Sensors. *Materials* vol 9 p 508

[33] Acoustic emission working group. Available online: www.aewg.org (accessed on 1 September 2017)

[34] Keshtgar A and Modarres M 2013 Detecting crack initiation based on acoustic emission, Chemical Engineering Transactions vol 33 pp 547-552 DOI: 10.3303/CET1333092

[35] Cajka R, Krivy V and Sekanina D 2011 Design and development of a testing device for experimental measurements of foundation slabs on the sub-soil *Transactions of the VŠB – Technical University of Ostrava, Civil Engineering Series* vol 11 iss 1 pp 1–5 DOI: https://doi.org/10.2478/v10160-011-0002-2

[36] Sucharda O, Smirakova M, Vaskova J, Mateckova P, Kubosek J and Cajka, R Punching Shear Failure of Concrete Ground Supported Slab *International Journal of Concrete Structures and Materials* vol 12 iss 1 Article number 36 DOI: 10.1186/s40069-018-0263-6

[37] Pazdera L, Cajka R, Topolar L, Mateckova P, Bilek V and Sucharda O 2019 Measurement and Utilization of Acoustic Emission for the Analysis and Monitoring of Concrete Slabs on the Subsoil *Periodica Polytechnica Civil Engineering* DOI:10.3311/PPci.12695

Effect of aggregate gradation on lime mortars rheology

M Vyšvařil[1], R Vozák[1]

[1] Faculty of Civil Engineering, Brno University of Technology, Veveří 331/95, Brno 602 00, Czech Republic

E-mail: vysvaril.m@fce.vutbr.cz

Abstract. The influence of basalt, limestone, and quartz aggregate and their various gradation on rheological properties of air lime mortars with constant water/binder ratio and binder/aggregate ratio is monitored in this work. It was found that the rheology of fine-grained lime mortars is influenced not only by the shape of the aggregate, but also by its granulometry, especially by the content of fines. The lime mortars with basalt, limestone or quartz aggregate are shear-thinning non-Newtonian liquids. Their shear stress, yield stress, and flow index decrease with increasing aggregate fraction. The viscosity of lime mortars is markedly influenced by the amount of lime matrix. The lime mortars with quartz and limestone aggregate are rheopectic suspensions.

1. Introduction

In the past few years, the results of research on the effect of aggregate type on lime mortar properties have been published. In particular, the properties of hardened lime mortars were monitored, and the different influence of quartz and limestone aggregates was observed especially [1–7]. Summarily, the limestone aggregate produced stronger mortars than quartz sand due to similarity between the calcitic binder matrix and the aggregate, and because of its angularity. Nevertheless, the results obtained for the individual aggregate cannot be applied generally for the similar type of aggregate. For example, using a limestone aggregate based on chalk leads to a reduction in strength and durability of lime mortars [3].

Previous works by Vyšvařil et al. [8–9] found that the use of limestone aggregate produces lower strengths and higher total porosity of dolomitic lime mortars and natural hydraulic lime mortars than quartz sand. Enhanced porosity of limestone mortars accelerated their carbonation and water absorption.

The influence of aggregate type (mineralogy, shape, surface roughness, porosity) on fresh state properties of lime mortars including a rheology has not been published yet despite the fact that the aggregate can affect a viscosity and thixotropic behaviour of lime mortars.

Studies of the effect of different types of aggregate on cement mortars rheology are more frequent. Essentially, the rheological properties of cement mortar are strongly influenced by the solid volume fraction, packing density, and amount of fines. The high amount of fines increases a yield stress and plastic viscosity by increasing the interparticle friction [10]. However, when the grain size of the fine aggregates is bigger than 0.7 mm, the yield stress is not affected by the size of the fine aggregate [11]. The use of solid content lower than 30% of volume does not have a significant effect on rheological properties of fresh concrete suspensions. The rheological behaviour of mixtures containing more than

IOP Conf. Series: Materials Science and Engineering **549** (2019) 012014 doi:10.1088/1757-899X/549/1/012014

30% of solid particles can be fairly describes by the Herschel-Bulkley model [12]. The increase in solid fraction results in higher yield stress, viscosity, and flow index.

In this work, the influence of three different types of aggregate and their various gradation on rheological properties of air lime mortars with constant water/binder ratio and binder/aggregate ratio is monitored. The knowledge of lime mortars rheology contributes to understand the behaviour of fresh lime materials that is important in a rendering.

2. Materials and methodology

The mortar mixtures were made of commercial hydrated lime CL 90-S (Carmeuse Czech Republic, Ltd.), crushed basalt aggregate (KÁMEN Zbraslav, Inc., the Dobkovičky quarry), crushed quartz sand (Filtrační písky, Ltd., Chlum u Doks), and crushed limestone aggregate (CEMEX Sand, L.P., the Štramberk quarry). The aggregates (0–2.5 mm) were sieved through 0.5 mm and 1 mm sieves and the individual fractions were used to make the respective mixtures. The granulometry of aggregates, loose bulk density, and experimental packing density of individual type and gradation of aggregate are shown in figure 1. The packing density were calculated as a ratio between the loose bulk density and the specific density of corresponding aggregate fraction [12]. The images of aggregates from optical stereoscope are shown in figure 2. It can be seen that the limestone aggregate (L) has the most angular shape and a rough surface.

Mortar mixtures were prepared using the constant lime:aggregate ratio of 1:1 by volume. In order to compare the effect of individual fraction of aggregate on the mortar rheological properties, all the mixtures were prepared with the same water/lime ratio of 1.4. This value was assessed by the reason of a good workability of fresh mortars and trouble-free measurements of rheological properties. This value of mixing water was also used in the foregoing research of lime mortars rheology [13]. The proportioning of the mortar mixtures is given in table 1.

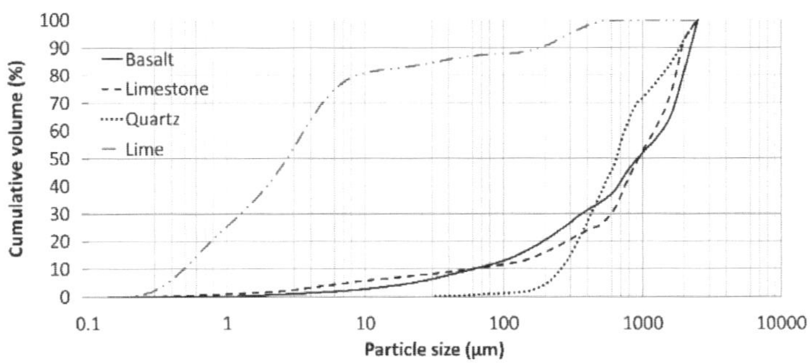

Type and gradation of aggregate (mm)	Loose bulk density (kg m^{-3})	Experimental packing density (–)
Basalt 0–0.5	1610	0.56
Basalt 0–1	1740	0.58
Basalt 0–2.5	1920	0.64
Limestone 0–0.5	1370	0.51
Limestone 0–1	1530	0.57
Limestone 0–2.5	1560	0.58
Quartz 0–0.5	1490	0.57
Quartz 0–1	1550	0.59
Quartz 0–2.5	1620	0.62

Figure 1. Grading curves of fine aggregates and lime; loose bulk density, and experimental packing density of individual type and gradation of aggregate.

B L Q

Figure 2. Images of basalt (B), limestone (L), and quartz (Q) aggregate obtained by optical stereoscope.

The rheological investigation including the characterization of flow properties and viscoelastic properties of mortars was carried out on the hybrid rheometer Discovery HR-1 (TA Instruments) and TRIOS 4.0.2.30774 software was used for the data evaluation. The Building Material Cell (BMC) and the paddle type rotor was adopted for the measurements. The gap thickness was chosen to be large enough (20 mm) for the mortar suspension. All the measurements have been done at 20 °C. The mortar mixture was introduced into the BMC after the 60 s long mixing procedure. The mortar was pre-sheared for 60 s at 100 s^{-1} to re-homogenize the sample and to eliminate its shear history. The rheological measurements were started after a period of rest of 60 s. The flow measurement comprised a shear rate increase (from 0.1 to 100 s^{-1}) applied through 30 steps with 15 s of measuring time at each shear rate followed by a decrease of shear rate on the same conditions. The results were expressed as shear rate vs. shear stress (flow curves) and the Herschel-Bulkley model [14] was applied to descending branches of the flow curves to analyze the experimental data. Thixotropy of the mortars was determined by the TRIOS software as a hysteresis area between the flow curves [14]. The viscoelastic properties of mortars were characterized by small amplitude frequency sweep tests with increasing the frequency from 0.1 Hz to 10 Hz. Experimental data were reported in terms of complex modulus (G^*) and loss tangent (tan δ) [14]. The strain of 0.003% used in the frequency sweep oscillation tests was lower than the critical strain for all studied mortars (table 2; γ_c) which was determined by varying strain amplitude oscillation test. All measurements were made three times and the results were averaged.

Table 1. Composition of mixtures.

Mixture	Lime (g)	Basalt (g)	Limestone (g)	Quartz (g)	Water (ml)
B-0.5	100	345.5	–	–	140
B-1	100	373.4	–	–	140
B-2.5	100	412.0	–	–	140
L-0.5	100	–	294.0	–	140
L-1	100	–	328.3	–	140
L-2.5	100	–	334.8	–	140
Q-0.5	100	–	–	319.7	140
Q-1	100	–	–	332.6	140
Q-2.5	100	–	–	347.6	140

3. Results and discussion

3.1. Flow properties

The type and gradation of the aggregate had a dramatic impact on the flow properties of mortars (figures 3–5). In all mortars, an increase in shear stress with increasing shear rate was observed and the shape of curves indicated the lime mortars as shear-thinning non-Newtonian liquids. The shear stress values, and flow index values, n, (table 2) decreased as the aggregate fraction increased. This is due to less internal friction in the mortars with a coarser aggregate fraction, and it is consistent with the results obtained with cement mortars [10]. Also, the shape of the aggregate grains played an important role in the flow properties of mortars. Limestone aggregate reached high shear stress values due to its angular grains, while quartz sand, as the most rounded aggregate, reached low shear stress values. Low packing density of limestone aggregate is one of the reasons for high shear stress values of limestone mortars (figure 1).

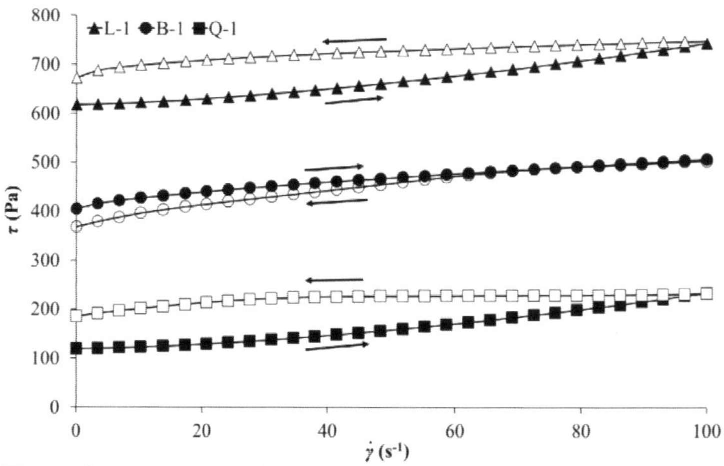

Figure 3. Flow curves of mortars with finest aggregate (0–0.5 mm).

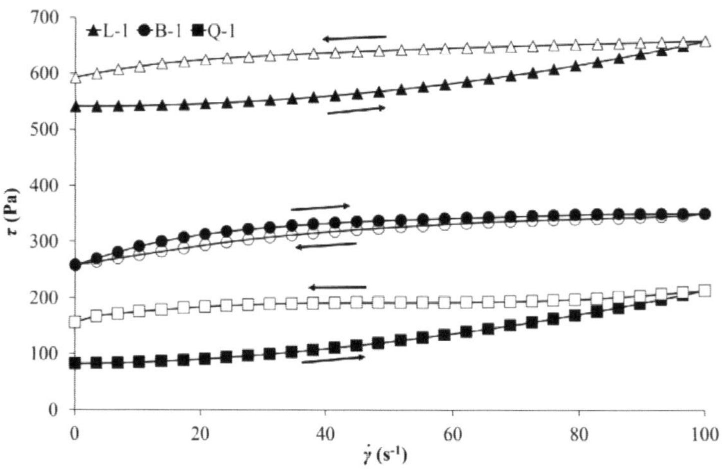

Figure 4. Flow curves of mortars with medium fine aggregate (0–1 mm).

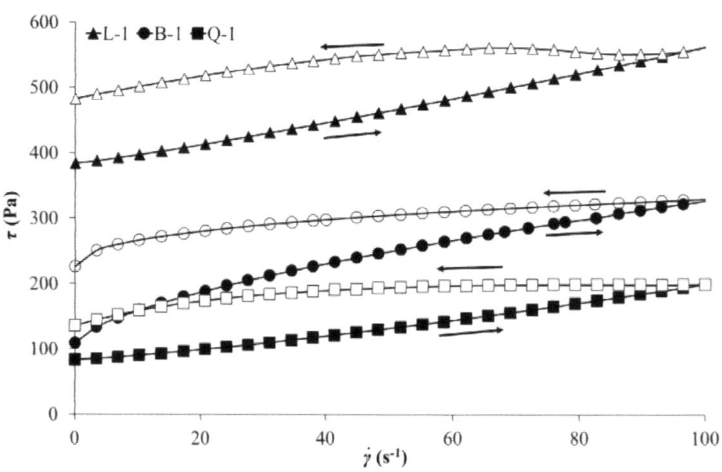

Figure 5. Flow curves of mortars with coarsest aggregate (0–2.5 mm).

The quartz sand mortars and the limestone mortars exhibited a rheopectic behaviour, meaning that they stiffened over time, which was reflected in higher shear stress values at decreasing shear rate. The lime mortars with basalt aggregate showed a gradual change from thixotropic to rheopectic behaviour with increasing aggregate fraction (table 2). The rheopectic behaviour of lime mortars and putties is commonly attributed to fragmentation of loose clusters of $Ca(OH)_2$ particles by fluid drag forces during the shearing process [15].

It is apparent from table 2 that the yield stress of mortars, τ_0, decreased, and the consistency coefficient, k, (comparable to the plastic viscosity) increased with increasing aggregate fraction. The rising viscosity is probably caused by the higher amount of lime binder matrix in the mortars with larger aggregate fraction, since less binder was used to coat the aggregate grains then in the mortars with fine aggregate with large surface area. The viscosity of lime mortars is dramatically influenced by the amount of continuous lime phase. Although this conclusion is inconsistent with a behaviour of cement mortars [16], it can be supported by different rheological properties of cement mortar and lime mortar due to the different viscosity of the binder putties. The lime putty obtained from the hard burnt lime achieves, immediately after slaking, high yield stress, viscosity and plasticity [17]. Hydrated lime also needs considerably higher amount of mixing water to obtain the same consistency as compared with the cement.

Table 2. Rheological parameters of fresh mortars.

Mixture	τ_0 (Pa)	k (Pa s)	n (-)	Thixotropy (Pa s)	γ_c (%)	G^* at 1 Hz (Pa)	tan δ at 1 Hz (-)
B-0.5	363.2	7.55	0.64	1351.9	0.016	4489	0.47
B-1	247.5	10.54	0.50	1161.3	0.008	2840	0.54
B-2.5	217.1	21.05	0.36	-5910.5	0.007	1745	0.59
L-0.5	668.8	11.07	0.43	-5576.3	0.015	10344	0.41
L-1	585.6	13.72	0.42	-5752.9	0.013	6602	0.44
L-2.5	465.1	19.52	0.36	-7227.7	0.006	2856	0.52
Q-0.5	180.8	7.62	0.47	-5709.1	0.006	1431	0.64
Q-1	152.8	9.74	0.37	-6045.9	0.005	1075	0.69
Q-2.5	125.2	15.85	0.37	-5149.6	0.005	1045	0.70

The rheopexy (time-dependent shear-thickening behaviour) of mortar mixtures increased with increasing aggregate fraction, in case of limestone especially. The lime particle disruption during the mixing of lime putty accelerates the hydrogel formation increasing viscosity and yield stress [18]. The rheopexy of lime mortars is supported by more efficient disintegration of lime clusters when mixing mortar with coarser angular aggregate. This was confirmed by monitoring the time evolution of shear stress of the limestone mortars (figure 6.). While the fine fraction mortar (L-0.5) was almost non-stiff during the measuring time, the coarse fraction mortar (L-2.5) showed after 5 min an increase in shear stress. At this time, the viscosity of the mortar also increased.

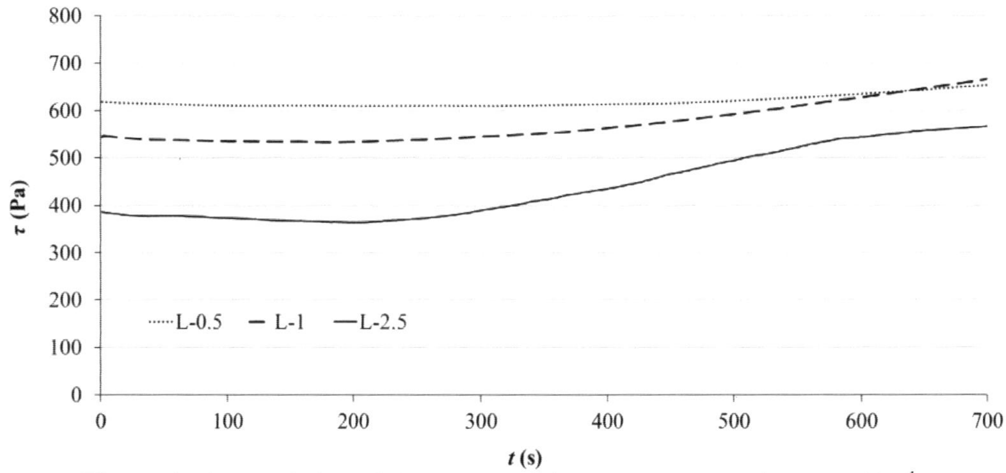

Figure 6. Time evolution of shear stress of limestone mortars (shear rate 10 s^{-1}).

3.2. Viscoelastic properties

The results of frequency sweep oscillation tests proved the relationship between the aggregate gradation and the stiffness of lime mortars; the ratio between dissipation and elasticity increased (growing G^*) with decreasing aggregate fraction and the mortars became more rigid (figures 7–9). Since the ratio between viscosity and elasticity (loss factor, tan δ) varies below 1 (table 2), the mortars show a higher proportion of the elastic component (G') than the viscous component (G''), suggesting that the mortars have a strong microstructure resistant to external intervention. The limestone mortars showed the lowest values of loss factor indicating the most uninfluenceable structures and their viscoelastic behaviour was closest to the solid matter. Increasing values of the loss factor at higher frequencies point to the beginning of liquid separation from the mortars structure. At such high oscillation frequencies, quartz mortars and B-2.5 mixture behave like liquids ($G'' > G'$, tan $\delta > 1$).

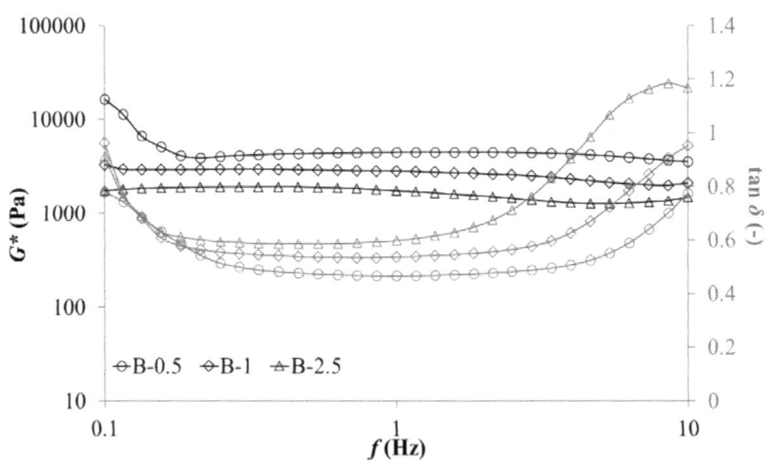

Figure 7. Evolution of complex modulus, G^*, and loss factor, tan δ, as a function of frequency of basalt mortars.

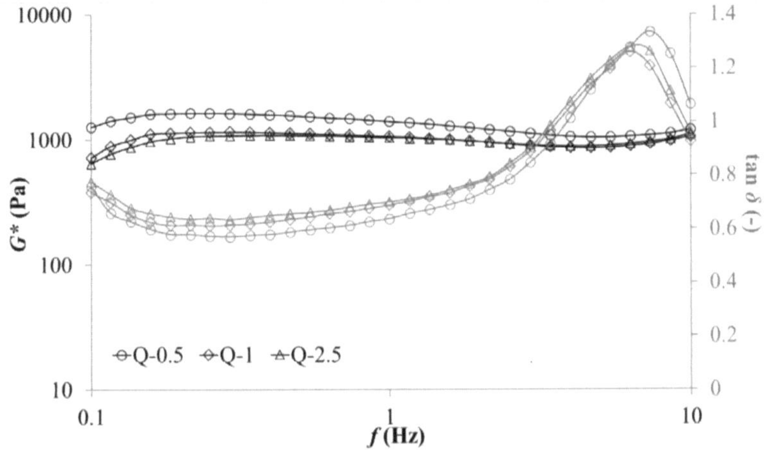

Figure 8. Evolution of complex modulus, G^*, and loss factor, $\tan\delta$, as a function of frequency of quartz mortars.

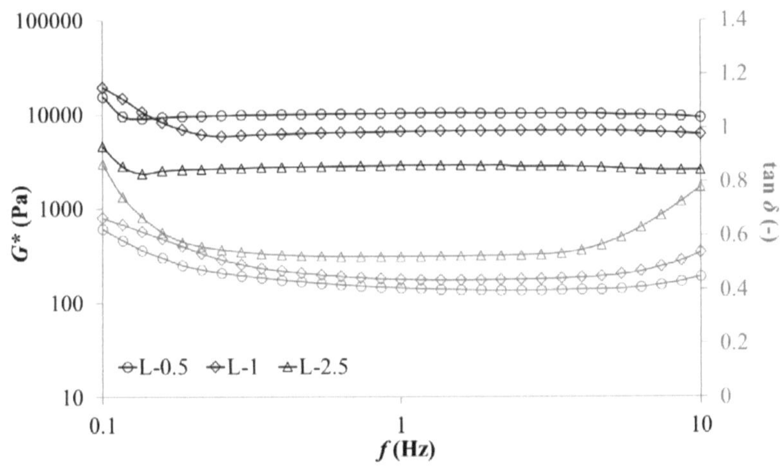

Figure 9. Evolution of complex modulus, G^*, and loss factor, $\tan\delta$, as a function of frequency of limestone mortars.

4. Conclusions

The influence of three various gradation of basalt, limestone, and quartz aggregate on rheological properties of air lime mortars were investigated. The main conclusions are as follows:

- The rheology of fine-grained lime mortars is influenced not only by the shape of the aggregate, but also by its granulometry, especially by the content of fines.
- The lime mortars with basalt, limestone or quartz aggregate are shear-thinning non-Newtonian liquids. Shear stress, yield stress, and flow index of mortars decreases with increasing aggregate fraction due to decrescent internal friction in the mortars. Angular grains of the limestone aggregate cause an increase in shear stress values, while the rounded grains of quartz decrease the shear stress of mortars.
- The influence of shape and type of aggregate on flow behaviour of mortars can be judged based on packing density of the aggregate. As the packing density decreases, shear stress and yield stress increase.
- The lime mortars with quartz and limestone aggregate are rheopectic suspensions, they stiffen over time. The lime mortars with basalt aggregate shows the change from thixotropic to

rheopectic behaviour with increasing aggregate fraction. The use of sharp-edged aggregate (crushed limestone) promotes the rheopectic behaviour of lime mortars in order to more efficiently break up the lime aggregates in the putty.

- The viscosity of lime mortars is dramatically influenced by the amount of continuous lime phase – the larger aggregate fraction, the higher amount of lime binder matrix in the mortar, and the higher viscosity of the mortar.
- The limestone mortars showed the lowest values of loss factor indicating the most uninfluenceable microstructures and their viscoelastic behaviour is closest to the solid matter.
- From a practical point of view, limestone or basalt aggregates are more suitable for manual plastering of vertical and ceiling surfaces, while rounded quartz aggregate, not increasing the viscosity of lime mortar to such an extent, is more suitable for machine thin-layer rendering.

Acknowledgments
This work has been financially supported by The Czech Science Foundation (GA CR) project No. 18-07332S and by Ministry of Education, Youth and Sports under the „National Sustainability Programme I" – the project No. LO1408 „AdMaS UP – Advanced Materials, Structures and Technologies".

References

[1] Lanas J and Álvarez J I 2003 *Cem. Concr. Res.* **33** 1867
[2] Pozo-Antonio J S 2015 *Constr. Build. Mater.* **77** 472
[3] Pavía S and Toomey B 2008 *Mater. Struct.* **41** 559
[4] Stefanidou M and Papayiannai I 2005 Cem. Concr. Compos. **27** 914
[5] Ellis S, Lawrence M and Walker P 2013 Chem. Mater. Res. **5** 97
[6] Santos A R, Veiga M do R, Silva A S, Brito J de and Álvarez J I 2018 *Constr. Build. Mater.* **187** 907
[7] Vyšvařil M, Žižlavský T and Bayer P 2016 *Appl. Mech. Mater.* **861** 141
[8] Vyšvařil M, Žižlavský T and Bayer P 2016 *Key Eng. Mater.* **722** 343–350
[9] Vyšvařil M, Žižlavský T, Zimmermann Š and Bayer P 2017 *Mater. Sci. Forum* **908** 35
[10] Westerholm M, Lagerblad B, Silfwerbrand J and Forssberg E 2008 *Cem. Concr. Compos.* **30** 274
[11] Han D, Kim J H, Lee J H and Kang S-T 2017 *IJCSM* **11** 627
[12] Kabagire K D, Diederich P, Yahia A and Chekired M 2017 *Constr. Build. Mater.* **151** 615
[13] Vyšvařil M., Hegrová M. and Žižlavský T 2018 *Solid State Phenom.* **276** 69
[14] Mezger T G 2014 *The Rheology Handbook* (Hanover: Vincentz Network)
[15] Kubátová D, Boháč M and Nečas R 2016 *Procedia Eng.* **151** 18
[16] Banfill P F G 1994 *Constr. Build. Mater.* **1** 43
[17] Ruiz-Agudo R and Rodriguez-Navarro C 2009 *Langmuir* **26** 3868
[18] Navrátilová E, Tihlaříková E, Neděla V, Rovnaníková P and Pavlík J 2017 *Sci. Rep.* **7** 17260

An optimal method of determining the position of bends on shear reinforcement as part of the diagnosis of reinforced concrete beam structures

O Anton[1], T Komárková[1], V Heřmánková[1]

[1] Brno University of Technology, Faculty of Civil Engineering, Veveří 331/95, 602 00 Brno, Czech Republic

E-mail: ondrej.anton@vutbr.cz

Abstract. The current period in the development of building industry can be characterized as a period of rapid development in this field, not only in the area of new constructions, but also in the area of a new uses of old constructions which find a new use after restoration, and as regards traffic facilities, especially bridge structures, in the area of verification of the existing functional characteristics and prediction of the lifetime of structures. This puts a strain on the field of diagnosis of structures, the development of which can be observed in the last 10 years in the form of a new wave of innovations of the existing NDT (non-destructive testing) methods as well as of the development of completely new devices and technologies. In spite of these facts, it can be said that in a number of cases, the diagnosis of structures is performed in traditional ways without using modern techniques and technologies. The aim of this paper is to present an optimal way of using NDT methods in the diagnosis of reinforced concrete beam structures, specifically in the determination of the position of shear bends on the main load-bearing reinforcement. The paper deals with the current state of the issue and illustrates, on a number of examples from practice, the applicability of individual methods as well as their limitations given by their physical principles and technological solutions.

1. Introduction to the issue

Today, there is a large number of reinforced concrete structures in the Czech Republic originating from various periods of the more than hundred-year-old tradition of manufacturing reinforced concrete. It is characteristic especially of the old structures that no project and construction documentation has been preserved. The reason of this can be found in the dramatic changes of political regimes in the past, in times of crisis such as World War Two, etc. These historic moments were often the cause of destruction or loss of the above-mentioned documents. A similar phenomenon is encountered, paradoxically, in the case of structures from the recent past, where the cause can be seen in the reluctance of the construction companies to provide construction documentation.

It is therefore a frequent task for the field of diagnosis of structural elements to determine characteristic properties of concrete in the selected elements of a reinforced concrete structure, and to detect the reinforcement of typical elements so that it is possible to perform a reanalysis of the actual load capacity of the structure, either to verify its functional characteristics or as a basis for restoration.

The present paper is focused on the specific field of diagnosis. Very frequent elements of monolithic reinforced concrete structures are reinforced concrete beams of various cross-sections and lengths. As regards the reinforcement survey aimed at static recalculation, the standard requirement is to detect the main load-bearing reinforcement in the middle of the element span near the bottom face and to detect the reinforcement near the upper face above the support. Apart from that, spacings of stirrups are detected. In those cases where reinforcement above the support is inaccessible for various reasons, it is possible to determine the number of reinforcing bars near the bottom face of the support, and to estimate the number of shear bends on the main reinforcement from the difference between the bars in comparison to the centre of the span. What is somewhat neglected is the possibility of determining the position and number of shear bends directly in the measurement on the side faces of the girder. And it is precisely this specific part of diagnosis that is the focus of this paper.

2. Selection of applicable diagnostic methods for determining the position of reinforcement

The field of diagnosis of building structures can utilize in fact four methods of localizing steel reinforcement in the reinforced concrete structure.

- **Chopped probes** – a method based on a mechanical removal of the covering layer of concrete using a hammer drill. It is a method widely used in the civil engineering surveys of aged structures, with a limited potential, involving the risk of frequent impossibility to detect all the reinforcement, local limitation of the method and the necessity of a destructive intervention. For a specific case of determining the position of shear bends of the main reinforcement, it is a completely unsuitable method [1, 2].
- **Electromagnetic indicators of reinforcement** – devices based on the principle of electromagnetic field, formation of eddy currents, and the magnetic properties of steel reinforcement. The major disadvantage of this method is a limitation given by the principle of the method, by the depth of the reinforcement under the structure surface, by the distance between individual bars and by the complexity of reinforcement in general. The applicability of this method is problematic in terms of detecting the position of the shear bends of main reinforcement, due to the fact that the detection depth of the probes is very limited, and it is also probable that there might be some stirrups or horizontal reinforcement between the concrete surface and shear reinforcement, which will not allow for localizing the reinforcement behind them [1, 2].
- **Georadar** – a method based on the principle of transmitting radio-frequency electromagnetic pulses into the material and the detection of their reflection from the inhomogeneities in the environment. With the appearance of the Hilti PS-1000 radar, this method started to be highly applicable also for the above-mentioned problem. Its advantages include the detection depth (Hilti PS1000 up to 400 mm, GPR Proceq up to 700 mm), immediate visualization of the result, and independence of other reinforcement closer to the surface. For the purposes of determining the position of bends on the main load-bearing reinforcement it is possible to use both areal scans and line scans, as is demonstrated in detail in the following chapter. Today, only two georadars are available on the market, designed primarily for the diagnosis of reinforced concrete structures. Hilti PS1000 X-Scan and Proceq GPR Live. Hilti PS1000 is equipped with three aerials in one probe with a frequency of 1.5 GHz and a detection depth of 300 mm, Proceq is equipped with one aerial of 0.9 – 3.5 GHz with a depth of up to 700 mm. Both devices allow for performing areal and line scans from the concrete surface, in both cases areal scans can be transformed into 3D pictures of the detected reinforcement [1, 2].

Figure 1. Both georadars available have a similar size of the probe.

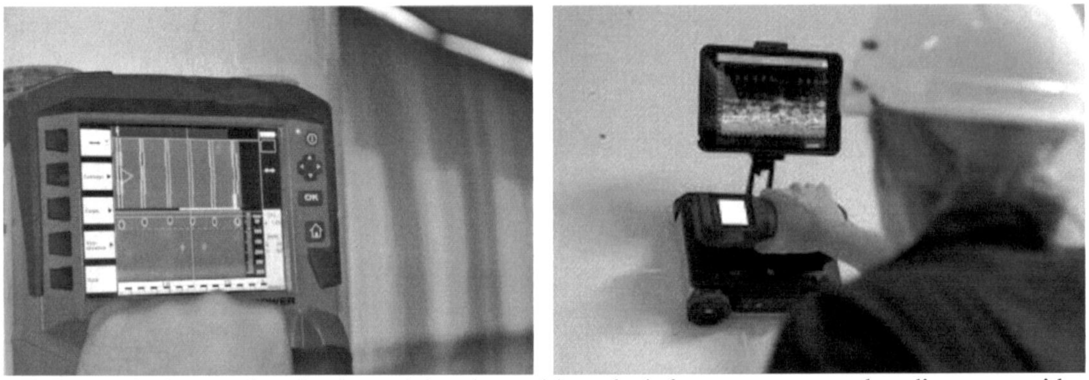

Figure 2. Both georadars for determining the position of reinforcement can produce line scans with online display of the results or a recording, on the left, Hilti PS1000 X-Scan, on the right, Proceq GPR Live [6].

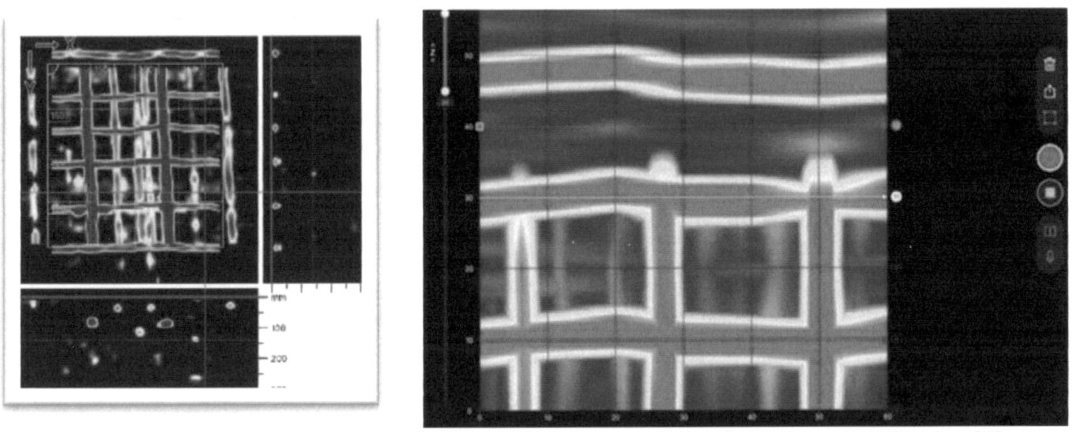

Figure 3. Areal scans, on the left, Hilti PS1000 X-Scan, on the right, Proceq GPR Live [6].

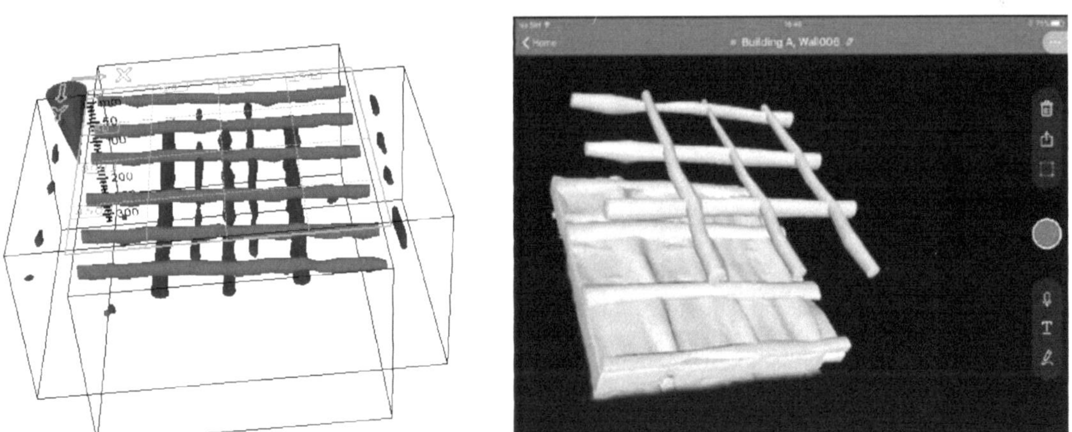

Figure 4. Both Hilti PS1000 X-Scan and Proceq GPR Live make it possible to display the detected reinforcement in a fully-fledged 3D projection [6].

- **Radiography** – a well-known traditional method utilizing the irradiation of the structure with γ-rays from the Co60 source, and subsequently the attenuation of the radiation when passing through the structure depending on the volume mass. Considering the fundamental difference between the volume mass of concrete and the density of steel, it is possible to represent the entire reinforcement on the output medium (radiographic film, electronic imaging media, etc.), and in the case of irradiation from several points of view, to determine precisely both the position of reinforcement, and its diameters. [1, 2].

3. Examples of practical application

3.1. Radiography

It is obvious from the above descriptions of the individual methods that the methods applicable for the solution of the given problem are radiography and georadar.

In the past, these tasks were performed by means of radiography but at the cost of accepting the expensiveness and time-consumingness of this method. Considering the necessary considerable irradiated width of the material given by the width of the reinforced concrete beam, we used, almost exclusively, Co60 radiography where the source of radiation was placed near one side face of the element examined, while the recording medium was situated near the opposite side face of the beam. With the appropriately set parameters of irradiation, the result was a radiogram with the reinforcement displayed in the irradiated place. The basic advantage was the 100% results in the determination of the reinforcement position while maintaining the principle of non-destructivity of this method. Its disadvantage was a long exposure time (given by the activity of the emitter and by the irradiated width of the concrete - in massive elements, tens of minutes to hours) and simultaneously a relatively small area of the element examined (given by the dimensions of the recording medium, usually 400 x 300 mm). Another limitation of this method is the fact that the method uses a high-activity gamma-ray source, which brings about a number of limitations resulting from the necessity to define and secure a temporary controlled area with all the requisites required by the regulations concerning health protection against ionising radiation. In practice it means that in order to apply radiography on the structure, it is necessary to prohibit completely the stay and movement of persons in the vicinity of the structure for the period of irradiation. Nevertheless, the method has been used successfully many times. Unfortunately, this method was abandoned in the Czech Republic at present due to its financial cost and the tightening safety regulations.

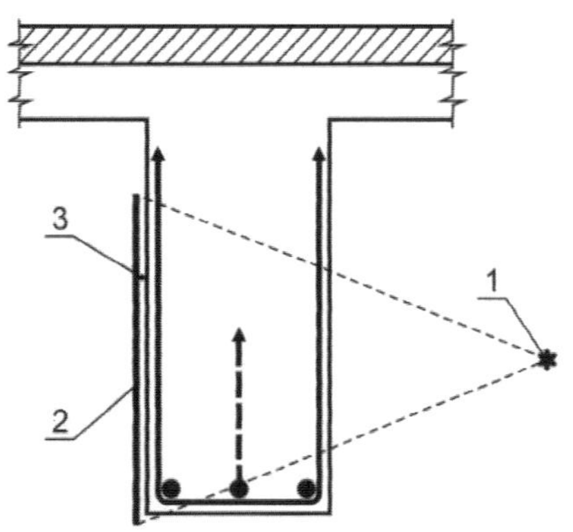

Figure 5. The left picture shows a diagram of the irradiation geometry of the reinforced concrete beam, where 1 – source of Co60 radiation, 2 – radiographic film, 3 – lead marks for precise identification of the position of the image The picture on the right shows a radiogram of the beam with a well apparent main load-bearing reinforcement in one layer, the last shear bend before the support, and one bar of the main reinforcement which bends simultaneously with the face of the beam haunch before the support [3].

3.2. Georadar - Example 1

Typical structures, in which the number and position of bends on the main reinforcement are important for the reanalysis of the structure, are reinforced concrete girder bridges. The first example of an appropriate use of the radar is the reinforced concrete girder bridge on the 1st class road near Holasovice. The bridge had already been diagnosed in 1990 and 2000 as part of its prepared restoration, and it was concluded that in the bridge girders, there are always 15 main reinforcements ROXOR R45 in three layers of 5, and it was supposed that part of them are shear bent in several places. Due to the budget constraints of the survey, the number and position of bends were not radiographically determined. In the subsequent survey in 2014, a series of successive areal scans were performed using the PS1000 radar, always on both side faces of the girders (considering the girder width of 600 mm and the radar's detection depth of 300 mm, the whole inner area was covered), which unequivocally detected the position of bends on the main load-bearing reinforcement, as is apparent from the pictures below.

Figure 6. Reinforced concrete girder bridge on the 1st class road near Holasovice [4].

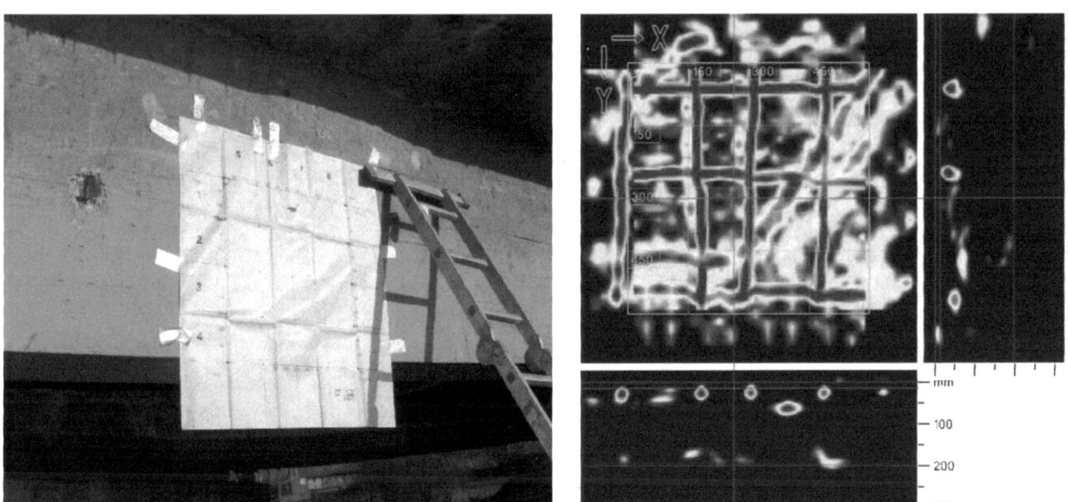

Figure 7. A grid for performing areal scans of 600 x 600 mm with the PS1000 probe, and on the right, a radarogram with the well apparent reinforcement – the main load-bearing reinforcement, its bends, stirrups and horizontal structural bars [4].

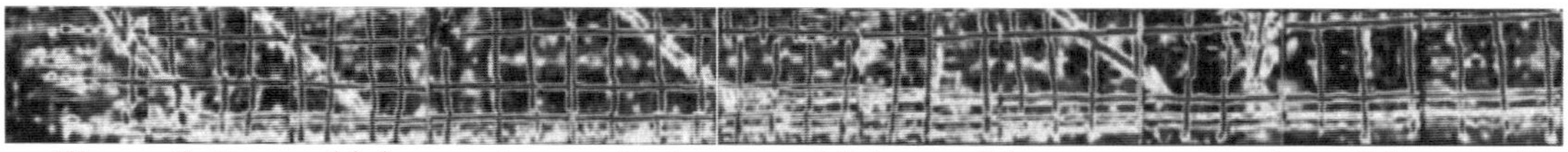

Figure 8. By combining areal scans (radarograms) it is possible to get an overview of the overall reinforcement of the element [4].

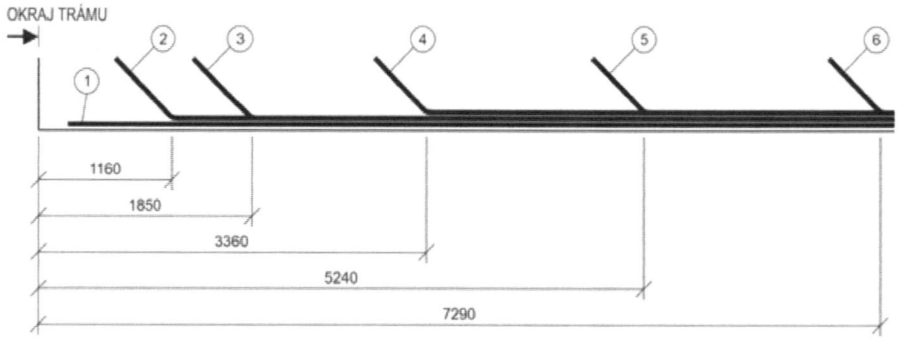

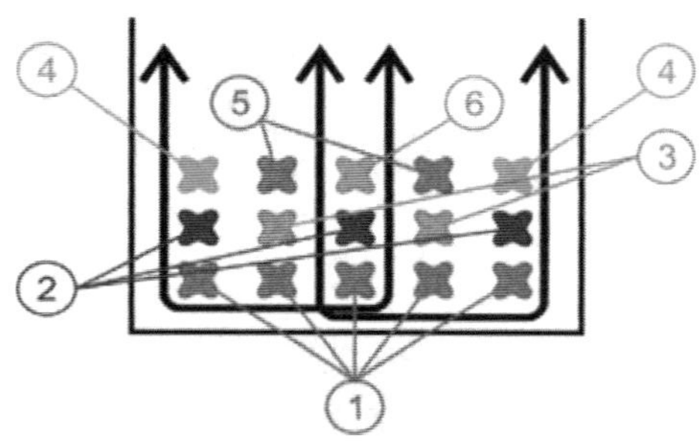

Figure 9. Diagram of the positions of bends on the main load-bearing reinforcement and a cross-section of the girder with the assignment of bend positions to individual bars; it is apparent that 10 of a total of 15 bars are bent in 5 positions [4].

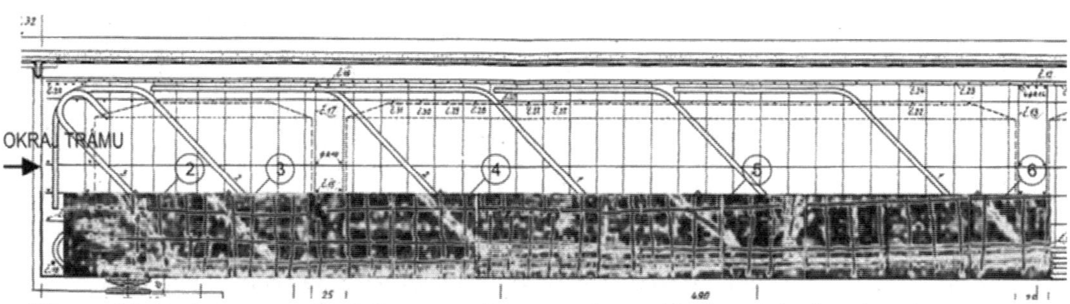

Figure 10. Radarograms of reinforcement in comparison with the original project documentation, which was found, paradoxically, in the possession of a private person only after performing the survey [4].

3.3. Georadar - Example 2

The second example is the determination of the position of bends on the main load-bearing reinforcement in the elements of the ceiling structure of the Pavilion Z at the Brno Exhibition Centre. On the subtle joists of 150 – 250 mm, it was possible to detect the bends by measurement from one side face of the joist only. Figure 11 documents the traditional way of determining the position of bends on the main reinforcement using the areal radar scan.

IOP Conf. Series: Materials Science and Engineering **549** (2019) 012015 doi:10.1088/1757-899X/549/1/012015

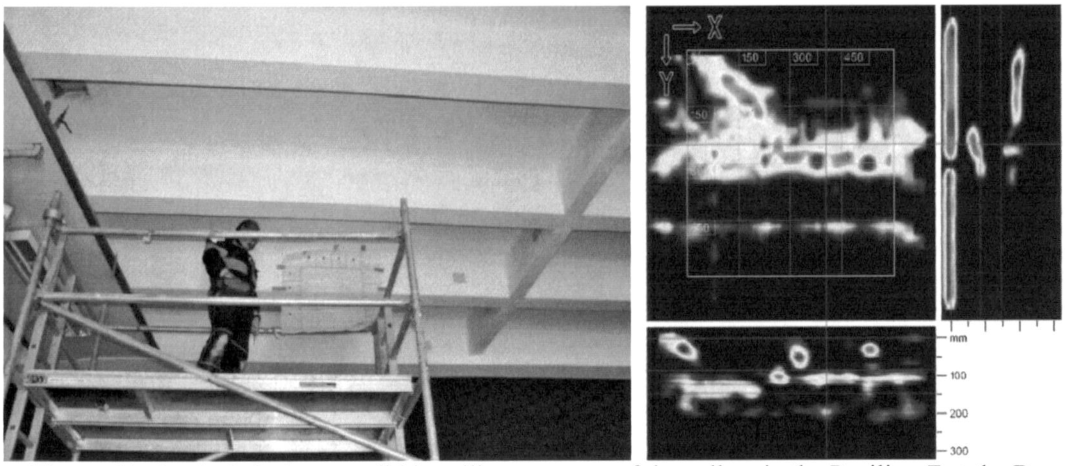

Figure 11. On the left, the monolithic ceiling structure of the gallery in the Pavilion Z at the Brno Exhibition Centre, the picture shows the already installed grid for areal measurement with the PS 1000 probe. The right picture shows a flat scan of 600 x 600 mm with a clearly visible bend of the main reinforcement [5].

On site it was also possible to test an alternative method based on the fact that the PS 1000 probe is equipped with three aerials, therefore, even when doing a line scan, we in fact get a picture made of three line sections of the measured environment. On the line scan (composed of two parts, a view of the measured line and a section), the "view" part clearly shows the position of bends, especially when we set the device to display in the "view" part only the reinforcement from the inner area of the joist and not the stirrups situated near the surface (see Figure 12). It is apparent from the measurement results that it is possible to make the measurement of the position of bends in the main load-bearing reinforcement even simpler by means of the basic line scans, although the areal scans have a higher information and documentation value.

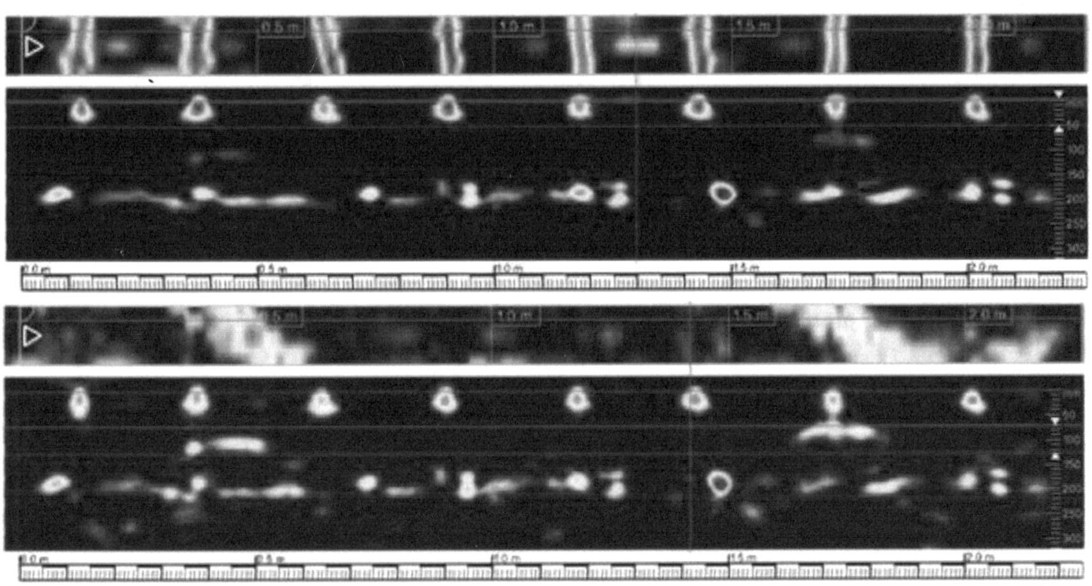

Figure 12. A line scan on the side face of the rib, in the picture above, only stirrups near the surface are displayed in the visible part, in the picture below, bends of the main load-bearing reinforcement bars are unequivocally identified in the visible part [5].

4. Conclusion

On the basis of three decades of the institution's experience with the use of radiography and several years of testing of georadar sets in the checks of the position of reinforcement in several hundreds of structures, it is undoubtedly possible to say that in most cases, the georadar method will be ideal for determining the position of bends in the main load-bearing reinforcement in the girders of reinforced concrete structures. It is currently the only NDT method with fully-fledged results in the desired extent. The only limitation could be the width of the element examined, if it exceeds double the detection depth of the radar set. This is, however, not very likely with the appearance of Proceq GPR Live (detection depth 700 mm, i.e. the limit width of the beam 1400 mm).

Acknowledgement

This paper has been worked out under the project No. LO1408 "AdMaS UP - Advanced Materials, Structures and Technologies", supported by Ministry of Education, Youth and Sports under the „National Sustainability Programme I".

References

[1] Anton O, Komárková T, Heřmánková V 2018 Conclusive Determination of Compliance with the Prescribed Reinforcement Elements of Concrete Structures - The Appropriate Methods and their Capabilities. Key Engineering Materials **776**:76-80. [accessed 2019Apr.10]. https://www.scientific.net/KEM.**776**.76

[2] Anton O, Komárková T, Heřmánková V 2018 Methods Appropriate for Determination of the Prescribed Reinforcement of the Elements of Reinforced Concrete Structures - Nowadays Used Methods. Key Engineering Materials **776**:41-45. [accessed 2019Apr.10]. https://www.scientific.net/KEM.**776**.41

[3] Hobst L 2007 Important Role of Radiography for determination of Conditions in Reinforced Concrete Structures. Engenharia Estudo e Pesquisa **9**:7.

[4] Anton O 2014 REPORT: Diagnostic Survey of Shear Reinforcement (Bends) of Beams of Bridge Structure on Road I / 57 in Holasovice, id. 57-016. ZPRÁVA: O diagnostickém průzkumu smykové výztuže (ohybů) trámů nosné konstrukce mostu na silnici I/57 v Holasovicích, ev.č. 57-016 (in Czech). Brno.

[5] Anton O 2014 REPORT: Determination of Reinforcement of Selected Elements of Reinforced Concrete Structure of Hall Z in the Area of BVV, Brno. ZPRÁVA: O stanovení vyztužení vybraných prvků železobetonové konstrukce pavilonu Z, v areálu BVV, Brno (in Czech). Brno.

[6] Products: Portable Ground Penetrating Radar - Proceq GPR Live. 2018. Proceq. [accessed 2019Feb.10]. https://www.proceq.com

IOP Conf. Series: Materials Science and Engineering **549** (2019) 012016 doi:10.1088/1757-899X/549/1/012016

Current possibilities of the diagnosis of prestressing steel in structures

O Anton[1], P Cikrle[1], V Heřmánková[1]

[1] Brno University of Technology, Faculty of Civil Engineering, Veveří 331/95, 602 00 Brno, Czech Republic

E-mail: ondrej.anton@vutbr.cz

Abstract. The field of diagnosis of structures has been experiencing a dynamic development in the last decades caused by the development and expansion of new NDT methods and instrumentation. At the same time, it is possible to observe changes in the focus of interest in this field, or more precisely, alternating waves of interest in specific types of diagnosis targeted at special types of structures. These waves of interest are always caused by a certain trend in the development of the society or building industry, and most often, by some major even in the field. In the last few years, particular attention has been paid to prestressed structures, especially to bridges, in connection with the recent accidents of bridges made from prestressed concrete in the Czech Republic and abroad. Considering the relatively large number of these bridges in the Czech Republic built in the second half of the 20th century, attention is paid to basic changes in the pre-construction condition surveys aimed at determining the condition and the remaining life of these structures. The paper focuses on the analysis of possibilities of assessing the condition of prestressing steel at the level of the present state of diagnosis of structures.

1. Introduction to the issue

In the Czech Republic, there is a large number of various types of structures made from prestressed concrete. The most frequent and most attention-drawing structures from the point of view of diagnosis targeted at determining the lifetime include bridges made from prefabricated prestressed girders of various types and bridges built e.g. as monolithic prestressed slabs, but also prestressed truss girders in the structures of industrial halls and other types of elements and structures are in the centre of interest. While hall roof girders, for example, are elements which can be replaced without problems if there are doubts of their functionality in the structure, as regards bridge structures, it is a fundamental problem complicated by general reluctance of the society and investors to consider their replacement, but also a problem at the level of public opinion, which has a tendency to attribute to the structures the character of "eternity" and endless durability.

Considering the recent accidents of prestressed bridge structures in the Czech Republic and abroad, extraordinary attention is paid to the pre-construction condition surveys of prestressed bridge structures with the aim of determining not only their technical condition but primarily of estimating on its basis the remaining lifetime of the structure in order to ensure the safety of its use. Immediately after the accident of the footbridge in Prague – Trója in 2017, attention was first paid to similar structures, which sometimes bordered on thoughtless hysteria oriented towards preventive demolition of structures. As regards the remaining prestressed bridges, fortunately, concerns have rather led to performing detailed pre-construction condition surveys and monitoring the condition of the structure.

There is a question, however, whether the contemporary state of diagnostic methods and devices available for their application correspond to the requirements for a perfect survey of the structure, or more precisely, for a survey with the results ideally relevant for determining the condition of the structure.

2. Diagnosis of prestressed structures

In order to assess the overall condition, lifetime and functional characteristics of a prestressed structure, it is essential to determine especially the condition of the prestressing steel. It makes no sense to discuss e.g. cables running freely outside the concrete cross-section, where the determination of the condition is easy, but above all, the prestressing wires and cables placed in the cable channels. The protection of reinforcement in cable channels from corrosion is largely dependent on timely and correct grouting of the cable channel with a suitable grout. Therefore, during the pre-construction condition surveys of prestressed structures and elements, extraordinary attention should be paid not only to the already existing corrosion of the prestressing steel, but also to the existence of correct grouting of cable channels itself, because the absence of grout is a defect which must be immediately repaired to reduce future risks.

Naturally, the diagnosis of prestressed structures pays attention to the other parameters as well, for example to concrete properties, but the basic parameter is the condition of the prestressing steel, or its endangering.

Diagnosis of the condition of prestressing steel can be divided into two phases. In the first phase, it is necessary to determine the precise position or course of prestressing wires or cables, in the second phase, to determine their condition and presence of grout.

3. Methods applicable for determining the position of prestressing steel in a structure

In spite of the fact that the precise position of prestressing steel or cable channels in the pre-tensioned and post-tensioned structures and elements is specified in the technical documentation, it is necessary to expect certain discrepancies arisen during the manufacture. Determining the precise position and course of prestressing channels and therefore also of reinforcement is thus logically the first step in their diagnosis. It must be noted, however, that in a number of cases, cable channels are delimited by steel protection pipes (seamless pipes, Sandrik) [2].

In general, there are four basic methods for determining the position of reinforcement in concrete.

- **Chopped probes** – a method based on a mechanical removal of the covering layer of concrete using a hammer drill. It is a method widely used in the civil engineering surveys of aged structures, with a limited potential. For a specific case of determining the position of prestressing steel, it is a completely unsuitable method [5, 6].
- **Electromagnetic indicators of reinforcement** – devices based on the principle of electromagnetic field, formation of eddy currents, and the magnetic properties of steel reinforcement. The major disadvantage of this method is a limitation, given by the principle of the method, by the depth of the reinforcement under the structure surface, by the distance between individual bars and by the complexity of reinforcement in general. The applicability of this method is problematic, as regards determining the position of prestressing steel, due to the fact that the detection depth of the probes is very limited, and it is also probable that there might be some soft reinforcement between the concrete surface and the prestressing steel, which will not allow for localizing other elements deeper in the concrete. [5, 6].
- **Georadar** – a method based on the principle of transmitting radio-frequency electromagnetic pulses into the material and the detection of their reflection from the inhomogeneities in the environment. With the appearance of the Hilti PS-1000 radar, this method started to be highly applicable also for the above mentioned problem. Its advantages include the detection depth (Hilti PS1000 up to 400 mm, GPR Proceq up to 700 mm), immediate visualization of the result, and

independence of other reinforcement closer to the surface. The position of prestressing steel can be determined by both linear and areal measurements [5, 6].

Today, only two georadars are available on the market and they are designed primarily for the diagnosis of reinforced concrete structures. Hilti PS1000 X-Scan and Proceq GPR Live. Hilti PS1000 is equipped with three aerials in one probe with a frequency of 1.5 GHz and a detection depth of 300 mm, Proceq is equipped with one aerial of 0.9 – 3.5 GHz with a depth of up to 700 mm. In both devices, it is possible to transform the measurement results into a fully-fledged 3D image.

Its disadvantage is, on the contrary, the necessity to perform the measurements from the flat surface of the structure, and the fact that certain minimum dimensions of the measured area are necessary. Therefore, it is not possible to determine the reinforcement e.g. in the intricately-shaped or inaccessible parts of the structure, which is especially the case of some types of prefabricated bridge girders.

- **Radiography** – a well-known traditional method utilizing the irradiation of the structure with γ-rays from the Co60 source, and subsequently the attenuation of the radiation when passing through the structure depending on the volume mass and irradiated width. Considering the fundamental difference between the volume mass of concrete and the density of steel, it is possible to represent the entire reinforcement on the output medium (radiographic film, electronic imaging media, etc.), and in the case of irradiation from several points of view, to determine precisely the position of reinforcement. Its disadvantage, on the contrary, is the relatively small area examined given by the size of the recording medium (usually 300 x 400 mm), and currently, the problematic availability of this diagnostic method [5, 6].

On the basis of the extensive experience of the author's institution, the choice of georadar is clear, it is an ideal method for determining the precise position of prestressing steel in the structure; in the case of elements for which the reinforcement is inaccessible for the radar measurement, it would be necessary to use radiography, which is not currently available in the Czech Republic for the generally known reasons (Co60 radiography) [3].

3.1. Examples of practical application of localizing the prestressing cables

The following examples illustrate the radar determination of the position of prestressing cables (Hilti PS1000 X-Scan device) on three common types of prefabricated bridge girders.

KA girder

The first example is a KA-67-type girder with a length of 12 m, designed as a closed profile of approximately rectangular cross-section with an inner hexagonal chamber. KA girders have been developed since the second half of the 1950s and standardized later by Dopravoprojekt Bratislava [1]. The girder is reinforced with a total of 13 prestressing cables made of 7 to 12 prestressing wires of Ø 4.5 mm. In the middle of the span, 9 cables are situated near the bottom face of the girder, two cables near the bottom face in the second layer, and two cables near the upper face, of which the two peripheral cables from the bottom layer and two from the second layer rise to the upper face towards the end of the girder. The position of cables was determined by means of Hilti PS1000 X-Scan radar.

IOP Conf. Series: Materials Science and Engineering **549** (2019) 012016 doi:10.1088/1757-899X/549/1/012016

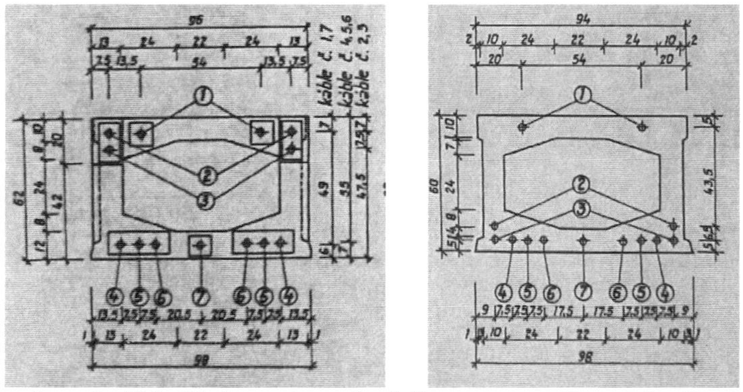

Figure 1. Type drawing of a KA-67 element, on the left: a cross-section at the front of the element, on the right: in the middle of the span [1].

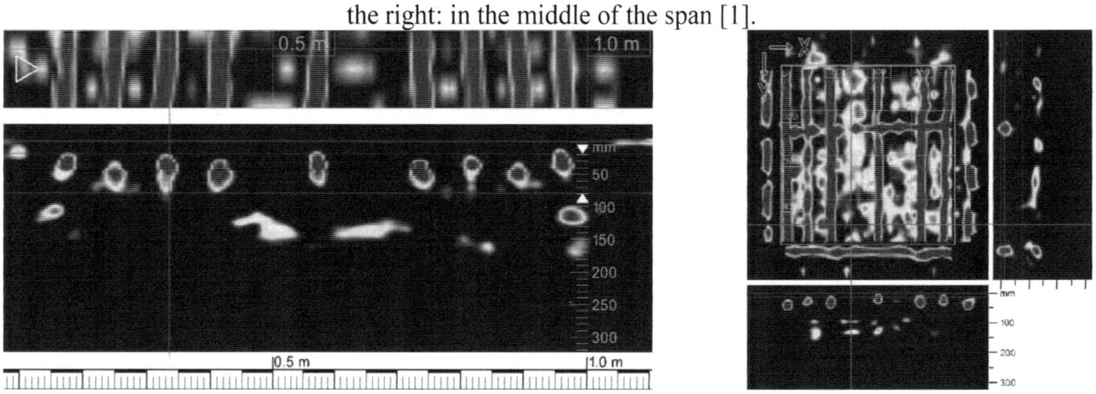

Figure 2. Radar scans (radarograms) taken on the lower side of the KA-67 girder, approximately in the middle of the span. On the left, a line scan across the girder captures all the 11 prestressing cables near the lower face including those in the second layer. The right picture shows an areal scan of 600 x 600 mm at the same place capturing the same cables and stirrups.

On this type of girder it is therefore possible to determine precisely the position of cables as long as they are situated near the bottom face of the girder, including the second layer. Cables near the upper face cannot be detected due to the hollow in the element. Cables which rise cannot be detected because of the depth limitation of the device. In case a radar with a higher depth range was used, it would probably be possible.

I girder

The second type of prestressed girder developed simultaneously with the KA girder is the so-called I girder. I girders were linked with the so-called latch connection at the protruding transverse reinforcement, the result of which was a slab with longitudinal holes [1]. The I girder variant used as an example is reinforced by six prestressing cables in the base of the vertical shaft and in the shaft itself (variant with a height of 1100 mm). The position of cables was determined by means of Hilti PS1000 X-Scan radar.

IOP Conf. Series: Materials Science and Engineering **549** (2019) 012016 doi:10.1088/1757-899X/549/1/012016

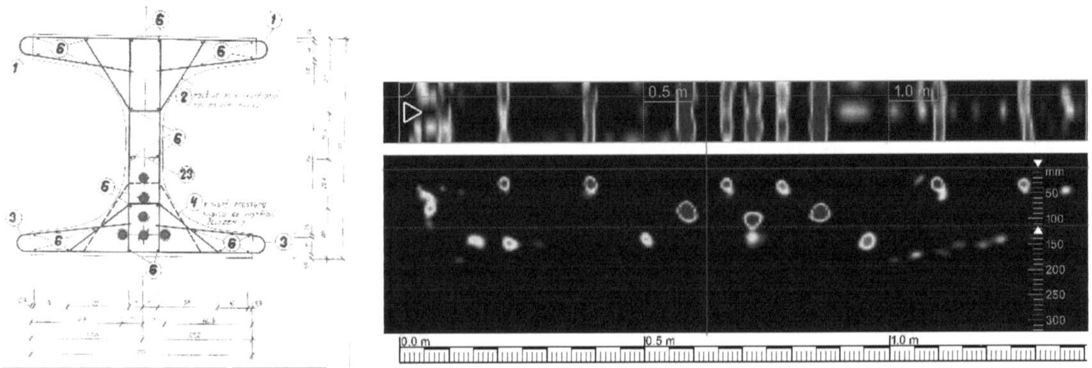

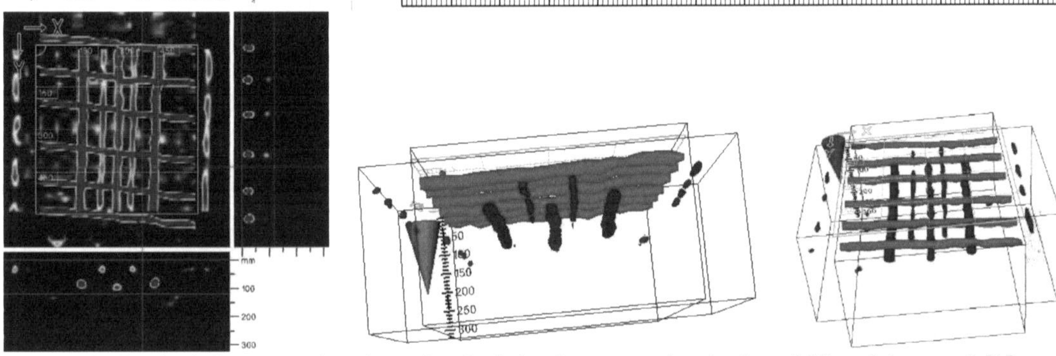

Figure 3. On the left, a type drawing of an I-girder (cross-section in the middle of the span) [1], on the right, a radar line scan from the bottom face capturing the bottom triple of prestressing cables and all the longitudinal soft reinforcement in the element.

Figure 4. A radar areal scan of 600 x 600 mm taken in the same place, for illustrative purposes in 2D, and in the middle and on the right, in 3D in various tilts. All the reinforcement is clearly visible.

Also in this type of girder it is possible to determine precisely the position of cables, but only for the bottom three ones. The remaining three cables situated in the shaft cannot be detected due to the low width of the shaft and their coverage by the lower cables. It might be possible to a certain extent from the inside of the chamber, but it is not accessible. The only exception is the peripheral girders with the accessible side surfaces.

DS-A girder with a strut frame structure and a central field

The last example is a DS-A girder with a strut frame structure and a central field from a motorway bridge. The girder is a chamber-type girder with prestressing reinforcement in the side walls [1].

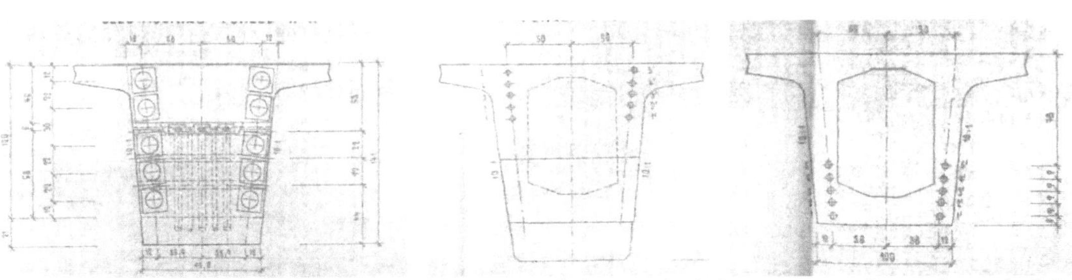

Figure 5. On the left, a type drawing of the DS-A girder, on the left, a cross-section at the end of the girder, in the middle, above the intermediate support (brace), and on the right, in the middle of the central field [1].

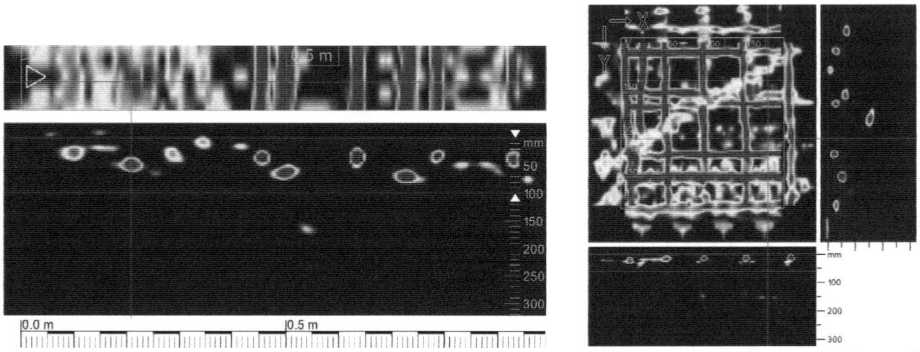

Figure 6. Radar scans (radarograms) taken on the side face of the DS-A girder approximately 3 m from the edge of the girder. Apart from soft reinforcement, the scans captured also 4 and 3 prestressing cables. The remaining two cables could not be detected due to the wider upper part of the girder. In the area above the intermediate supports, it was possible to capture only one or two bottom cables on each side, for the same reason.

Also in this type of girder it is possible to determine precisely the position of cables, but only if they are situated below the flat side of the girder, where the probe can move.

In conclusion of this part it can be said that radar can be used for determining the precise position of cables but, depending on the type of girders, part of the cables are always undetectable (approx. 30 – 50% of the total number or length of cables).

4. Subsequent methods for the detection of grouting of cable channels and condition of cables (corrosion)

4.1. Visual assessment

Unfortunately, the standard method for the determination of cable channel grouting is currently a destructive method, which consists in detecting the position of cables, making a core drill up to the edge of the cable channel (ideally covered by a steel pipe), breaking off the drill, impact drilling to the contact with the prestressing cable, and assessing the grouting, corrosion, or carbonation of the grout.

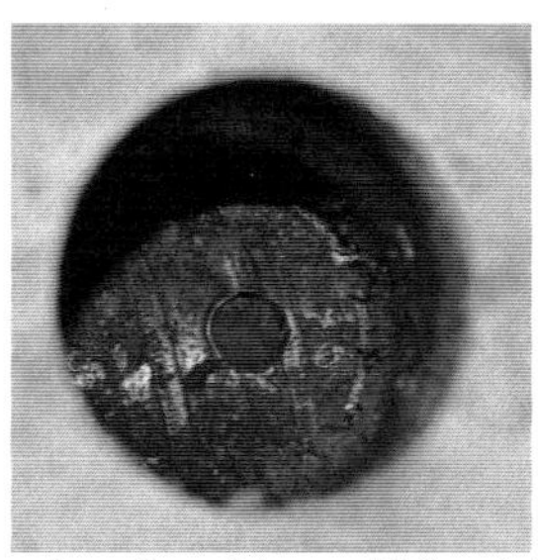

Figure 7. A drill to the steel protection pipe, breaking off the drill and puncturing the pipe to the cable. Currently, standard practice of checking.

This method has significant shortcomings – when we disregard the risk of damaging the cable when working carelessly, the major disadvantage is only local (point) evaluation of the condition which cannot capture the truly essential part of the structure.

4.2. Radiography

An alternative method is to use Co60 radiography (gammagraphy). With this method it is possible to evaluate, with a suitable irradiating configuration, the grouting (or its absence) without damaging the cable channel or its protection pipe. The problem is that nobody deals with this type of radiography in the Czech Republic at present, but this can change over time. Nevertheless, considering the complicated realization of gammagraphy, it is again a limited, local measurement not capturing the whole structure.

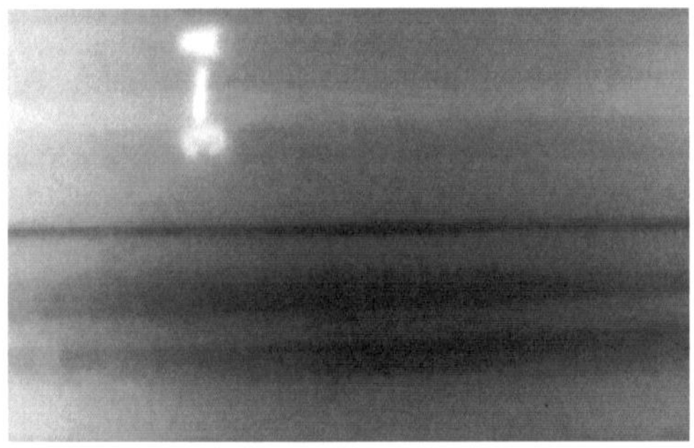

Figure 8 An example of a gammagram of a prestressed structure model with two prestressing cables, thanks to the difference in blackening it is possible to determine unequivocally the grouted and ungrouted parts (the sample below is ungrouted).

4.3. Development of new methods

The present can be characterized by a dynamic development of new diagnostic methods and devices. Hopefully, there will be alternative methods also for the diagnosis of prestressing steel. On the basis of the experiments made in cooperation with the Proceq company, it seems that new radar and ultrasonic devices could be one of the possibilities. On the premises of the Faculty of Civil Engineering, BUT in Brno, tests were performed on model blocks, which also included a model of the grouted and the ungrouted cable channel with reinforcement, by means of the Proceq GPR Live and the ultrasonic reflection device Proceq Pundit PL-200 PE. Preliminary results achieved on this model suggest the possibility of further development and appropriateness of other experiments in this respect.

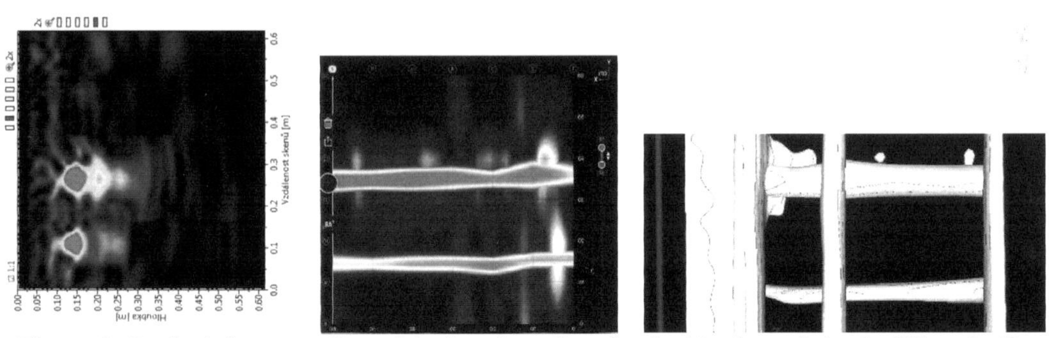

Figure 9. On the left, a recording of the location of a pair of cable channels by the US probe Proceq Pundit PL-200 PE, in the middle and on the right, a recording of the areal scan by the Proceq GPR Live radar. The results show the apparent difference between the grouted and the ungrouted channel (above), and a possibility of using a modification of these methods and devices (naturally with regard to e.g. the presence of steel pipe etc.) [4].

5. Conclusion

On the basis of the long-lasting experience of the institution in the area of pre-construction condition surveys and on the basis of a number of experiments it is possible to say that, as regards examples of structures made of prestressed concrete presented here, the structural systems are far ahead of the level and possibilities of the diagnostic methods. At the current level of knowledge, it is possible to perform diagnosis of the condition of prestressing steel only to a limited extent given by the accessibility of cables and to evaluate the state of reinforcement only at certain points. It is only the development of new methods that could allow for at least expanding this diagnosis to a larger scope of reinforcement volume in the structure, and thus provide a more relevant picture of the condition and the remaining life of the structure.

Acknowledgement
This paper has been worked out under the project No. LO1408 "AdMaS UP - Advanced Materials, Structures and Technologies", supported by Ministry of Education, Youth and Sports under the „National Sustainability Programme I".

References

[1] Vítek J 2016 History of Pre-stressed Concrete. Historie předpjatého betonu (in Czech). Praha: ČKAIT (Betonové stavitelství).
[2] Cikrle P, Anton O 2015 Evolution of Non-destructive Methods for Concrete Testing in the Last 25 Years. Vývoj nedestruktivních metod pro zkoušení betonu za posledních 25 let (in Czech). Beton TKS **15**:5.
[3] Hobst L 2007 Important Role of Radiography for determination of Conditions in Reinforced Concrete Structures. Engenharia Estudo e Pesquisa **9**:7.
[4] Downloads: Pundit Live Array Presentation. 2018. Proceq. [accessed 2018Aug.14].
[5] Anton O, Komárková T and Heřmánková V 2018 Conclusive Determination of Compliance with the Prescribed Reinforcement Elements of Concrete Structures - The Appropriate Methods and their Capabilities. Key Engineering Materials **776**:76-80. [accessed 2019Apr.10]. https://www.scientific.net/KEM.**776**.76
[6] Anton O, Komárková T and Heřmánková V 2018 Methods Appropriate for Determination of the Prescribed Reinforcement of the Elements of Reinforced Concrete Structures - Nowadays Used Methods. Key Engineering Materials **776**:41-45. [accessed 2019Apr.10]. https://www.scientific.net/KEM.776.41

IOP Conf. Series: Materials Science and Engineering **549** (2019) 012017 doi:10.1088/1757-899X/549/1/012017

Experimental investigation of the influence of the plasticizing agent on the properties of cement mortars

V Gregorová[1], Z Štefunková[1] and M Ledererová[1]

[1] Slovak University of Technology in Bratislava, Faculty of Civil Engineering, Department of Material Engineering, Radlinského 11, 810 05 Bratislava, Slovakia

E-mail: valeria.gregorova@stuba.sk

Abstract. Negative properties of concrete include the volume changes occurring during their solidification and hardening process. These deformations and cracks lead to reduce the strength and the durability of structures. The most important volume change of concrete is shrinkage resulting from the ongoing physical and chemical processes. The concrete with a high water to cement ratio is especially prone on the drying shrinkage. This article deals with the verification of the properties of cement mortars using selected kinds of plasticizing agents developed for self-levelling cement screeds. Applied agents (MS 3-12, MS 4-12, MS 5-12 and reference Berascreed 101) are based on modified polycarboxylates. The dosage of these agents was determined by the previous research in the amount of 1.5% by weight of the binder. The use of agents had a positive impact on the consistency of the fresh cement mortars and increased values of the initial and the long-term compressive strength. The negative impact of these agents was noted at the shrinkage and bulk densities of fresh and hardened cement mortars. Based on the achieved results can be concluded, that the plasticizing agent MS 4-12 may be considered a suitable substitute for the reference agent.

1. Introduction

Cement composites, especially concretes are currently the most widely used building materials in the world. The resulting concrete properties in plastic or hardened state are influenced not only by the concrete composition itself, i.e. properties of cement (mineralogical composition, quantity and fineness of grinding), water to cement ratio, properties of aggregates (dosage, granulometry), properties of additives (kind, dosage and composition), as well by a method for dosing of the individual components, mixing them, transport and compacting of fresh concrete and especially by the treatment method.

The origins of cement composites are connected with the knowledge of technology of hydraulic binders. Composites development progressed rapidly and reflecting the increasing demands of civil engineers on the building materials in the implementation of increasingly demanding constructions and often in very complicated conditions [7].

Self-Compacting Concrete (SCC) passed shortly rapid development since the creation of the concept of desired characteristics to the normal use of the free market construction work [7]. The important progress in the development of concrete with bigger fluidity occurred in Japan in the 80 years of the twenty century, the first significant applications were realized early the 90s. It is characterized by very good workability, stability and texture uniformity and resistance to component segregation. The hardened SCC generally achieves higher strength and durability compared to conventional concrete [4, 5]. The desirable properties of fresh SCC such as e.g. fluidity, low blocking flow between the reinforcement, the resistance to agitation, and segregation, sweating, good pump ability and mobility can be achieved by suitable composition [7]. The mobility of the concrete is ensured by the addition of a plasticizer (super plasticizer).

Shrinkage may be a serious problem, even if self-levelling cement screeds in which the water to cement ratio is relatively high, even when using high-quality super plasticizers. The basic technological measure for its reduction is the treatment of fresh composite. Often times, the correct method of treating or reducing the water to cement ratio using appropriate additives, do not have to be the prevention of crack formation [10]. This applies especially in the large-scale structures (screeds, floors, ceilings), where there is a very high risk of excessive evaporation of water from the structure and the subsequent formation of shrinkage cracks.

The volume changes of concrete and other cement composites occur as a result of physical and chemical processes in the cement matrix. Reduction of volume causes tensions that may result in cracks or other deformations. We distinguish several kinds of shrinkage. The shrinkage associated with the changes of the water content (moisture content) in the composite structure may be a plastic, which occurs at the initial stages of hydration, or so-called drying shrinkage manifested in composites in hardened condition [9].

A special kind of shrinkage is carbonation shrinkage that occurs in the surface layer of the composite in the long-term reaction of atmospheric carbon dioxide with cement components. The value of total shrinkage of concrete ranges from 0.3‰ to 0.7‰ and mortars shrinkage from 0.7‰ to 1.5‰ [8]. There are many ways of reducing the shrinkage of cement composites. The one of the ways is suitable reinforcement of concrete. Different types of dispersed fibres can be used to eliminating the effects of shrinkage of self-levelling cement screeds [3]. The next option is to use such an additive inorganic material as the fly ash and the shrinkage reducing agent (SRA) [2, 6].

Plasticizing agents are substances that affect the dosage of water. They can reduce the amount of mixing water with unchanged consistency or contrariwise, at the same dosage of water are able to improve the workability, i.e. the consistency of fresh concrete. Reducing the water to cement ratio has the effect on increasing the compressive strength and improving hardened concrete properties. Reducing the water to cement ratio using plasticizing agents leads to a significant reduction in shrinkage and creep of the concrete. The application of plasticizing agents with constant water to cement ratio has not significant influence on the shrinkage of concrete. Special kind of plasticizing agents are superplasticizers characterized by a very strong fluidity effect. They significantly affect the consistency of fresh concrete. Increasing dosage of plasticizing agents results in a lower consistency (better workability). The dosage of agents ranges from 0.6 to 1% by weight of agents solutions from the weight of cement. Superplasticizers have not direct impact on the shrinkage, creep or durability of concrete [1].

2. Materials and methodology

2.1. Materials

Cement CEM II A/S 42.5 R, in according with STN EN 197-1 [15], with following properties: compressive strength after 2 days 21.6 MPa, after 28 days 49.7 MPa, initial setting time 200 min, setting time 250 min and soundness (expansion) by Le Chatelier 7.2 mm.

Aggregate. Natural quarried aggregate meeting the requirements of STN EN 13139 [11] divided by sieve analysis according to STN EN 12620+A1 [12] on fraction 0/1, 1/2 and 2/4 mm.

Fly ash is characterized as the grey powder with apparent density 700 to 900 kg.m^{-3}; activity index after 28 days min 75%; activity index after 90 days min 85% and other characteristics declared by producer. The chemical compositions of fly ash (wt. %) are: SiO_2 (38.76); Al_2O_3 (16.08); Fe_2O_3 (3.21); CaO (21.28); MgO (1.98); TiO_2 (0.51); SO_3 (13.99).

Water supplied by the public water supply company was used in the mixtures, in according with STN EN 1008 [13]. All cement mortars had w/c ratio = 0.55.

Plasticizing agents. To improve the flow ability of the cement mortars were applied four kinds of super plasticizing agents based on modified polycarboxylates, which differ from each other by the type (molecule) of polycarboxylate. The agent *Berascreed 101,* characterized as the yellow-brown homogeneous solution by density at 20°C 1 070 kg.m^{-3}; pH value at 20°C 6.7±0.5; dry matter content 24% ± 1.2%, was used for reference mortar (labelled as REF). The agent *MS 4-12* was used for mixture labelled as P1; for mixture labelled as P2: *MS 3-12* and for mixture labelled as P3: *MS 5-12.* These agents have the same properties declared by producer: density at 20 °C 1 060 kg.m^{-3}±20 kg.m^{-3}; pH value at 20°C 6.0±2.0; dry matter content 20%±1%; chloride content max 0.10% by weight of agent. Furthermore, the stabilizing agent *BR stabilizer 101* with density 1 026 kg.m^{-3}, 3.6% dry weight and pH 9.5 was used to reduce the segregation of the cement mortars.

2.2. Mixture design

The design of cement mortar composition was based on commercially produced self-levelling cement screeds (mortars). The general composition of cement mortar is given in table 1. The composition of the mortars were each differed only in the kind of plasticizing agents, wherein the dose was estimated from previous research in an amount of 1.5% by weight of the binder for super plasticizing agent and 1.6% of the weight of the binder for stabilizer.

Table 1. The overview of cement mortars mixing ratio to cubic meter.

	Dosage (kg m^{-3})
Cement	243.15
Fly ash	154.00
Aggregate	
- fraction 0/1	518.72
- fraction 1/2	207.49
- fraction 2/4	311.23
Agent	
- Plasticizing	6.00
- Stabilizing	6.24
Water	
(w/c=0.55)	218.84

2.3. Measuring methods

Cement mortars were mixed according to STN EN 196-1 [14] in a normalized laboratory mixer. After that, the consistency and the bulk density of the fresh cement mortars were determined. The consistency was determined by a flow test on the Haegermann flow table as the average value of two perpendicular measurements of the cake diameter resulting from the spill of mortar after lifting the mould. Then was followed the change in the consistency of the mortar with a time of maturing. The consistency was monitored every 30 minutes from the completion of mixing of the mortar to the time, that the mortar do not longer show signs of flow ability, i.e. it was not possible to measure its spill.

The test samples to determine compressive strength were cylindrical with diameter and heights of 30 mm and were cured in humid environment. The compressive strength of hardened mortars was measured after 2, 28 and 90 days. The test samples to determine shrinkage were beams of 40 x 40 x 160 mm with glass contacts in front. They were cured for 24 hours in humid environment. After this time, samples were unmoulded and the initial length measurements were done. Subsequently, the samples were kept in a laboratory environment (temperature 20 °C±5 °C, relative air humidity φ = ca 50%). The drying shrinkage of cement mortars was determined as loss of length compared to the initial length determined after 24 hours of curing in moist environment. Shrinkage was measured after 2, 4, 7, 14, 28, 56 and 90 days of curing in the laboratory environment. The length measurement was done using the Graf-Kaufmann device. The weight loss was measured on the same samples used for the shrinkage measurements.

3. Results and discussion

Achieved results of monitored parameters of fresh and hardened cement mortars are given in table 2.

The application of four kinds of plasticizing agents had various impacts on physical and mechanical properties of cement mortars, which can be seen in figure 1 to 3.

Table 2. Physical and mechanical properties of testing cement mortars.

	Fresh cement mortars		Hardened cement mortars					
	Bulk density ($kg\ m^{-3}$)	Consistency (mm)	Bulk density ($kg\ m^{-3}$)			Compressive strength (MPa)		
			2 days	28 days	90 days	2 days	28 days	90 days
REF	2 030	275.0	1 960	1 920	1 920	6.77	19.26	20.80
P1	2 120	282.5	2 160	2 010	1 990	9.85	24.38	26.00
P2	2 070	229.5	2 170	2 010	2 030	7.55	17.80	23.10
P3	2 080	245.5	2 125	1 980	1 970	8.86	21.64	25.55

3.1. Consistency

The use of plasticizing agents resulted in a high fluidity of cement mortars. Flow diameter ranged from 229.5 mm to 282.5 mm. The most liquid initial consistency was achieved using the MS 4-12 plasticizer (P1). Its value was higher by about 2.7% compared to the reference mortar.

The consistency of cement mortars were gradually changed with a time of their maturing. All of the observed mortars showed even bigger fluidity during the first 60 minutes. The further course of the change in consistency depended on the type of used plasticizer, which was resulted to different rates of loss of workability (Figure 1). The fastest loss of workability was achieved for reference mortar after about 210 minutes. Mortars tested for new plasticizing agents were still workable after 4 hours. Mortar with the plasticizing agent MS 5-12 (P3) has maintained the longest ability for workability.

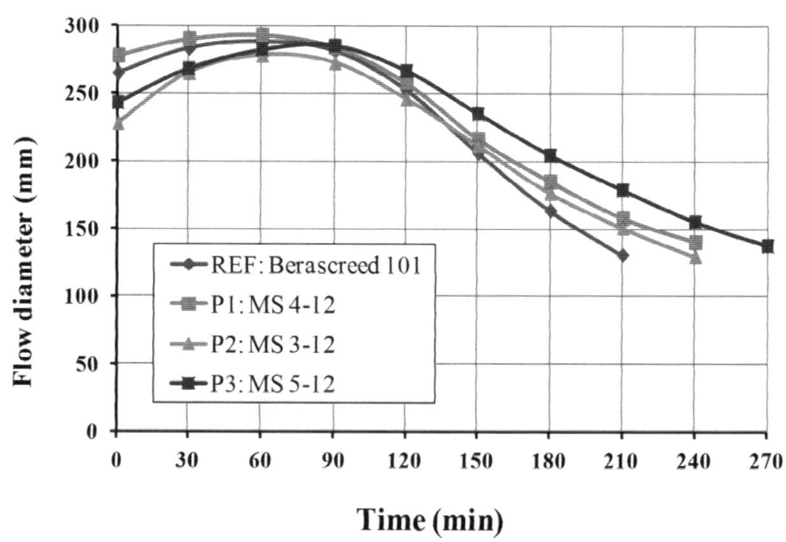

Figure 1. The change in the consistency of the cement mortar with a time of maturing.

3.2. Bulk density and weight loss

The values of density of fresh mortar were almost identical to the reference mortar, which reached a value of 2030 kg.m^{-3}. Some difference was observed for sample P1, whose bulk density was 2120 kg.m^{-3}. Bulk density of hardened cement mortars was gradually declined with a time of keeping the samples in humid environment. The reference mortar reached the lowest value of the bulk density, i.e. 1920 kg.m^{-3} (90 days of hardening). The values of bulk density of cement mortars with new plasticizing agents in the same reference period ranged from 1970 kg.m^{-3} to 2030 kg.m^{-3}.

Volume changes in cement mortar had started to show from the start of drying. The biggest weight loss occurred in the first days of setting and subsequent hardening of the mortars. After 28 days of hardening, there was observed only a minimal weight change of mortars and after 56 days the weight has not changed practically. The biggest weight loss compared to the reference mortar was reached on the cement mortar (P3), i.e. 8.8%.

3.3. Shrinkage

Course of shrinkage of cement mortar (Fig. 2) corresponded with the weight loss of the sample. The biggest shrinkage was reflected in the first days, gradually decreased and after 28 days was changed minimally. The weight of reference mortar increased in the first days, i.e. there was a swelling and not shrinkage of the mortar. The overall lowest shrinkage was found just at the reference mortar, i.e. 0.66‰ after 90 days of hardening. The lowest shrinkage values of the new tested plasticizing agents reached cement mortar P1 with a value of 0.76‰ (90 days).

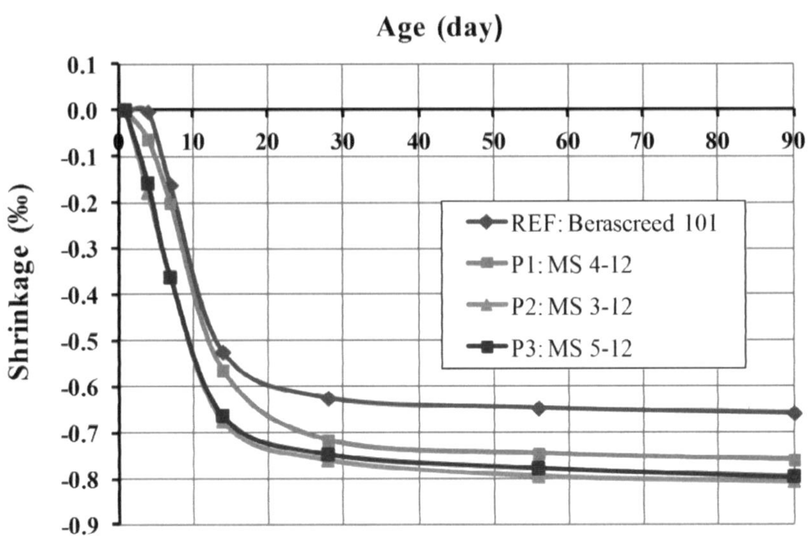

Figure 2. The shrinkage of cement mortars in the application of four kinds of plasticizing agent.

3.4. Compressive strength

The influence of plasticizing agents on the compressive strength of cement mortars can be seen in figure 3. Using these agents had only a minimal effect on an increase the initial (2-day) and the long term (28 and 90 days) compressive strength of cement mortars compared to the reference mortar.

The resulting values of compressive strength in the monitoring period ranged from 7.55 MPa to 26.00 MPa. A positive increase of strength characteristics of testing mortars was observed by the application plasticizer MS 4-12 (P1). Due to the achieved lower compressive strengths, the plasticizing agent MS 3-12 (P2) appears to be the least suitable.

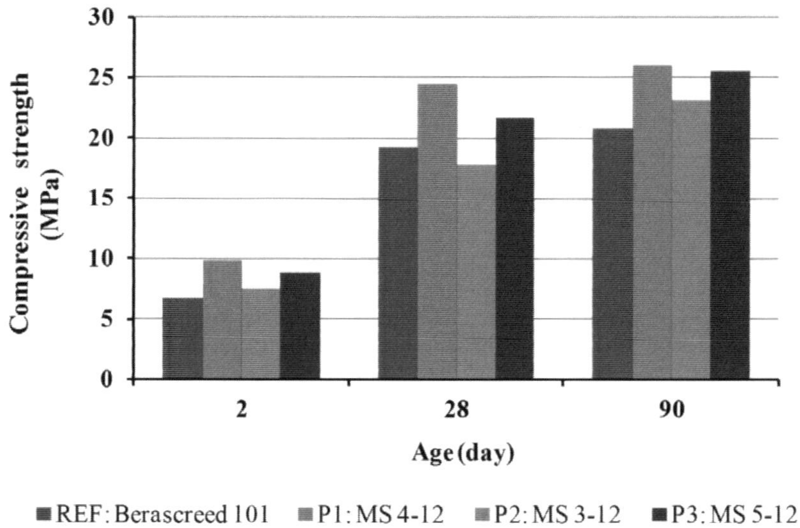

Figure 3. The compressive strength of cement mortars in the application of four kinds of plasticizing agent.

4. Conclusion

Based on the results of the experiments can be concluded that the use of new types of plasticizing agents had only a slightly different effect on the observed properties of cement mortars compared to the reference additive.

Application of plasticizing agents resulted in a high fluidity of cement mortars. Flow diameter of testing mortars ranged from 229.5 mm to 282.5 mm. In the case of use the plasticizer MS 4-12, the initial consistency was slightly improved, the effectiveness of other ingredients (MS 3-12 and MS 5-12) was lower. Changing the mortars consistency with time of maturing was during 210 minutes almost identical in all cases. Another course of the change in consistency depended on the kind of used plasticizing agent, which was resulted to different rates of loss of workability. Mortar with the plasticizer MS 5-12 has maintained the longest ability for workability.

The values of fresh mortar´s densities were almost identical with the reference mortar, which reached a value of 2030 $kg.m^{-3}$. Some difference was observed on mortar with plasticizer MS 4-12, whose bulk density was 2120 $kg.m^{-3}$. Bulk density of hardened cement mortars was gradually declined with a time of keeping the samples in the humid environment. The reference mortar reached the lowest value of the bulk density, i.e. 1920 $kg.m^{-3}$ (90 days). The values of bulk density of cement mortars with new plasticizing agents in the same reference period ranged from 1 970 $kg.m^{-3}$ to 2 030 $kg.m^{-3}$.

The biggest shrinkage of mortars was reflected in the first days. It was decreased gradually and after 28 days was changed minimally. The overall lowest shrinkage was observed at the reference mortar, i.e. 0.66‰ after 90 days. The lowest shrinkage values of new plasticizing agents reached a mortar with the plasticizer MS 4-12, i.e. 0.76‰ (90 days).

The results showed a positive effect of the tested plasticizing agents on the initial and the long-term compressive strength of mortars. All new plasticizers resulted to higher values of compressive strength as a reference sample. The one exception is mortar P2 (plasticizer MS 3-12) at the age of 28 days, the strength value of which was higher as a reference sample. However, this difference was minimal. The compressive strength values ranged from 7.55 MPa to 9.85 MPa after 2 days of hardening and from 23.10 MPa to 26.00 MPa after 90 days of hardening.

Due to the achieved results, the plasticizer MS 4-12 seems to be a suitable replacement for plasticizing agent Berascreed 101.

Acknowledgment
This article was supported by the Slovak Research and Development Agency under the contract No. APVV-15-0681 and by the Science Grant Agency - project VEGA No. 1/0501/17.

References
[1] Bajza A and Rouseková I 2006 *The concrete technology* (Bratislava: Jaga Group) pp 190
[2] Collepardi M 2005 Effects of Shrinkage Reducing Admixtures in Shrinkage Compensating Concrete under Non-Wet Curing Conditions *Cem. Concr. Compos.* **27** pp 704-708
[3] Ďubek M, Makýš P, Ďubek S and Petro M 2018 The evaluation of the content of fibers in steel fiber reinforced structures and image analysis *Journal of Civil Engineering and Management* **24** (Vilnius: VGTU Press) pp 183-192 https://doi.org/10.3846/jcem.2018.1642
[4] Hayakawa M 1993 Development and Application of Super Workable Concrete. *Special Concretes – Workability and Mixing* (London: E & FN SPON) pp 183-190
[5] Hela R and Hubáček A 2005 Influence of extreme climatic conditions on properties of aerated self-compacting concrete *The 3rd Conference - Special Concretes.* (Malenovice: Sekurkon Ostrava) pp 29-34
[6] Matulová K and Unčík S 2017 Effect of shrinkage-reducing admixture on the properties of cement composites (London: Taylor & Francis Group) pp. 329-334.
[7] Schutter G D, Bartos P J M, Domone P, Gibbs J and Hela R 2008 *The self-compacting concrete* (Praha: ČSB Servis, s.r.o.) pp 344
[8] Svoboda L 2005 *Building Materials* (Bratislava: Jaga Group) p 470
[9] Tazawa E and Miyazawa S 1995 Chemical and Autogenous Shrinkage of Hydrating Cement Paste *Cem. Concr. Res.* **25** pp 288-292
[10] Unčík S and Matulová K 2012 *Using an expansion cements to reduce the shrinkage of concrete* https://www.tzbportal.sk/stavebnictvo/vyuzitie-expanznych-cementov-na-redukovanie-zmrastovania-betonu.html [accessed March 12 2019]
[11] STN EN 13139/AC (72 1503) 2004 *Aggregates for mortar*
[12] STN EN 12620+A1 (72 1502) 2008 *Aggregates for concrete*
[13] STN EN 1008 (73 2028) 2003 *Mixing water concrete.* Specification for sampling, testing and assessing the suitability of water, including water recovered from processes in the concrete industy, as mixing water for concrete
[14] STN EN 196-1 (72 2100) 2016 *Methods of testing cement* Part 1 Determination of strength
[15] STN EN 197-1 2012 *Cement* Part 1 Composition, specifications and conformity criteria for common cements

IOP Conf. Series: Materials Science and Engineering **549** (2019) 012018 doi:10.1088/1757-899X/549/1/012018

An experimental preparation of fibre concrete to software´s detection of fibres

V Gregorová[1], **M Ďubek**[2], **S Ďubek**[2] **and Z Štefunková**[1]

[1] Slovak University of Technology in Bratislava, Faculty of Civil Engineering, Department of Material Engineering, Radlinského 11, 810 05 Bratislava, Slovakia
[2] Slovak University of Technology in Bratislava, Faculty of Civil Engineering, Department of Building Technology, Radlinského 11, 810 05 Bratislava, Slovakia

E-mail: valeria.gregorova@stuba.sk

Abstract. Fibre concrete requirements are only met if the fibres in its volume are evenly distributed. Their direction and orientation should be equal around each axis in a three-dimensional space. The requirement is that the fibres to be evaluated in the final hardened fibre concrete are cut from fibre concrete sections. Homogeneity can also be verified on fibre-concrete cross-sections. Fibre Position Evaluation is a mathematical task that is not presented here but is currently made up of the results of this experiment. In the first step, it was necessary to prepare samples to obtain the necessary cross-sections. In the next step, it was necessary to create conditions by creating a contrast between the fibres and the concrete to create software capable of detecting fibres. In this paper, the principle of detection of fibres based on the detection of cut section images is described. There are several methods used to highlight the concrete matrix over the fibres. Also presented are the results of adding and detecting fibres.

1. Introduction

The basic requirement for full-fledged effect of fibres in steel fibre reinforced concrete "SFRC" is even distribution of fibres called homogeneity. SFRC refers to composite material exhibiting higher strength characteristics, consisting of concrete, and fibres. SFRC is a cement composite, which is characterized by high tensile and flexural strength. These are the qualities of demanded and utilized in modern architectural elements of structures.

The deployment of fibres in fresh or hardened concrete has been the subject of several studies, for example [3, 5, 7]. SFRC has been investigated by different devices through metal detectors to X-ray machines [8, 9, 11, 16, 19, 20] or by electromagnetically methods [9]. To ensure homogeneity of fibres in concrete is not a straightforward process. The purpose of measuring and experiments was to determine how the fibres in concrete are actually arranged and define the uniformity of fibre arrangement, i.e. homogeneity. A number of factors come into play during the process of manufacturing fibre reinforced concrete structures that influence the final layout and orientation of the fibres. The ideal approach would be to orient the fibres according to the load direction of the structure. As part of the research it is also necessary to follow the direction of fibre rotation according to the direction of filling of the sample [12, 17]. The choice of fibres in the production of SFRC structures plays an important role. As a matter of fact, the shape of these structures during the production affects the final position and arrangement of the fibres. The orientation factor of the fibres affects the direction of the fibres and therefore the dimensional characteristics of the beams are crucial to choosing the length of the fibres [6, 10]. All experimental research was conducted in laboratory conditions. In practice, however, inspection directly on the site with the results of the right steps in the implementation of SFRC structures is required.

2. Materials and methodology

2.1. Materials

Material used. Mixed Portland cement CEM II / B-M (S-V-LL) 32.5 R, mined aggregate and steel fibres 60 mm long and 1 mm in diameter were used to produce the six fibre-concrete samples (beams). As for the concrete production, a fresh reference concrete formula was used, according to EN 14845: 2007 [21- 24] with water to cement ratio of 0.55. The cement dose amounted to 320 kg/m^3, water dose 176 kg/m^3, aggregate fractions 0/4 mm and 8/16 mm aggregate fractions [21] were loaded both with 952 kg/m^3. The dosage of the fibres [1, 13] was determined according to the draft for floor structures - 30 kg/m^3.

Experimental beams. For research purposes and to verify dispersion ratings of the reinforcement, beam samples with dimensions of 150x150x550 mm have been produced (Figure 1). The beams were cut into smaller parts with dimensions of 100 mm. The cut surfaces of the fibre-reinforced concrete beams (150x150 mm) were photographed and analysed with the help of a software program. The developed software was supposed to identify the number of fibres, their direction and spatial orientation taken from the images of fibre-reinforced sample cuts.

Cutting samples and photography. The cutting of experimental beam samples was conducted on a bench saw with a high-speed cutting disc. The beam has been slid on a sliding table (runner) at an appropriate speed so that the fibres on the surface of the cut would not have been deformed. Diamonds on the cutting disc could damage the surface of the transacted fibre, which may result in lower light reflection during photography. This may cause an incomplete loading of fibres by the developed software. In the case of damaged fibres, the surface of the cut may additionally be sanded, achieving a straight cut surface of the fibres.

The fibres used in the experimental beam sample production were selected with respect to its dimensions and were identical with the fibres used to produce the SFRC structure. The direction of detected fibres is dependent upon the indicative factor of the structure and therefore the dimensional characteristics of the beams crucial to the choice of the fibre's length. The orientation factor has been the subject of the research [6, 10, 15] and indicates the direction and position of the fibres at the edges of the sample. The casting walls have a big impact on the direction of the fibres and therefore the indicative factor is in the element's middle $\alpha 1 = 0.5$; $\alpha 2 = 0.6$ and at the edges $\alpha 3 = 0.84$. Depending on the shape of the casting the "α" is calculated as a total. The fibres were thus selected with a length of 60 mm and a diameter of 1.06 mm. Their slenderness ratio equals to 55. Slenderness ratio characterizes the shape of the fibre and determines the ratio of fibre length to its thickness. The same formula was used in the concrete implementation of SFRC floors [18, 4] in industrial buildings.

2.2. Design of the contrast

Chemistry contrast. According to the available chemicals to enhance the contrast of colours between concrete and fibres, we chose primarily the use of concrete surface properties. Specifically, it is the use of cementitious alkalinity in concrete, which can be used to stain the surface of a fractured concrete surface with acid-base indicators. Indicators are substances that change its characteristic features indicate the change or the current status of the chemical composition of the system. Colour acid-base indicators change colour depending on the pH value of the material [2, 14].

The cement stone in the concrete is alkaline (the pore liquid has a pH value about of 12 to 13.5 depending on the type of cement used). Acid-base indicator solutions change a pH value dependent staining. The bromothymol blue used has a yellow colour at a pH value less than 6.0 and turns blue at a pH value greater than 7.6. Colour of "A" represented the wet surface of concrete with a pH value about 12 to 13 using the obtained blue colour indicator (Figure 1 on the left). A similar change occurs even with the use of an experimental substance - phenolphthalein. For this experiment, the acid-base indicator phenolphthalein was chosen as its alcoholic solution. It stained the concrete surface to red, that can be seen in figure 1 on the left side labelled as colour "B". In the latter case, substances were used to colour the surface of the concrete using other added chemicals. By rubbing the surface of the concrete with potassium ferrocyanide and adding a solution of ferric salts (in my case ferric nitrate) gives us a blue precipitate, called as *the Berlin Blue,* labelled as colour "C" in figure 1 on the left side. The concrete surface is coloured only if it is still not carbonated.

Figure 1. Left sample with chemical tests, right sample with pen ink. A: bromothymol blue; B: phenolphtelein; C: potassium ferrocyanide and ferric nitrate.

Mechanics contrast. The sawing speed is signed in the grooves formed on the exposed fibre surfaces. This is an undesirable effect for surface shooting because the fibre surfaces do not form a single-coloured surface, but several shades. Elimination of this deficiency would consist in additional grinding. For grinding, we chose a water-cooled electric grinder and grinding wheels used for grinding and polishing ornamental stones and marble. It is a powerful grinder with a smooth blade that reaches a smooth and flat concrete and fibre surface at high speeds (Figure 4 and 5). In this way, the formation of a better surface was achieved and the fibres had a flat surface capable of reflecting enough of the same light in the image.

2.3. Photography

The camera was placed on a tripod to achieve higher image quality during imaging (Figure 2). Additional flash has not been used during imaging, so the software is also able to process the reflection of natural light from the fibres. Shiny surfaces of the cut fibres with higher reflectivity than concrete reflected more light. Thus, they appear on images as bright areas (areas with high brightness). Imaging is conducted on all surface cuts. The cut surfaces have not been additionally treated.

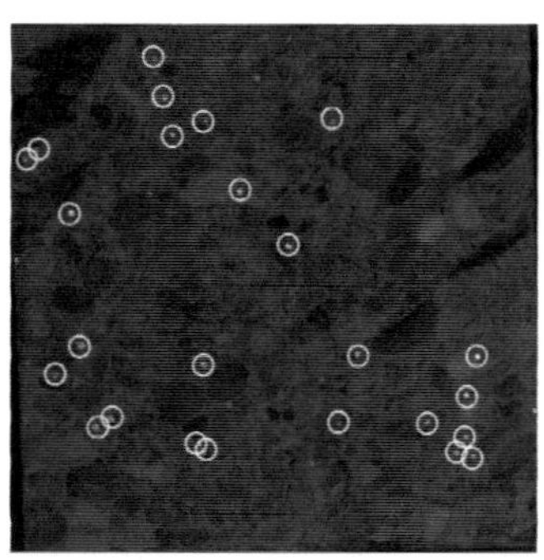

Figure 2. Manual fibre counting.

2.4. Analysis software

To search for fibres a specific application has been suggested (Figure 3). The basic application is designed in C ++. The application architecture is built to make use of external libraries OpenCV. OpenCV is an open-platform library designed for image processing. It is focused on the computer vision and image processing in real-time [25]. Graphical application environment is created in the library of the Microsoft Foundation Class Library (MFC), which encapsulates the Windows API in C ++ classes, ensuring full compatibility with most Windows platforms. Used technologies ensure scalability of the application and expandability of the new features in the future.

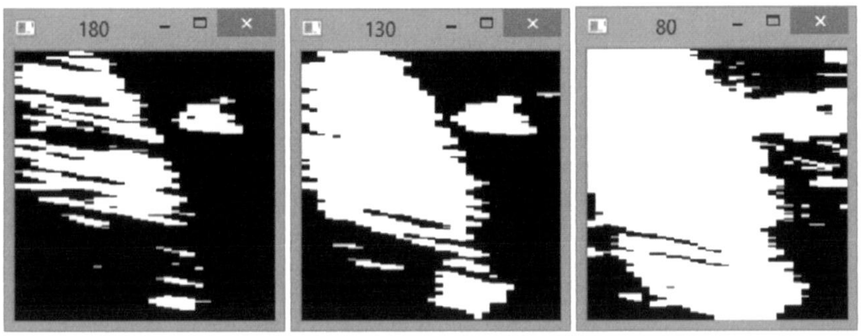

Figure 3. An exemplary software thresholding and search for contours of the fibre.

On to the loaded input image different filters of the OpenCV library are applied. If the input image is large and it contains dead spots on the edges, the software uses thresholding (image processing). Subsequently contours are searched for, while convexity defects are combined to form one solid object. The result of this step is the cut-out image of the concrete sample. The image of the concrete sample is then processed using morphological reconstruction to eliminate local maximums. Subsequently, methods of erosion and dilation were applied to improve the quality of bright parts. Along with the binary thresholding bright parts were obtained, which in our case represents the found fibres.

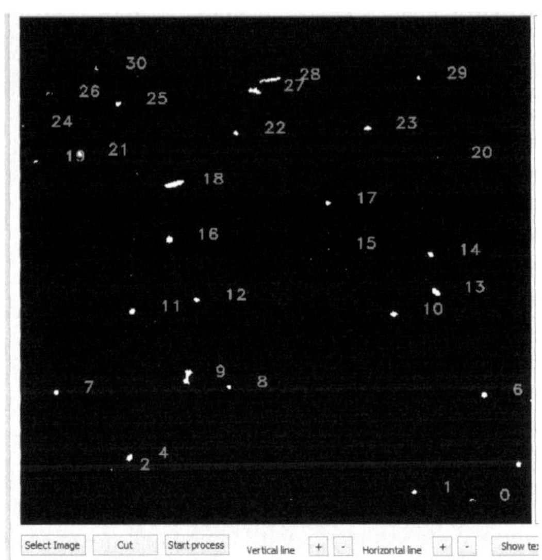

Figure 4. The trimmed image with fibres detected.

3. Results

To confirm the reliability of the software, it was necessary to detect multiple images. Each image analysed by the software was also analysed manually. The software was able to identify all the threads from the sample sharpened image. The success rate was 99.97% of the found fibres compared to the actual state. This was 2.32% more for inkjet images. The reason could be the creation of high contrast between grey and concrete fibres in weak to strong blue (Figure 1, right). In the picture, we can see a strongly blue "puddle" with a grey border, causing extra fibres to load. For further imaging, samples were prepared by cutting with additional surface grinding for higher accuracy of detected fibres. Table 1 shows the results from sample number 1 in sections B to K.

Table 1. Number of fibres with chemical and mechanical contrasting, and deficiency on sample 1.

	Chemical contrast			**Mechanical contrast**		
	Manual counting	Analysis software	Variation (%)	Manual counting	Analysis software	Variation (%)
1B	31	30	103.23	31	31	100.00
1C; 1D	44	42	95.45	44	44	100.00
1E; 1F	38	38	107.89	38	37	97.37
1G; 1H	54	52	100.00	54	54	100.00
1I; 1J	63	63	100.00	63	63	100.00
1K	41	38	107.32	41	42	102.44
Average			102.32			99.97

Multiple samples had to be processed to obtain the relevant output. All resulting measurements showed the need to scan mechanically modified fibre concrete surfaces. The software showed the most accurate results after mechanical surface treatment. Its success rate was 99.97%. At the same time, the number of fibres in single sections represents fibre density and homogeneity along the sample.

4. Conclusion

The research presented in this article deals with the evaluation of uniformity of dispersed fibres in fibre reinforced concrete structures. Its focus lies on the basic characteristics of fibre concrete. It is a composite material with ideally dispersed reinforcing elements in the form of fibres.

The process of automatic image processing is already used in practise. However, it is at the beginning in the field of fibre concrete. Its benefits are needed for homogeneity research and the speed of evaluation of results. It saves time and its accuracy is sometimes higher than manual counting and identification of fibres. The correct algorithm will be able to determine the homogeneity in the final fibre concrete.

The chemical process was best demonstrated in increasing contrast. However, the software detected a higher number of threads in the section. This could be due to wet stones in the concrete matrix. At that time, the software also included these small stones. The result is an average of 2.32% more fibre-identified software than the actual amount of fibre detected. The lower contrast was reflected in the sample ground. The fibres get a higher gloss, which is more accurate for the developed software in its identification spectrum. The results showed an accuracy of 99.97% on 6 samples. This process is useful for further processing images on samples. Thanks to automated image processing, fibre counting is possible efficiently.

The tool creates a basis for further processing of fibre homogeneity issues and serves as a control tool for the evaluation of manufacturing fibre reinforced concrete structures. It can also be applied to production of fibre reinforced concrete structures with controlled orientation, where the assumption of acting internal forces is known. It also creates a basis for the development of evaluation software and more detailed expression of decomposition of forces into individual axes. Currently the conducted research is aimed at verifying the possibility of evaluating the images obtained from the CT-Scan device - see picture below.

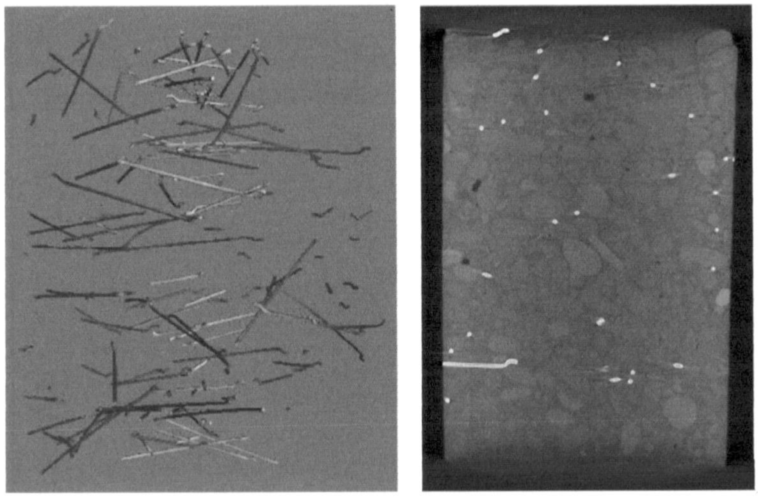

Figure 5. Output images from the CT-SCAN left image sample and right radiographs.

Acknowledgment
This article was supported by the Slovak Research and Development Agency under the contract No. APVV-15-0681.

References

[1] Anton O, Komarkova T, Lanik J, Hermankova, V 2018 Development of new methods for determining the distribution of steel fibres in the hardened steel fibre reinforced concrete - Possibilities of production and verification of test specimens *IOP Conference Series: Materials Science and Engineering*, **385** (1), art. no. 012002,

[2] Bína J et al.: Malá encyklopédia chémie, Bratislava: Obzor, 1968,Číslo povolenia 1040/I-1967

[3] Bui V K, Geiker M R, Shah S P 2003 Rheology of fibre reinforced cementitious materials. *Naaman AE, Reinhardt HW (eds) Proceedings of HPFRCC4 workshop, Ann Arbor*. RILEM Publications, Cachan, pp 221–231

[4] Ďubek M, Makýš P, Ďubek S and Petro M 2018 The evaluation of the content of fibres in steel fibre reinforced structures and image analysis *Journal of Civil Engineering and Management* (Vilnius: VGTU Press) pp 183-192 https://doi.org/10.3846/jcem.2018.1642

[5] Ferrara L, Meda A 2006 Relationships between fibre distribution, workability and the mechanical properties of SFRC applied to precast roof elements. Mater Struct **39**:411–420

[6] Gettu R, Gardner D R, Saldivar H, Barrgán B E 2005, Study of the distribution and orientation of fibres in SFRC specimens, RILEM, *Materials and Structures* **38**.

[7] Katzer J, Domski J 2012 Quality and mechanical properties of engineered steel fibres used as reinforcement for concrete. *Constr Build Mater* **34**:243–248

[8] Kim KY, Yun TS, Choo J, Kang DH, Shin HS 2012 Determination of air-void parameters of hardened cement-based materials using X-ray computed tomography. *Constr Build Mater* **37**:93–101

[9] Komarkova T, Lanik J, Topolar L, Stoller J, Stonis P 2018 Experimental assessment of steel fibre reinforced concretes with different concentrations of fibres *IOP Conference Series: Materials Science and Engineering*, **385** (1), art. no. 012028, .

[10] Lee B Y, Kim J K, Kim J S, Kim Y Y 2009, Quantitative evaluation technique of Polyvinyl Alcohol (PVA) fibre dispersion in engineered cementitious composites, *Cement & Concrete Composites* **31**, 408–417

[11] Liu JP, Li CF, Liu JZ, Cui G, Yang Z 2013 Study on 3D spatial distribution of steel fibres in fibre reinforced cementitious composites through micro-CT technique. *Constr Build Mater* **48**:656–661

[12] Martinie L, Roussel N 2011 Simple tools for fibre orientation prediction in industrial practice. *Cem Concr Res* **41**:993–1000

[13] Motyčka V, Henková S, Kantová R, 2014 Considers the Opinions of Construction Companies When Planning Study Programme Content, *Conference Proceedings of 3rd Cyprus International Conference on Educational Research*, ISSN 1877-0428,

[14] Minarovičová K, Antošová N Interaction between pathology and technology of buildings. *MINAROVIČOVÁ, K. -- URBÁN, D. -- RYCHTÁRIKOVÁ, M. ATF 2016: book of posters from 4th International Conference on Applied Technology*. 15-17 September 2016, Leuven, Belgium: The Katholieke Universiteit Leuven, 2016, s. 26. ISBN 9789086497973

[15] Nečasová B, Liška P, Novotný M 2019 Ageing of Adhesive Joints for Façade Applications – Comparison of Artificial and Real Weathering Conditions, *MATEC Web Conf.* Volume 279, 2019, 10th International Scientific Conference Building Defects (Building Defects 2018), ISSN 2261-236X, ISBN 978-2-7598-9032-3, EDP Sciences,

[16] Ponikiewski T, Gołaszewski J, Rudzki M, Bugdol M 2015 Determination of steel fibres distribution in self-compacting concrete beams using X-ray computed tomography. *Arch Civ Mech Eng* 15:558–568

[17] Şanal İ, Ozyurt N 2013 To what extent does the fibre orientation affect mechanical performance? *Constr Build Mater* **44**:671–681

[18] Švajlenka J, Kozlovská M, Posiváková T 2018 Analysis of selected building constructions used in industrial construction in terms of sustainability benefits *Sustainability* (Switzerland), 10 (12), art. no. 4394, .

[19] Wong RCK, Chau KT 2005 Estimation of air void and aggregate spatial distributions in concrete under uniaxial compression using computer tomography scanning. *Cem Concr Res* 35(8):1566–1576

[20] Yang, C. J. Jin, L. B. Chen, D. C. Qi, J. Practical Measurement for Steel Fibre Distribution of the SFRC Beams, *Applied Mechanics and Materials*, Vols. **256-259**, pp. 840-843, 2013

[21] STN EN 12620+A1 (72 1502) 2008 *Aggregates for concrete*

[22] STN EN 14889-1 2006 *Fibres for concrete - Part 1: Steel fibres - Definitions, specifications and conformity*

[23] STN EN 14845-1 2007 *Test methods for fibres in concrete - Part 1: Reference concretes*

[24] STN EN 14845-2 2006 *Test methods for fibres in concrete - Part 2: Effect on concrete*

[25] MS Company HTML Color Codes http://www.recenzie.com/kody-farieb [accessed April 16 2019]

Experimental determination of freeze-thaw resistance in self-compacting concretes

D Kocáb[1], M Lišovský[2] and P Žítt[1]

[1] Brno University of Technology, Faculty of Civil Engineering, Veveří 331/95, Brno, Czech Republic
[2] Svoboda a syn, s.r.o., Jahodová 524/62, Brno, Czech Republic

E-mail: dalibor.kocab@vutbr.cz

Abstract. The paper describes the determination of freeze-thaw resistance in self-compacting concretes and compares several test methods used for this purpose. The basic principle of testing freeze-thaw resistance is the observation of changes, in this case the loss of certain properties, in concrete attacked by freeze-thaw cycles. This paper documents measurements of tensile strength, tensile splitting strength, and dynamic Young's modulus measured by the ultrasonic pulse velocity test and resonance method. The outcome of the experiment is an evaluation of how suitable each test is for determining the freeze-thaw resistance of self-compacting concrete.

1. Introduction

Self-compacting concrete (SCC) possesses properties thanks to which it quickly received popularity in civil engineering. The first SCC was manufactured in Japan approximately 30 years ago [1] and has entered widespread usage since then. Unlike ordinary concretes, SCC is more sensitive to the quality of feedstock (mainly aggregate and admixtures), composition (variations in particle size distribution, too many fine particles), and the actual placement and curing [2,3].

Environmental matters of building design and construction have recently come into focus as well, whether they pertain to service life or sustainability [4]. Resistance to various adverse effects is therefore important for self-compacting concretes as well [5,6,7]. One of the main factors that may substantially reduce the durability of a concrete structure is the combination of freezing temperatures and high moisture. When concrete is subjected to repeated freezing and thawing, it begins to degrade if water is present. Dry concrete suffers practically no damage from this type of attack [8]. When water comes in contact with concrete, it enters its pore structure and when it freezes (at roughly -0.5 °C in the capillary system), problems may occur. The volume of ice is approximately 9 % greater than the volume of water, which is what causes internal stress in the pore structure. The ensuing damage to the structure may bring about an irreversible reduction in the basic material properties of the concrete [9,10].

It is therefore the objective of this paper to assess the possibilities of determining the freeze-thaw resistance of self-compacting concretes with an emphasis on non-destructive tests (NDT).

2. Experiment
The experiment was performed with specimens made of four self-compacting concretes, which were subjected to freeze-thaw (F-T) cycles. Besides that, all the concretes were tested for basic material properties using both destructive and non-destructive tests.

2.1. Material
For better navigation, the four SCC formulas were identified by the letters A through D. All SCC were made with the same cement CEM I 42.5 R, the same 0-4 mm sand, and the same coarse aggregate of the fraction 4-8 mm and 8-16 mm. They only differed in the amount of plasticiser and the addition of micro-milled limestone and fly ash at varying ratios. Because these concretes were being developed for the commercial market, the paper will not disclose their composition in any more detail.

2.2. Test specimens
Each SCC was made into 9 prisms with the nominal dimensions of 100×100×400 mm. Once cast in steel moulds, the specimens were left to set for approximately 1 hour at standard laboratory conditions and afterwards covered by a PE sheet. At the age of 24 hours the specimens were removed from the moulds and stored at a temperature of (20 ± 2) °C and relative humidity over 95 %. When they reached the age of 25 days they were placed under water at (20 ± 2) °C, where they remained for three days until beginning of the tests.

2.3. Test methods
At the age of 28 days, all the specimens were tested for dynamic Young's modulus. Given the rather low result variability (there were only minor differences between the specimens), see table 1, the specimens of all the concretes were divided into three sets as follows – prisms no. 1 to 3 were reference, 4 to 6 were tested for F-T resistance after 75 cycles and prisms no. 7 to 9 were tested for F-T resistance after 150 cycles.

As stated earlier, the first property to be determined was dynamic Young's modulus; it was measured by the ultrasonic pulse velocity test and resonance method. Young's modulus measured by the ultrasonic pulse velocity test bears the designation E_{cu} and its value was calculated in compliance with ČSN 73 1371 [11]. The pulse velocity was measured using a Pundit PL-200 with 150 kHz transducers. Dynamic Young's modulus measured by the resonance method, in keeping with ČSN 73 1372 [12], is identified as E_{crL} in the case of longitudinal vibration and E_{crf} in flexural vibration. The natural vibration frequencies were determined using a Handyscope HS4 oscilloscope and a software capable of the fast Fourier transform. The reference specimens were then tested for flexural strength f_{cf} according to EN 12390-5 [13] and tensile splitting strength f_{ct} according EN 12390-6 [14] (this test used prism fragments left after the f_{cf} test).

Specimens no. 4 to 9 of all four concretes were then placed in a KD-20 test chamber that allows cyclic freezing and thawing. One F-T cycle consisted of 4 hours of freezing at an air temperature of -18 °C and 2 hours of thawing in water at 20 °C in compliance with ČSN 73 1322 [15].

3. Results and discussion
After 25, 50, and 75 F-T cycles the specimens were tested for dynamic Young's modulus to see how this property changes depending on the number of cycles. The outcome is the relative dynamic modulus of elasticity (RDM), which is relative to the reference value measured prior to the commencement of the tests (as well as the relative strength development – "f-ratio" in the graphs). The relative development of Young's modulus calculated from the data obtained by the ultrasonic pulse velocity test is identified as RDM (U), resonance method with longitudinal vibration RDM (FL), and resonance method with flexural vibration RDM (FF). Figure 1 shows plots of these results.

IOP Conf. Series: Materials Science and Engineering **549** (2019) 012019 doi:10.1088/1757-899X/549/1/012019

Table 1. Average values of the properties before testing including sample standard deviations (s.s.d.).

Concrete	E_{cu} [GPa]	E_{crL} [GPa]	E_{crf} [GPa]	f_{cf} [MPa]	f_{ct} [MPa]
A: average	**43.8**	**41.3**	**42.0**	**8.6**	**5.2**
A: s.s.d.	0.45	0.34	0.41	0.35	0.24
B: average	**41.1**	**38.7**	**39.4**	**8.0**	**4.6**
B: s.s.d.	0.32	0.21	0.12	0.22	0.20
C: average	**40.2**	**38.1**	**38.6**	**8.5**	**5.2**
C: s.s.d.	0.42	0.45	0.49	0.17	0.34
D: average	**45.1**	**42.4**	**43.3**	**7.5**	**5.2**
D: s.s.d.	0.50	0.44	0.48	0.54	0.28

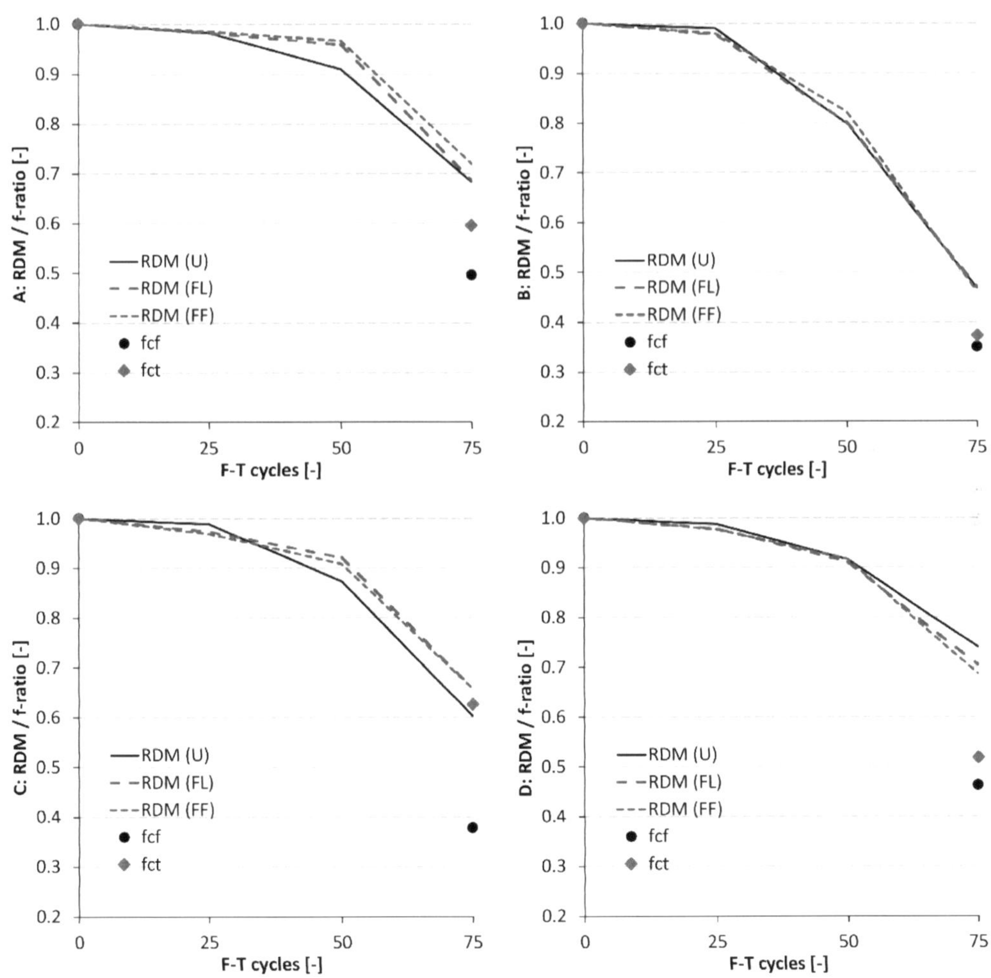

Figure 1. Relative development of the properties of all 4 concretes depending on the number of F-T cycles.

Prisms no. 4 through 6 were also tested for flexural strength and tensile splitting strength. The development of the parameters (see figure 1) and visual examination of the prisms (figure 2) clearly show that the prisms failed to resist the planned 150 F-T cycles. At 50 F-T cycles there is already a visible drop in RDM, especially in concrete B (down to almost 80 % of the original value) and after 75 F-T cycles it is clear that all the concretes suffered severe degradation. The high degree of damage to the internal structure was also manifested in the unorthodox failure that occured during the flexural strength test (figure 3) in several prisms. The test was therefore ended after 75 F-T cycles and the strength values were tested in all the specimens. Table 2 shows the results of dynamic Young's modulus and strength properties after 75 F-T cycles.

Figure 2. Fracture (across the label) in concrete B after exposure to 75 F-T cycles.

Figure 3. Unusual failure in prism no. 9 of concrete B during the flexural strength test.

Table 2. Average values of the properties including sample standard deviations (s.s.d.) after 75 F-T cycles.

Concrete	E_{cu} [GPa]	E_{crL} [GPa]	E_{crf} [GPa]	f_{cf} [MPa]	f_{ct} [MPa]
A: average	**29.9**	**28.3**	**30.2**	**4.3**	**3.1**
A: s.s.d.	5.02	6.19	4.72	1.14	0.72
B: average	**19.1**	**18.1**	**18.1**	**2.8**	**1.7**
B: s.s.d.	6.37	6.53	6.37	0.80	0.35
C: average	**24.2**	**25.0**	**25.4**	**3.2**	**3.2**
C: s.s.d.	4.82	3.51	3.85	1.19	0.73
D: average	**33.6**	**29.9**	**29.8**	**3.5**	**2.7**
D: s.s.d.	2.09	1.98	2.47	0.75	0.37

The massive increase in internal cracking, most of which occurred between the 50th and 75th F-T cycle (in concrete B already between the 25th and 50th cycle), is well illustrated by the value and especially amplitude shape of the natural frequencies of longitudinal vibration (figure 4).

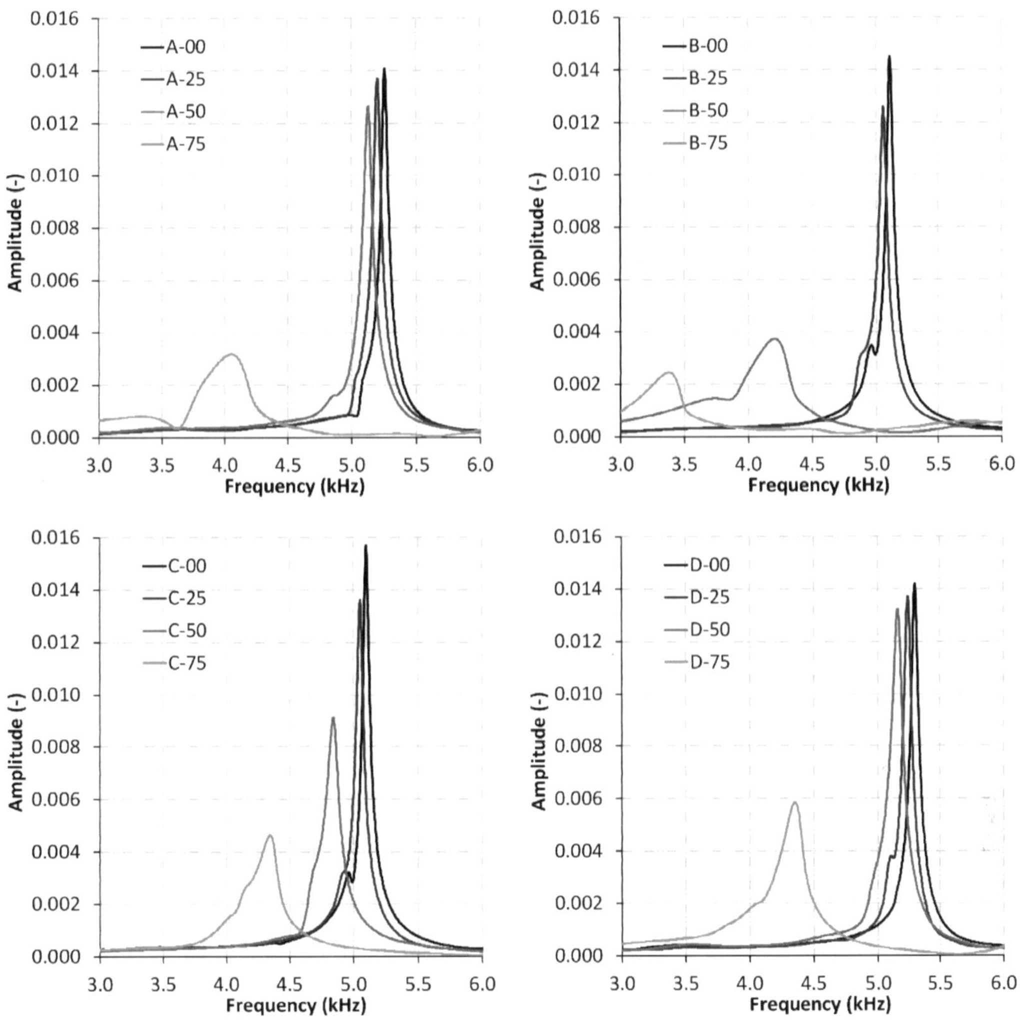

Figure 4. Natural frequencies of longitudinal vibration in a specimen (prism no. 7 of all the concretes was chosen for this demonstration) in dependence on the number of F-T cycles; the designation of the concrete is indicated in the graph legend.

The results show that the flexural strength test is the best method for assessing the degree of damage in self-compacting concretes after a freeze-thaw resistance test. After 75 F-T cycles, the flexural strength of all four concretes dropped below 50 % of its original value (i.e. the flexural strength of the reference specimens which were not subjected to F-T cycles). The second most sensitive method for determining the freeze-thaw resistance appears to be the tensile splitting strength test; with the exception of concrete C it showed the second highest decrease (in the case of concrete B it was nearly identical to the relative decrease in flexural strength). A critical factor at play during this test is the moment when it is ended. The photos in figure 5 show that the maximum force applied need not correspond to the force at which the SCC specimen damaged by F-T cycles suffered a failure. The first image “a” shows the beginning of microcracks, mainly near the loading strips. The second image, “b”, shows microcracks that span almost the entire height of the specimen – the concrete is compressed and the microcracks close. Image "c" shows a vertical crack that indicates damage by tensile splitting stress. Given the fact that the self-compacting concrete was badly damaged by internal microcracks, it had a low modulus of elasticity (it lacked "stiffness") and as the test continued and the loading stress increased, the crack became worse (images "d" through "f" at the bottom of figure 5). The test was ended once the loading stress dropped and the specimen suffered a complete failure, see image "f". The calculation was therefore made with the force being applied when the vertical crack first appeared, because if maximum loading stress were used, the tensile splitting strength of the prisms would be only slightly lower than that of the reference prisms. This is a potential problem with using this method and should be therefore taken into account.

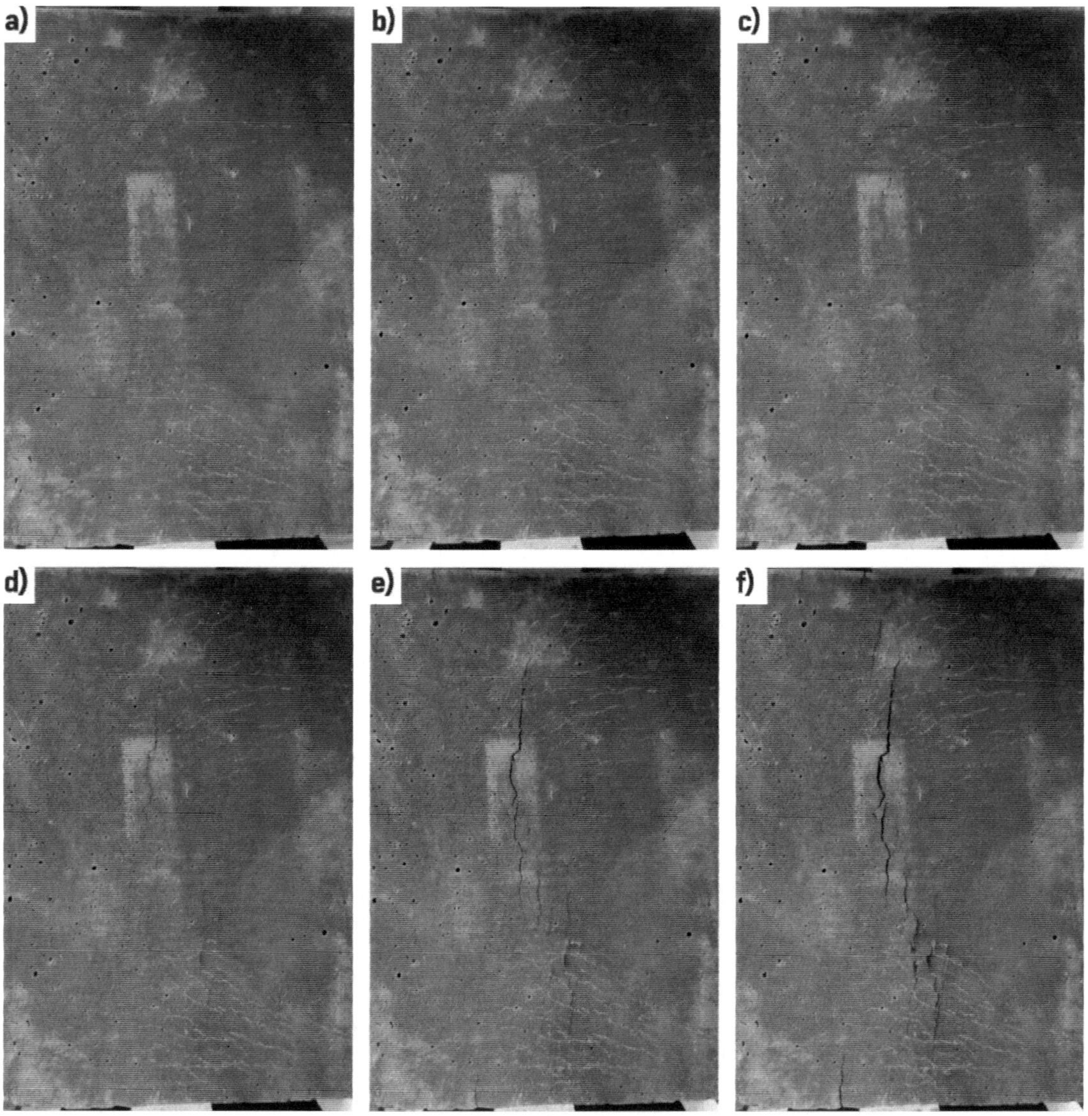

Figure 5. Progress of the tensile splitting strength test performed with SCC after 75 F-T cycles demonstrated on the failure of a test specimen no. 8 from concrete C.

4. Conclusion

The measurements show that all the test procedures assessed herein were quite able to reflect the decrease in the quality of the self-compacting concretes after exposure to cyclic freezing and thawing.

Differences between the methods are only in their sensitivity to internal structure damage. The greatest decrease of all the observed properties was in all cases found in flexural strength. Testing tensile splitting strength is also rather accurate, however, attention should be paid to the specific loading stress at which the specimen suffers a fracture. The relative dynamic modulus of elasticity determined by the two non-destructive tests showed almost identical progress. The decrease in dynamic Young's modulus may have been the lowest of all the parameters being observed, but it still reflects the damage rather reliably, and the same goes for the resonance method, as documented by the data in figure 4. In case of RDM it was only necessary to determine a stricter criterion of its maximum decrease if the concrete in question could be dubbed frost-resistant. A great advantage of the NDT methods is using the same specimens during every stage of the test.

Acknowledgement

This paper has been written as a part of project No. 19-22708S, supported by the GAČR - Czech Science Foundation and project No. FAST-S-19-6002, supported by Brno University of Technology.

References

[1] Okamura H and Ouchi M 2003 Self-Compacting Concrete *Journal of Advanced Concrete Technology* **1**(1) pp 5-15
[2] Zhao H, Sun W, Wu X and Gao B 2015 The properties of the self-compacting concrete with fly ash and ground granulated blast furnace slag mineral admixtures *Journal of Cleaner Production* **95** pp 66-74
[3] Tuyan M, Mardani-Aghabaglou A and Ramyar K 2013 Freeze–thaw resistance, mechanical and transport properties of self-consolidating concrete incorporating coarse recycled concrete aggregate *Materials and Design* **53** pp 983-991
[4] fib Bulletin No. 28 2004 *Environmental Design* (Lausanne, Switzerland: fib) 80
[5] Sideris K K and Anagnostopoulos N S 2013 Durability of normal strength self-compacting concretes and their impact on service life of reinforced concrete structures *Construction and Building Materials* **41** pp 491-497
[6] Ryan P C and O'Connor A 2016 Comparing the durability of self-compacting concretes and conventionally vibrated concretes in chloride rich environments *Construction and Building Materials* **120** pp 504-513
[7] Kanellopoulos A, Petrou M F and Ioannou I 2012 Durability performance of self-compacting concrete *Construction and Building Materials* **37** pp 320-325
[8] Collepardi M 2006 *The New Concrete* (Grafiche Tintoretto: Treviso, 1. ed)
[9] Öz H Ö, Yücel H E and Günes M 2016 Freeze-Thaw Resistance of Self Compacting Concrete Incorporating Basic Pumice *International Journal of Theoretical and Applied Mechanics* **1** pp 285-291
[10] Neville A 2011 *Properties of concrete* (New York, Pearson, 5. ed.)
[11] ČSN 73 1371 *Non-destructive testing of concrete – Method of ultrasonic pulse testing of concrete (in Czech)* (Prague: UNMZ)
[12] ČSN 73 1372 *Non-destructive testing of concrete – Testing of concrete by resonance method (in Czech)* (Prague: UNMZ)
[13] EN 12390-5 *Testing hardened concrete – Part 5: Flexural strength of test specimens* (Brussels: CEN)
[14] EN 12390-6 *Testing hardened concrete – Part 6: Tensile splitting strength of test specimens* (Brussels: CEN)
[15] ČSN 73 1322 *Determination of frost resistance of concrete (in Czech)* (Prague: UHM)

Acid attack on hardened cement paste by acids forming low soluble calcium salts

V Pavlík[1]

[1]Slovak University of Technology, Faculty of Civil Engineering, Department of Material Engineering, Radlinského 11, 810 05, Bratislava, Slovak Republic

E-mail: vladimir.pavlik@stuba.sk

Abstract. Solutions of acids have very aggressive effect on the cement matrix in concrete. Aggressiveness of the solutions increases with concentration of an acid but also depends on the acid type. The corrosion effect of solutions of various types of acids and their mixtures on hardened cement pastes is described in the paper. The results confirmed, that the rate of corrosion of hardened cement paste in solutions of strong acids (HCl, HNO_3) was faster than in solutions of weak acids. However, the rate of corrosion of hardened cement paste was lower in solutions of acids that produce low soluble calcium salts, because calcium salts filled the pores in the layer of corrosion products and thus increased its diffusion resistance.

1. Introduction

Solutions of acids represent a specific and very aggressive environment for buildings and constructions based on cement composites. Acid solutions attack hardened cement matrix. Carbonaceous aggregate is likewise a subject of acid attack. The occurrence of acid solutions in the natural environment is relatively rare and acidity of the environment is usually relatively low. Waters with increased natural acidity include peat waters that contain humic acids and acids produced by the plants root system. Atmospheric acid precipitation with pH values in the range of 3 to 4 also may occur. Solutions of various acids with increased concentrations may occur mainly in agriculture, industry or in waste waters. A special case is an acid attack in the sewage pipes and sewage systems that is caused by sulfuric acid originated from biochemical oxidation of hydrogen sulfide by the action of aerobic sulfur bacteria of the genus Thiobacillus.

There exist various types of acids. By their composition we distinguish between inorganic and organic acids. By their strength acids can be divided into "strong" and "weak" acids. "Strong acids" are very highly, or totally ionized in the water solutions into hydrogen cations H^+ and corresponding anions (e.g., HCl, HNO_3, H_2SO_4). "Weak acids" are only partially ionized and establish equilibrium between the unionized molecules and ions in the solution (e.g. organic acids and carbonic acids). Acids also can be classified by solubility of their calcium salts in water (Tab. 1), because this property strongly affects the rate of an acid attack on concrete.

Specimens made of hardened cement paste (HCP) or cement mortar are often used to study the corrosion resistance of a cementitious material. All compounds of cement hydration are unstable in an acidic environment. Acids react with both, hydrated and unhydrated compounds in HCP and decompose them. The end products of the chemical reactions between the acid solution (H^+ ions) and the cement hydration products are the calcium salts of the acid and a decalcified residue of the cement hydration products [1-7] (Fig. 1). If the calcium salts of the acid are well soluble in water, the ions of Ca^{2+} gradually diffuse from the reaction zone into the aggressive solution (Fig. 2). Other, less soluble or insoluble decalcified corrosion products remain on the surface of the sample where they form a porous layer of decalcified residue [3, 8-15]. The layer of corrosion products in this case consists of amorphous hydrated oxides - SiO_2.aq., Al_2O_3.aq. and Fe_2O_3.aq. At a sufficiently low pH values, the hydrogels Al_2O_3.aq. and Fe_2O_3.aq. also gradually dissolve. The porosity of the corrosion product layer in this case is up to 80% for the paste with *w/c* = 0.4 [6, 8, 9].

Table 1. Examples of various acids with different solubility of their calcium salts in water.

Acid type	Solubility of Ca-salts	(g /100 g of water)
HCl, HNO_3, HCOOH, CH_3COOH	High solubility	1 - 300
Citric, tartaric or sulfuric acid	Middle solubility	0.01 – 1 or 2
Oxalic or hydrofluoric	Low solubility	< 0.01

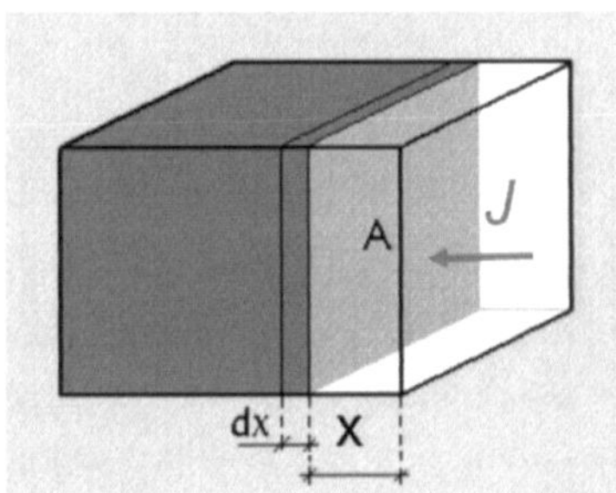

Figure 1. Scheme showing formation a layer of corrosion products on the surface of the cement composite with thickness X. A is the reaction area and *J* is diffusion flow [9].

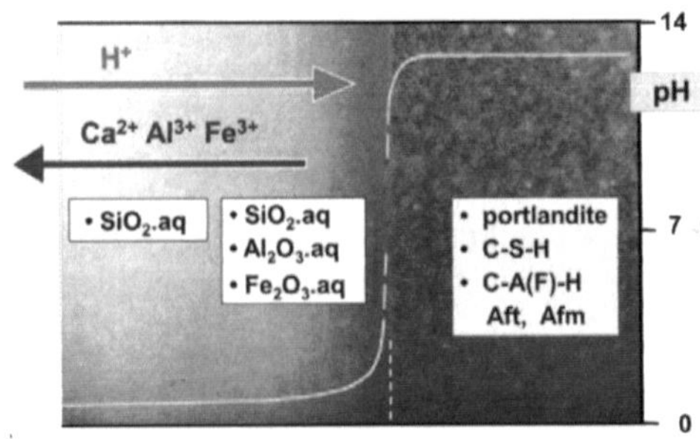

Figure 2. Diffusion flows of acid ions through the layer of corrosion products to the reaction zone and opposite flow of soluble products into the aggressive solution. The figure corresponds to an action of a strong acid with pH that allows solubility of hydroxides of Al and Fe [9].

The aggressive effect of various acids depends on many factors, predominantly on type and concentration of an acid and also on water-to-cement ratio (and neutralization capacity) of HCP. Diffusion of aggressive substance across the layer of corrosion products to the reaction zone is the slowest stadium of the corrosion process and it is a factor that determines the corrosion rate. It is affected by the depth and diffusion resistance of the corrosion layer. Thickness of the layer of corrosion products progressively increases in time and follows a √t function. There are two types of mathematical relationships that express progress of acid corrosion as a function of time. The first are derived theoretically from the diffusion models (Fick I. and II. law). Rombén [1], for example presented a mathematical model in the form:

$$x = \sqrt{2kDct} \tag{1}$$

where x is the thickness of the diffusion layer, D is diffusion coefficient the acid in the corrosion layer, k is a non-specified constant, and c is the concentration of acid and t is duration of acid attack. Other corrosion models were published by numerous authors, e.g. [16-19]. Modification of some simpler models can also be written in a form [9]:

$$\boldsymbol{x} = \sqrt{\frac{\boldsymbol{2Dct}}{\boldsymbol{\beta_c}}} \tag{2}$$

where β_c is a total neutralising capacity of hardened cement paste. Neutralisation capacity corresponds to the amount of acid (mol) required to completely decompose a volume of HCP (m^3) and transform it into a soluble or insoluble corrosion products (mol/m^3). Neutralisation capacity of HCP increases with decreasing value of water-to-cement ratio of the paste.

Other (empirical) relationships were derived from evaluation of experimental results by the method of the least squares. These empirical relationships were presented usually in the form (3) or (4)

$$x = k_1 \, . \, t^{n} \tag{3}$$

$$x = k_2 \, . \, c^{m} . \, t^{n} \tag{4}$$

where again, x is depth of corrosion, c is concentration of acid (in mol/l), t is time of corrosion, and k_1, k_2, m and n are constants derived from experimental results; values of m, n are near to 0.5.

2. Experimental

2.1. Materials and methods

a) Portland cement CEM I 42.5. Chemical composition of cement was as follows: CaO 63.43 %, MgO 1.37 %, SiO_2 19.10 %, Fe_2O_3 2.13 %, Al_2O_3 5.87 %, SO_3 1.85 %, insoluble residue 2.13 % and ignition loss 2.22 %.
b) Pure acids: (a) Hydrochloric acid (38%); (b) Acetic acid (96%), (c) Formic acid (98%), (d) Sulfuric acid (98%), Benzoic acid (crystalline, solubility about 0.29 g/100g of water at 20 °C).

Two types of test specimens were prepared from cement pastes with water/cement ratio (w/c) of 0.4:

a) Glass tubes that were opened on only one side were filled with the cement paste and then cured initially in water for 28 days. The tube samples were then immersed in diluted aggressive solutions of different acids (Fig. 3). The acid solutions were kept at nearly constant concentration during the test. Weight of samples and depth of corrosion were determined at predetermined time.
b) Test samples of cubical and cylindrical shape. The cubes with side of 20 mm and/or cylinders with diameter and height of 30 mm were prepared in steel molds and after demolding were initially cured in water for 60 days. Thereafter they were immersed in aggressive solutions of acids. In both experiments the concentration of aggressive solutions was kept constant by changing the solution during the experiment. Compressive strength of samples was determined after predetermine time periods.

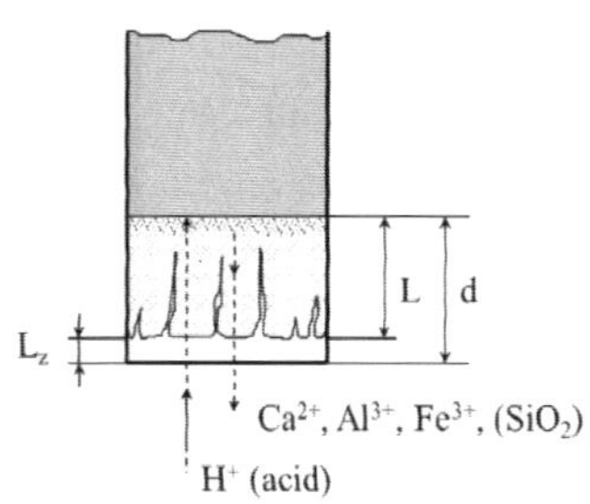

Figure 3. Method of testing the corrosion rate of HCPs in glass tubes immersed in acid solutions, where: L, L_z and d is the thickness of a corrosion products layer, shrinkage of the corrosion products layer and corrosion depth, respectively.

Analytical techniques: Mineralogical composition of the pastes was studied by powder X-ray diffraction using Bruker D2 PHASER diffractometer (results are mentioned but not presented here). Scanning electron microscope (SEM) Vega 3, Tescan was used for observing samples morphologies. The SEM was equipped with x-act SDD Oxford Instruments detector for EDX/EDS microanalysis.

2.2. Results and discussion

2.2.1. Acid attack by strong and weak acids

Effect of solutions of different acids with equal concentration on the development of depth of corrosion (d) of hardened cement paste (HCP) is presented in Fig. 4. Concentrations of acids were kept constant by changing aggressive solutions. The highest rate of corroded layer development was obtained in hydrochloric and nitric acids that represented strong acids giving a highly soluble calcium salts. Effect of weak acids (formic and acetic acid) was lower. Increase in depth of corrosion of HCP was the slowest in the sulfuric acid solution. The reason was a relatively low solubility of calcium sulfate dihydrate that formed in the reaction and accumulated in the corrosion products layer, making the layer less porous and permeable.

Effect of solutions of hydrochloric acid with different concentrations on depth of corrosion (*d*) of HCP with *w/c* = 0.4 is presented in Figure 5. By applying the method of least squares on the experimental results the relation between depth of corrosion and two variable parameters, concentration (*c*) and time (*t*) was found (equation (5)).

$$d = 2.19\, c^{\,0.69} .\, t^{\,0.63} \tag{5}$$

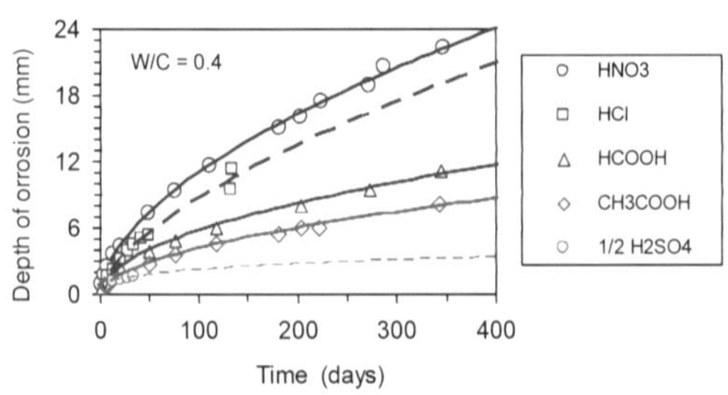

Figure 4. Increase in depth of corrosion in HCPs (*w/c* = 0.4) in glass tubes in the aggressive solutions of different acids with constant concentration ($c = 0.1$ mol.dm^{-3}).

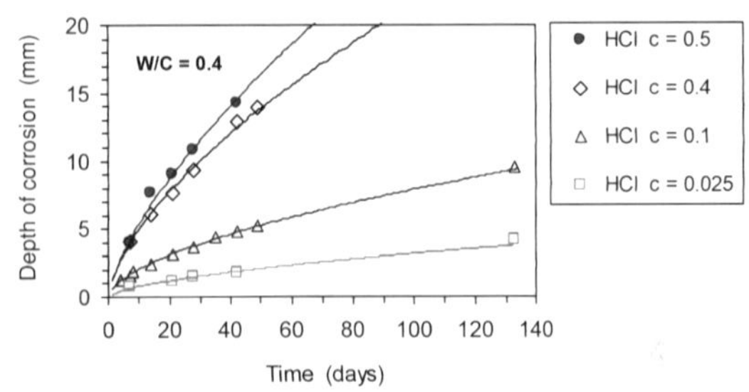

Figure 5. Corrosion of HCP with *w/c* = 0.4 in solutions of HCl with various concentration (mol.dm^{-3}).

2.2.2. Acid attack by a solution of hydrochloric and sulfuric acids and their mixtures

Aggressive effect of solutions of HCl, H_2SO_4 and equimolar solutions of a mixture of both acids was studied using cubic samples of HCP and also samples in glass tubes. Calcium salt of hydrochloric acid is well soluble, calcium salt of sulfuric acid is much less soluble. Solubility of $CaCl_2.6H_2O$ is 279.3 g/100g of water and $CaSO_4.2H_2O$ is 0.22 g/100g of water at 0°C.

The layer of corrosion products on the surface of HCP increased faster in the solution of HCl than in the solution of H_2SO_4. Compressive strength decrease of cube samples was also faster in the solution of HCl (Fig. 6). The rate of corrosion in the solution of a mixture of both acids was intermediate. Visual appearance of test specimens is in Fig. 7. Corrosion layer formed on the surface of HCP in the solution of HCl was mostly composed from the decalcified SiO_2.aq gel. It was weak and porous. When it dried it shrank and cracked, its color was lighter and it could be easily peeled off.

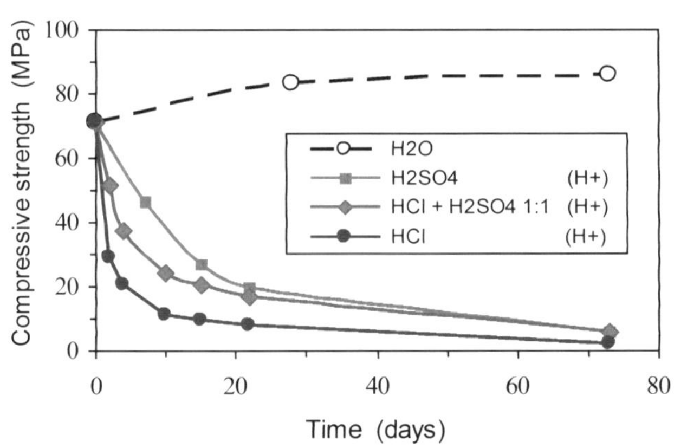

Figure 6. Compressive strength of HCP`s specimens subjected to attack of hydrochloric acid, sulfuric acid and mixed solution of both. Concentration of H^+ ions in each solution was equal ($[H^+] = 0.25$ mol/l).

In the case of the H_2SO_4 attack, the layer of corrosion products was less porous and was composed from the mixture of low soluble gypsum ($CaSO_4.2H_2O$) and SiO_2.aq gel (Figs. 7 - 9). After drying the layer was relatively hard and solid. The layer of corrosion products was less porous and generated greater diffusion resistance for the penetrating sulfuric acid. The corrosion rate, therefore, was slower (Figs. 4, 6 and 10). The results are in agreement with those of other authors [20].

In the case of an acid attack of a solution composed of mix of HCl and H_2SO_4 the calcium sulfate precipitated not only inside the corrosion layer but also in the close vicinity of the sample in the form of a porous wad of irregular shape (Fig. 7). The reason was higher concentration of H^+ ions originating from both acids and resulting Ca^{2+} ions than sulfate anions in the solution ($[Ca^{2+}] > [SO_4^{2-}]$).

Figure 7. Visual appearance of HCP samples with *w/c* = 0.4 after curing for 25 days in water (top left) and after their exposure in the aggressive solution of hydrochloric acid, sulfuric acid and a mixture of both acids, with molar proportion of HCl : H_2SO_4 = 1 : 1. All the acid solutions had equal concentration ($c(H^+) = 0.25$ mol/l).

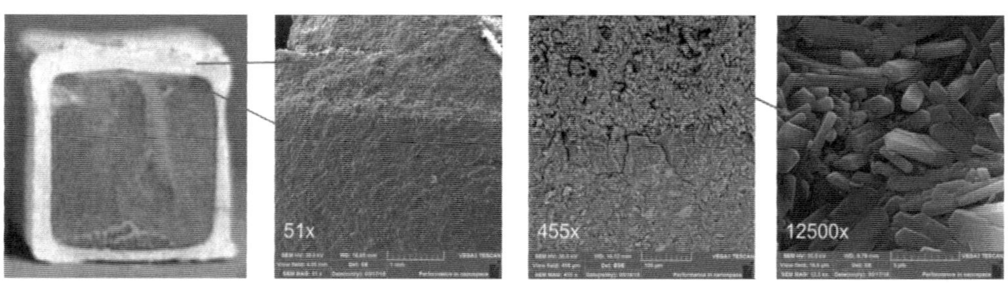

Figure 8. Cut section of a HCP cube sample attacked by the sulfuric acid solution with SEM images of areas near the reaction zone and gypsum crystals in the layer of corrosion products.

Figures 10 (a, b) show corrosion of HCP pastes (tubes) subjected to an attack by solutions of hydrochloric and sulfuric acids and their mixes. Figure 10 (a) shows that thickness of the layer of corrosion products increased more rapidly in the solution of HCl than the solution of H_2SO_4. In the cases of an attack of aggressive solutions containing a mixture of HCl and H_2SO_4 the corrosion rate was lower than in the HCl alone despite the fact that the solutions had nearly equal concentration of H^+ ions. The rate of corrosion decreased with the increasing ratio of sulfuric to hydrochloric acid. The corrosion rate in solutions of H_2SO_4 slightly increased with increasing concentration of sulfuric acid.

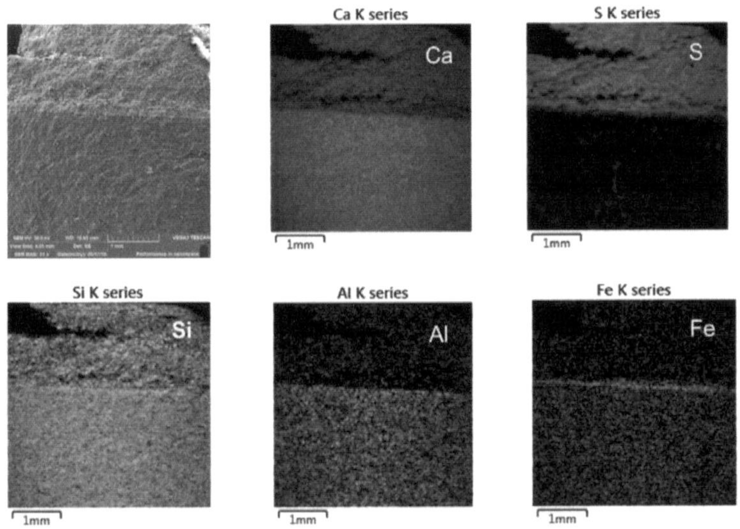

Figure 9. SEM image of areas near reaction zone in the sample attacked by the sulfuric acid solution (Fig. 8) and EDX distribution maps for Ca, S, Si, Al and Fe. Products of cement hydration were decomposed. Ions of Al, Fe and partly Ca were leached out from the corrosion layer. S and Ca are present in the layer in the form of a low soluble gypsum.

(a)

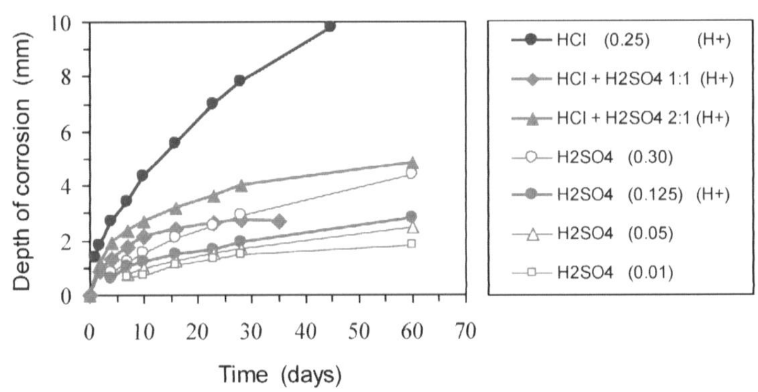

(b)

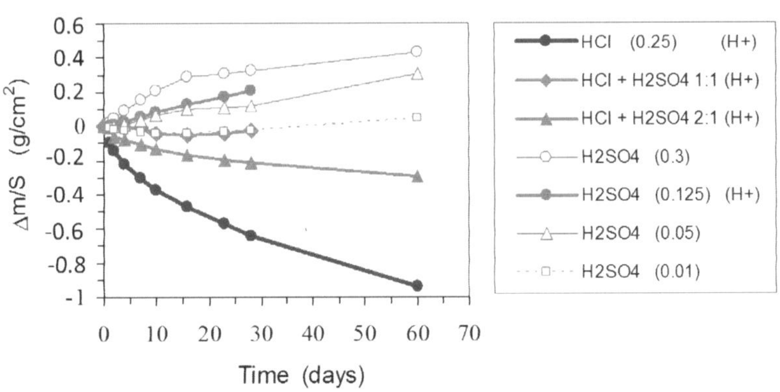

Figures 10 (a, b). Corrosion of hardened cement pastes in solution of HCl, H_2SO_4 and mixtures of both acids. Concentrations of solutions in mol/dm^3 is in parentheses. Symbol (H^+) refers to the solutions with equal concentration of H^+ ions ($[H^+] = 0.25$ mol/dm^3). The results correspond to tests performed on samples in glass tubes.

Weight of the samples stored in the HCl solution gradually decreased due to dissolution of Ca^{2+} ions from the corrosion layer. However, the weight of the samples in the solution of H_2SO_4 increased or did not change (Fig. 10 b). The changes confirm accumulation of gypsum in the corrosion product layer, corresponding to Figs. 7-9. In the samples immersed in a mixture of hydrochloric acid and sulfuric acid, gypsum precipitated in the layer of corrosion products and also around the surface of the sample as was described previously. The low soluble gypsum precipitate in the corrosion layer and spreading around the sample slowed the diffusion of the aggressive component to the reaction zone.

2.2.3. Acid attack by solutions of acetic and benzoic acids

Corrosion effect of solutions of two weak acids, acetic and benzoic acids was studied using cylindrical samples of HCP (Fig. 11 a, b). The acids differ in size of their molecules and solubility of their Ca-salts (the solubility of calcium acetate hydrate at 0 °C is about 43.6 g/100 g water; the solubility of calcium benzoate trihydrate is about 4.5 g /100 g water). Concentration of both acids was about 0.025 mol/l (saturated solution in the case of benzoic acid).

(a) 

(b) 

Figure 11. (a) Appearance of samples of HCPs after storage for 120 days in the solution of acetic (left) and benzoic acid (right). (b) A cross-sectional area of the cores of samples in the solution of benzoic (left, B) and acetic acid (right, A) with a visible "core" layer on the edges.

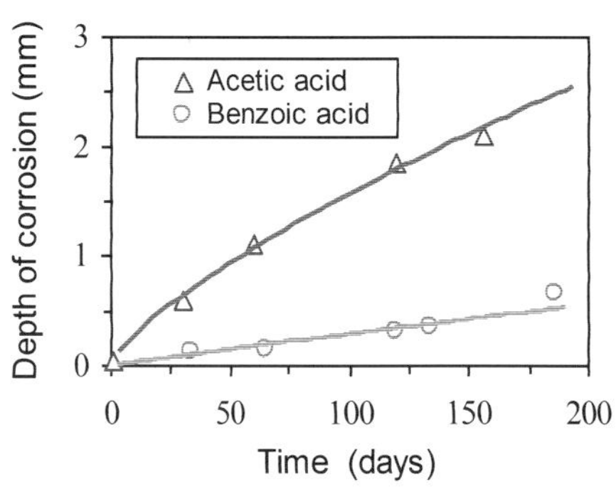

(a)

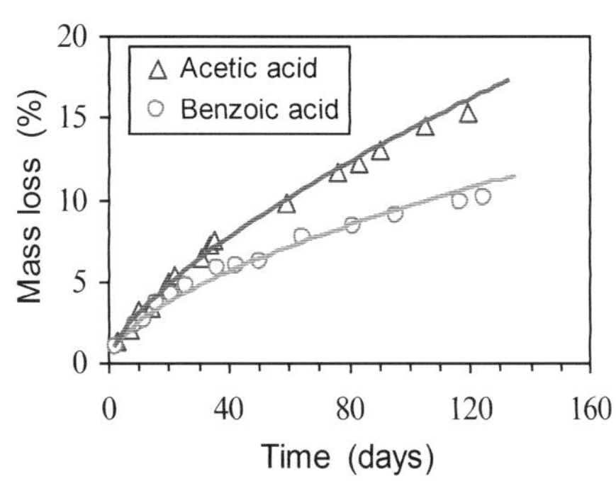

(b)

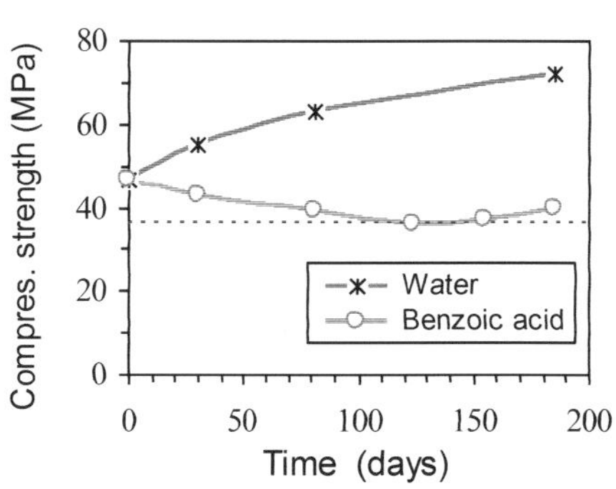

(c)

Figure 12 (a-c). Corrosion of HCP samples immersed in acetic and benzoic acid solutions (c = 0.025 mol/l):
(a) Depth of corrosion,
(b) Mass loss, and
(c) Compressive strength changes.

Corrosion layer that formed on the surface of samples in the solution of acetic acid was porous. It shrank and cracked after drying (Fig. 11 a, b). The layer of corrosion products that formed in benzoic acid was relatively compact and it could not be separated from the "core" of samples. A thin "core" layer also was formed under the layer of corrosion products during the action of both acids (Fig. 11 b).

Corrosion of HCP samples in the acid solutions proceeded at different rates (Figs 12 a-c). The thickness of a layer of corrosion products increased much more rapidly in the solution of acetic acid (Fig. 12 a). It was difficult to determine thickness of the corroded layer in corroded sample from benzoic acid as there was no defined interface boundary between the layer of corrosion products and uncorroded core. The weight of samples immersed in the benzoic acid solution decreased more slowly than in acetic acid solution (Fig. 12 b). The compressive strength of HCP samples immersed in the benzoic acid was reduced by about 24% over the 120 days and then began to increase slightly (Fig. 12 (c)). This was probably the result of two conflicting effects. The first was slight increase of the compressive strength of the core, the second was a very slow decrease of a cross-section area of uncorroded core. The rate of both effect changes differently in time.

3. Conclusions

Solutions of acids have a significant corrosion effect on cement based materials. Aggressiveness of the solutions increases with concentration of acids but also depends on the acid type. The results presented confirmed, that the rate of corrosion of hardened cement paste in solutions of strong acids (HCl, HNO_3) was faster than in solutions of weak acids (acetic and formic acid). However, the rate of corrosion of hardened cement paste was much slower in solutions of acids that produce low soluble calcium salts (sulfuric and benzoic acid), because calcium salts fill the pores in the layer of corrosion products and thus increase its diffusion resistance. Formation of low soluble salts can affect corrosion rate also in the cases of attack by mixture of different acids with high and low Ca-salts solubility.

Acknowledgment

This article was supported by the Slovak Research and Development Agency under the contract No. APVV-15-0681 and by the Science Grant Agency - project VEGA No. 1/050/17.

References

[1] Rombén L 1978 *Aspects on Testing Methods for Acid Attack on Concrete*. Stockholm. BI forskning research 1:78 (und 9:79), Cement-och betong instituted
[2] Moskvin V M, Ivanov F M, Alexejev S N, Gusejev E A 1980 *Corrosion of Concrete and Reinforced Concrete, Methods of Their Protection*. Strojizdat Moscow (in Russian)
[3] Alexander M, Bertron A, De Belie N 2013 *Performance of Cement-Based Materials in Aggressive Aqueous Environments*. State-of-the Art Report, Rilem, TC 201-PAE
[4] Živica V, Bajza A 2001 *Constr. Build. Mater.* 15 pp 331–340
[5] Živica V, Palou M T, Križma M, Bagel L 2012 *Constr. Build. Mater.* 36 pp 623–629
[6] Unčík S, Bajza A, Rouseková I 1994 *Slovak Journal of Civil Engineering* 2 pp 26-34
[7] Pavlík V 1994 *Cem. Concr. Res.* 24 pp 551–562
[8] Pavlík V 1994 *Cem. Concr. Res.* 24 pp 1495–1508
[9] Pavlík V 2005 *Edition of Scietific Works*, No 7, Slovak University of Technology, Faculty of Civil Eng.
[10] Gutberlet T, Hilbig H, Beddoe R E 2015 *Cem. Concr. Compos.* 74 pp 35-43
[11] De Ceukelaire L 1992 *Cem. Concr. Res.* 22 pp 903-914
[12] Chandra S 1988 *Cem. Concr. Res.* 18 pp 193-203
[13] Goñi S, Macías A, Madrid J, Díez M 1999 *J. Am. Ceram. Soc.* 82 pp 1275-1280
[14] Israel D, Macphee D E, Lachovsky E E 1997 *J. Materials Science* 32 pp 4109-4116
[15] Macías A, Goñi S, Madrid J 1999 *Cem. Concr. Res.* 29 pp 2005-2009
[16] Grube H and Rechenberg W 1987 *Beton* 37 pp 446-451
[17] Hinsenveld A 1992 *PhD Dissertation*, University of Cincinnati OH
[18] Baker P G, Bishop PL 1997 *J. Hazardous Materials* 52 pp 311-333
[19] Beddoe R E 2016 *Cem. Concr. Res.* 88 pp 20–35
[20] H. Yuan H, Dangla P, Chatellier P, Chaussadent T 2013 *Cem. Concr. Res.* 53 pp 267–277

Practical applications of electroosmotic system using electrodes with conductive plastics and advanced osmotic pulse system

G Pavlendová[1], P Šín[1], J Šujanová[2]

[1] Slovak University of Technology in Bratislava, Radlinského 11, 810 05 Bratislava
[2] University of Ss. Cyril and Methodius in Trnava, Hajdóczyho 1, 917 01 Trnava

E-mail: qpavlendova@stuba.sk

Abstract. Practical applications of the two commonly used electroosmotic systems in Slovakia were monitored and described in the paper. Emerged problems were analysed, to test the relevance of the system used for moisture removal from the object. Own installation of electroosmotic system was done. Porous structure and salt content of treated masonry was analysed and the influence of advanced osmotic pulse technology on salts' precipitation and moisture field homogenisation was tested. Best results were achieved using electrodes with conductive plastics and the power unit with pulse system.

1. Introduction

The usage of electro osmosis in Civil Engineering has longer data as it is one of the least destructive methods for removing noxious moisture from masonries. Excess moisture in a building is a serious problem that affects the quality of life and the durability of construction. Electro osmotic utilisation can be divided according to the applied power supply, connecting wire and electrode's shape and material. In the paper experiences with two, in Slovakia often used, electroosmotic systems are described.

2. Experimental

Electrodes with conductive plastics, Kerasan consist of electrical wire- special alloy of titanium ∅ 3 mm, resistant to salts, electrically conductive nonpolar plastic (1st layer of the shielding) that ensures that the material does not change its dense structure and so protects the wire and electrically conductive polar plastic (2nd layer of shielding). The 2nd layer loses its closed structure which extends into a slightly larger volume. The resultant porosity of outer layer increases the effective surface area of the electrode.

The electrodes are preferably laid diagonally when the wall is accessible from both sides. If the walls are only accessible from one side, the positive and negative electrodes are placed over each other at a distance of 2 times the wall thickness. Direct current to the electrodes is applied through the control unit. Voltage and the current flow (mA) is set up according to the conditions in masonry. The unit is directly connected to 230 V power source. Power consumption of the unit is up to 0.8 kWh per year per 1 meter of the masonry [1].

IOP Conf. Series: Materials Science and Engineering **549** (2019) 012021 doi:10.1088/1757-899X/549/1/012021

2.1. Monitoring practical applications of Kerasan system and troubleshooting

We were monitoring installations in the old part of Bratislava at: Chapel Corpus Christi at Panská Street, Building of the Slovak Philharmonic, Palackého 100/2, Honorary Consulate of the Principality of Monaco in the Slovak Republic at Gorkého 7A in Bratislava and a family house at Holubyho 10. At the Chapel Corpus Christi, electroosmotic circuit has been functioning since 2010, without any problems and the place is much drier and healthier than ever before. Restoration work at the Building of the Slovak Philharmonic was done by several companies, for one part of the building 6 electroosmotic circuits were installed in 2011. In 2013 problems occurred with one of them. The measurements in situ did not prove problems in the circuit, so the control unit was changed, one anode added and for some time the polarity changed to get the uniform distribution of the water content close to the anode and decrease the value of electrical resistance close to it. No other problems occurred but since the owner has changed we do not have any further information. At the Honorary Consulate of the Principality of Monaco two electroosmotic circuits were built in 2008 to solve moisture problems, one was set into function. The restoration of the building continued. On 29.7.2014 a control was done and moist places and deteriorated masonry with clearly visible salt signs appeared. Twelve control places were specified for the moisture control, but the tests and further inspection revealed that the conductive mortar was replaced with nonconductive, one of the wires was slightly damaged, and conductive material was found in the field of electroomosis. After replacing of the damaged parts, electroosmotic circuit started to function again. Family house was another example what can happen if there were more companies involved without properly documentation left. While installing the electroosmotic circuit unprofessionally done undercutting was revealed, the outer wall was covered with cement plaster, what meant that water could not get out and all the water was kept in. Because of all these two circuits were installed after putting in order the chaos done before.

2.2. Own installation at Beblavého street

The installation procedure at Beblavého Street, in the cellar of a company house.

Figure 2.2.1.
Wall with rising moisture.
Source: (Author).

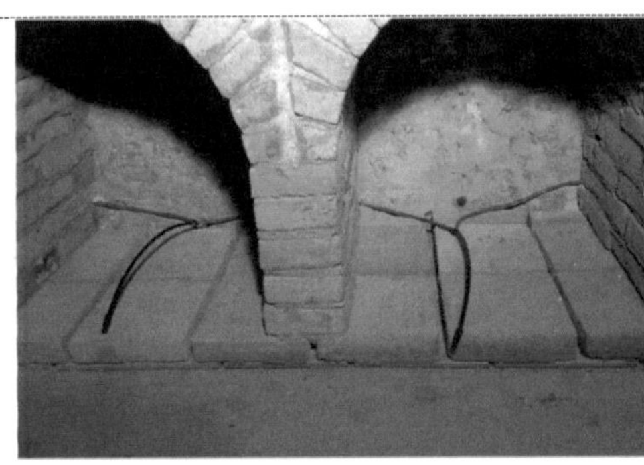

Figure 2.2.2.
Cathode's positioning. Opening for inserting the cathode (length 40 cm, angle 45°). The distance between the electrodes is 80 - 100 cm, the diameter of the drilled hole is 16 mm.
Source: (Author).

Figure 2.2.3.
Slot for inserting the anode, the slot is 3 cm high and 3-5 cm deep.
Source: (Author).

Figure 2.2.4.
Filling in with conductive mortar.
Source: (Author).

The device started to operate on 27.8.2015. Although the measurements have shown that electroosmosis improved the condition in the wall we spotted other problems that needed understanding and solving. We observed salts 'precipitation and wet mortar in the anode vicinity so material testing and salt analysis were done. See Figures *2.2.*5 and 2.2.6)

Figure 2.2.5. Salts' precipitation.
Source: (Author).

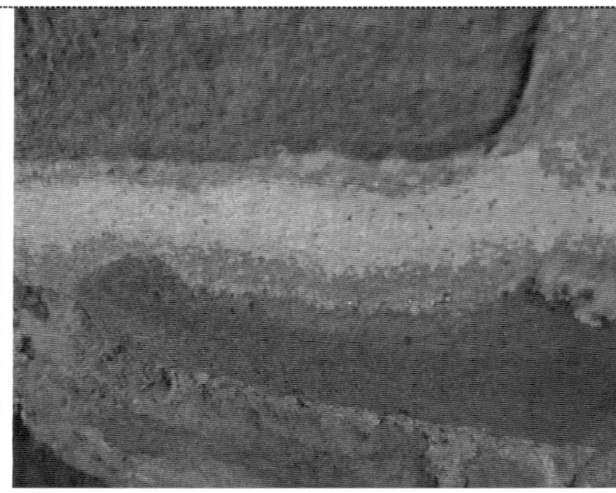

Figure 2.2.6. Wetting of the mortar in the vicinity of the anode.
Source: (Author).

2.2.1. Testing the material characteristics and salt content. Thermogravimetric Analysis TGA and Differential Scanning Calorimetry DSC of historical brick and plaster was performed by Mettler Toledo TGA/DSC 2 at the heating rate 10°C per minute from room temperature to 1000 °C. DSC and TGA curves of brick and plaster looked very similar. Until 300 °C it was possible to see evaporation of physically bounded water. This was represented by the decrease of both pairs of curves. From about 400 to 800°C removing of chemically bounded water (dehydroxylation) was present which was represented by the further decreasing of TGA curves and "plateau" of DSC curves. This meant that fired brick and plaster were certainly historical once because in the middle ages it was not possible to achieve higher temperatures than 700 °C because the firing was made by wood and therefore only part of dehydroxylation process occurred. At 800 °C the calcination was present which was represented by the at DSC curves. The fact that both TGA curves and both DSC curves were very similar lead to the conclusion that probably some portion of this brick was added into the plaster.

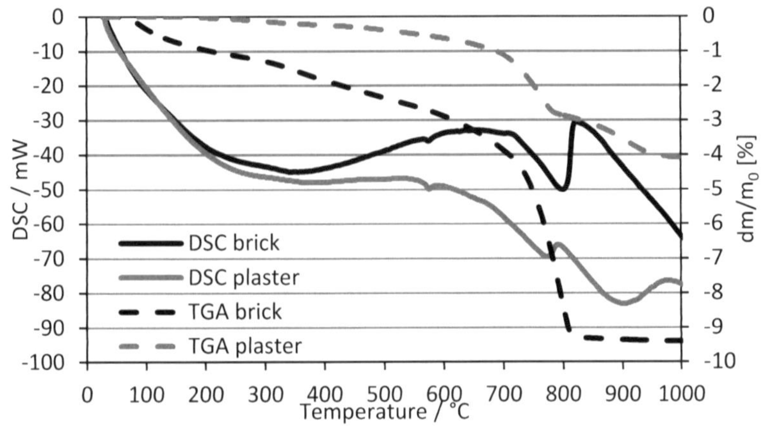

Figure 2.2.1.1. Thermogravimetric Analysis TGA and Differential Scanning Calorimetry DSC.

Pore size distribution of historical brick and plaster was performed by the porosimeter Pascal 140 Series Thermo Scientific. The maximum number of pores was at 1 μm. This also added to our presumption that part of the brick was added into the plaster and they were manufactured by the same process from similar types of materials with very similar pore size distribution.

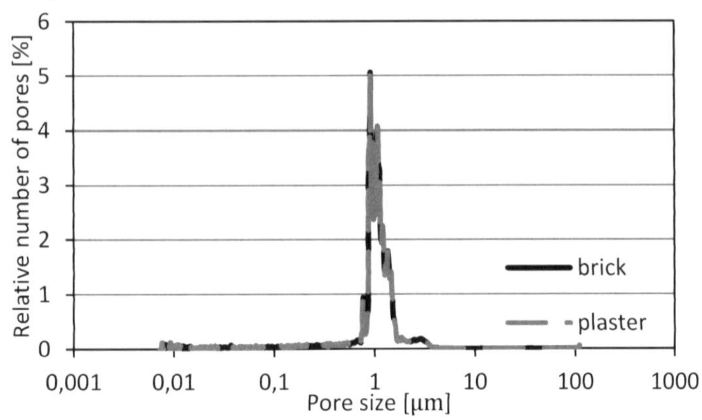

Figure 2.2.1.2. Pore size distribution of historical brick and plaster.

The studies of salt taken from the vicinity of electrodes were carried out by means of the scanning electron microscope Vega 3 from Tescan company (SE Detector, 30 kV, high vacuum 9×10^{-3} Pa) working with EDX Link 300 ISIS from Oxford Instruments (Detector Au(Pd), 30 kV, low vacuum 100 Pa, resolution 128 eV). The samples were prepared by fixing the powder particles to microscope holder, using a conducting carbon strip.

From the point of morphology the microstructure was formed by polyhedral grains in all cases. The structure of salts close to the anode showed also calcium phosphate (most probably gypsum) crystals.

The spectra obtained during EDX studies were used for carrying out the quantitative analysis. Quantitative analysis showed that the elements with the highest percentage were oxygen: 47.55 mw. %, sodium: 28.89 mw. % and sulphur present in amount: 17.07 mw. %.

IOP Conf. Series: Materials Science and Engineering **549** (2019) 012021 doi:10.1088/1757-899X/549/1/012021

Figure 2.2.1.3. SEM pictures of salts in the vicinity of the second anode deposited on a carbon strip 4020× magnification. Source: (Author).

Figure 2.2.1.4. SEM pictures of salts in the vicinity of the second anode deposited on a carbon strip 1990× magnification. Source: (Author).

We tested the salts in the area of the fifth cathode.

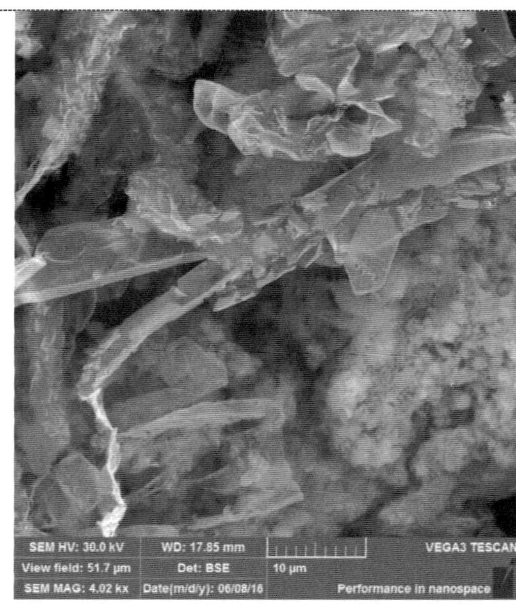

Figure 2.2.1.5. SEM pictures of the salts in the vicinity of the fifth cathode deposited on a carbon strip 4020× magnification. Source: (Author).

Figure 2.2.1.6. SEM pictures of the salts in the vicinity of the fifth cathode deposited on a carbon strip 1990× magnification. Source: (Author).

The elements with the highest percentage were oxygen: (49.98 -64.63) mw. %, sodium: (1.94- 14.16) mw. %, calcium: (13.71 - 34.01) mw. %, carbon: (4.67 - 8.65) mw. %, and other elements like Mg, Al, Si, Cl, K and Fe were within the range (0.13-5.16) mw. % .

IOP Conf. Series: Materials Science and Engineering **549** (2019) 012021 doi:10.1088/1757-899X/549/1/012021

From the results it was not clear, what kind of salts we were dealing with. In the old part of Bratislava the level of groundwater is high and the building is just under the castle hill, which means the influence of heavy rains, as well. In the last years increasing temperature of ground water was documented [2] - a new factor, that appeared recently. We decided to ask the company working with AOP (advanced osmotic pulse) technology [3], whether we could test their control unit consisting of a power supply and main controller board connected to a rectifier output stage. AOP also uses titanium wire for anodes, but without plastics and cathodes are regular copper clad ground spears The control unit (AKJI ECBS)works through administering a series of low voltage pulsating charges, which means technology uses pulses of electricity to reverse the flow of water seepage. Removal of moisture decreases the conductivity of materials, especially in the area closest to the anode. Eventually it comes to the point when no more dewatering is possible since the passage of current is hindered by this dry area. By changing the direction of flow more uniform water profile during dewatering can be achieved. The technology was invented and tested in the old USA military premises and some old concrete dam chambers [4, 5]. The power unit was used from January 2018 to January 2019. The current increased 4 times, from the average 102 mA to the average 419 mA (see Fig. 2.2.1.7 – 2.2.1.10). For the better resolution the current values in the plot have been divided by 5. The unit of the current is therefor in the plot presented as [mA/5], so the value of the current in the plot represented by 20 units was in the reality current with the value 100 mA. Precipitation of the salts was only in the vicinity of the cathodes and the mortar in the area of anode dried. The moisture amount decreased as well. Now the previous power unit is used and so far without changes. As mentioned in [4, 6] the reason was not only higher current, but changing the direction as well. We did not change the electrodes, it would mean drilling in the walls again and from experiences we believed that electrodes with conductive plastics were the better choice. We do not sell electroosmotic systems, so we are completely free to choose the best from both mentioned systems. The combination of the electrodes with conductive plastic and AOP control unit provided so far the best results.

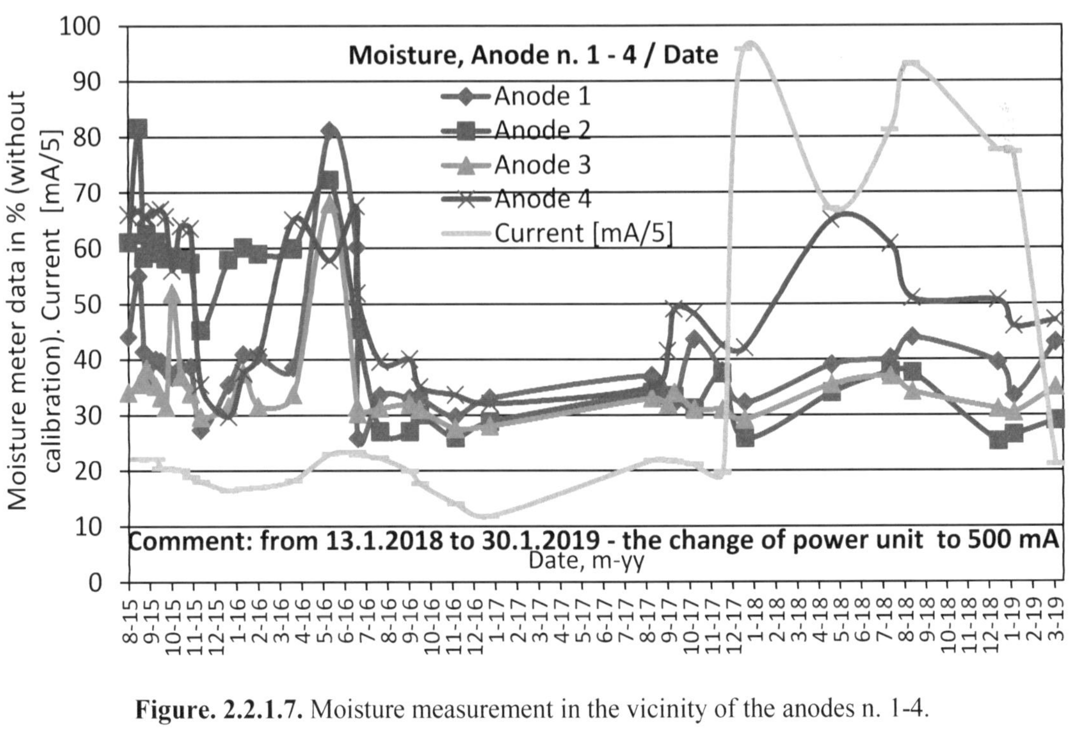

Figure. 2.2.1.7. Moisture measurement in the vicinity of the anodes n. 1-4.

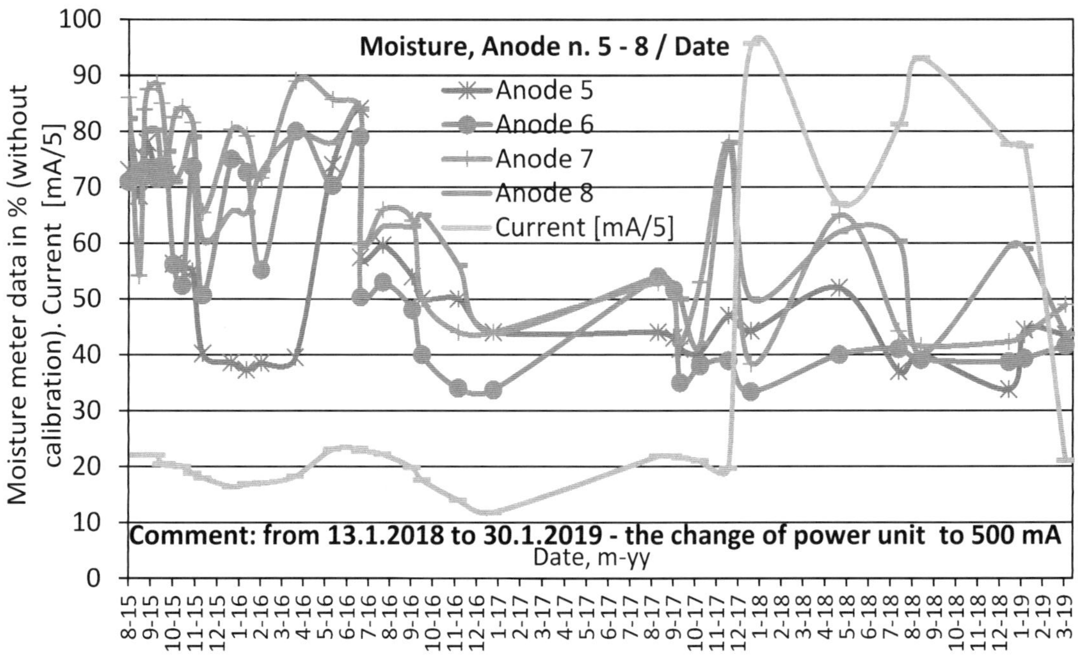

Figure. 2.2.1.8. Moisture measurement in the vicinity of the anodes n. 5-8.

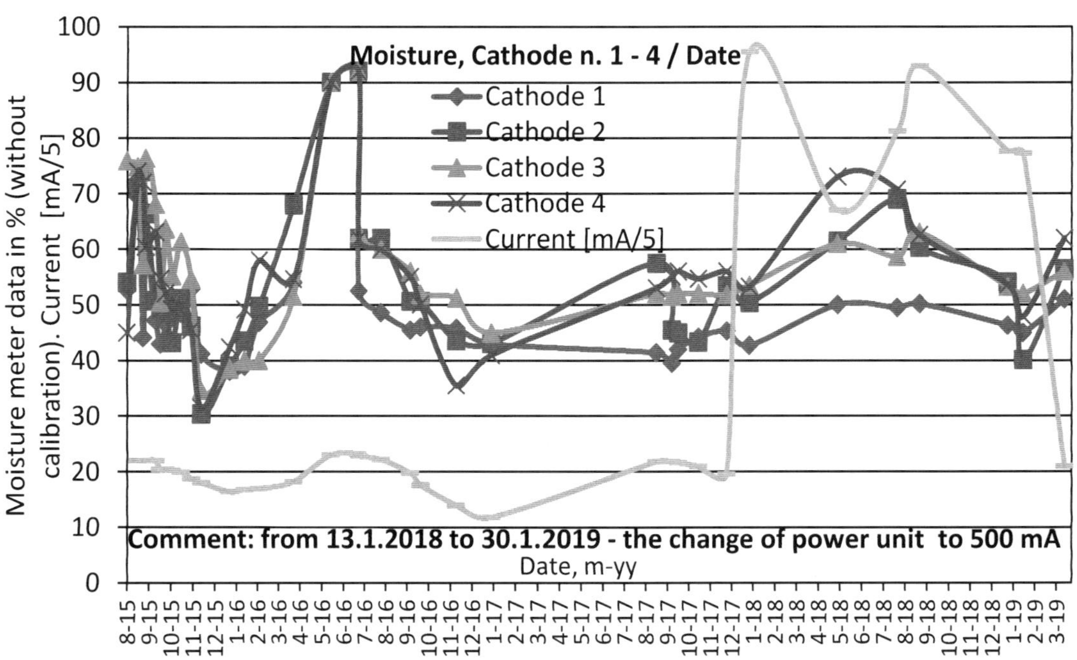

Figure. 2.2.1.9. Moisture measurement in the vicinity of the cathodes n. 1-4.

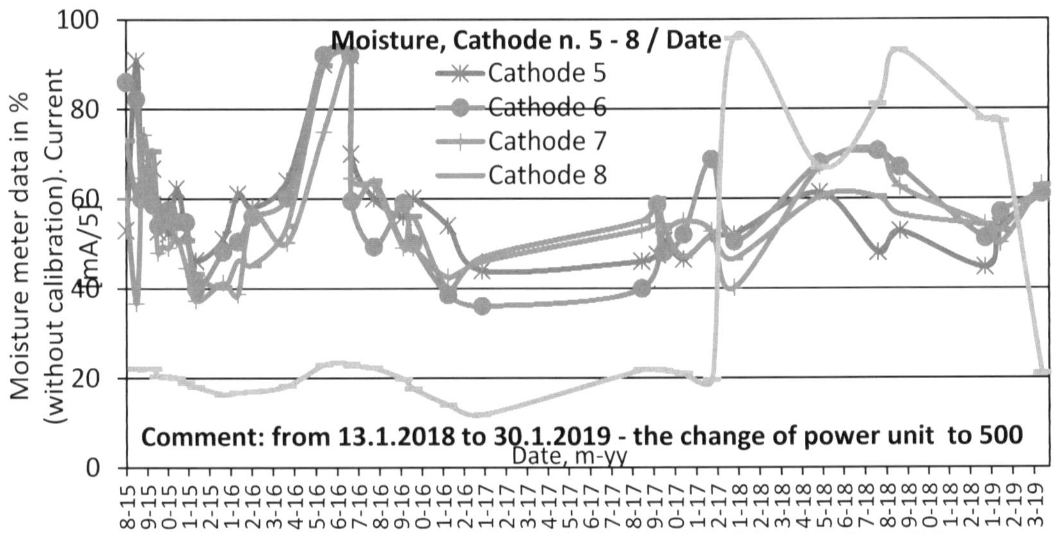

Figure. 2.2.1.10. Moisture measurement in the vicinity of the cathodes n. 5-8.

3. Conclusion

Unwanted moisture in building is a big problem mainly for historical buildings. Electroosmosis is for sure a solution because of its least destructive application; however its popularity decreased, mainly because of many systems of passive electroosmosis that are on the market. Companies claim that the system works without any external source. They have always been a controversial issue. On theoretical grounds, there is no evidence that they can work, we have done research in this field [6], but there are many ways they use to persuade the customer and they enlarge their mistrust even to the functioning systems. We should also keep in mind that electroosmosis does not stop water entering the capillaries from the ground, but it keeps the amount entering the wall in acceptable amount. Another important thing is that masonry is complicated porous structure and there are always many factors involved, so before choosing the final decision on how to deal with moisture thorough inspection and analysis should be done.

Acknowledgments The research was funded by the grant APVV-15-0681.

References

[1] Hydrobeton 2014 [online]. Available from: http://www.hydrobeton.sk/fileadmin/Sanacia_vlhkeho_muriva.pdf

[2] Krčmář D, Rusnáková D, Tóthová E, Fľaková R & Ženišová Z 2019 *Podzemná voda* **23**, **2** 113

[3] Triton 2016 [online]. Available from: http://www.tritonnorway.com/technical.html

[4] Little D A and Von Fay K F 2013 Electro-Osmotic Pulse Leak Repair Method: Evaluation in Trinity Dam Bonnet Chamber - Central Valley Project, Trinity River Division – California [online]. California. Available from: http://www.google.sk/url?sa=t&rct=j&q=&esrc=s&source=web&cd=1&ved=0CC0QFjAA&url=http://www.usbr.gov/research/projects/download_product.cfm?id=734&ei=QCQbU__HAYnesga12YCwAg&usg=AFQjCNGjmIfYxnreimh-2c6JKkRcDiV3vA&sig2=ZwGGelsrunMYJtQ_p-pbVg&bvm=bv.62578216,d.Yms

[5] Marshall O S, Morefield S, McInerney M K, Hock V F, & Femmer P D 2011 Electro-osmotic pulse (EOP) treatment method [online]. US007935236B2. United States. Available from: http://www.google.sk/url?sa=t&rct=j&q=&esrc=s&source=web&cd=1&ved=0CCcQFjAA&url=http://patentimages.storage.googleapis.com/pdfs/US7935236.pdf&ei=VccZUvXKsuS7AaZl4HAAQ&usg=AFQjCNFduokBHqDavJkO-9u9TEJ6DOj37w&bvm=bv.62578216,d.ZG

[6] Pavlendová G 2015 *Habilitation thesis: The Effective Usage Of Electro Osmosis*

Influence of glass and polymer fibres on physical and mechanical properties of cement composites

Z Štefunková[1], M Ledererová[1], S Cápayová[2] and V Gregorová[1]

[1] Slovak University of Technology, Faculty of Civil Engineering, Radlinského 11, 810 05 Bratislava, Department of Material Engineering, Slovakia
[2] Slovak University of Technology, Faculty of Civil Engineering, Radlinského 11, 810 05 Bratislava, Department of Transport Construction and Traffic, Slovakia

E-mail: zuzana.stefunkova@stuba.sk

Abstract. This paper focused on the effects of using glass and polypropylene fibres in cement mortars. Were used glass fibres Cem-FIL Anti-Crak HP 12mm long and polymer fibres BeneSteel 55 (shortened to 12mm) in 1wt.%, 3wt% and 5wt% of cement. The water and filler dose were constant. Tests were performed on 40x40x160mm beams. From the results we have found that by the addition of fibres, the shrinkage of hardened cement composites is eliminated, with the use of polymeric fibres for better results. The filaments also positively influenced tensile strength at bending. By adding the fibres, the consistency of the fresh mortar deteriorates, and the bulk density of fresh and hardened mortars decreases, resulting in less compressive strength.

1. Introductions

In the design and manufacture of fibre-reinforced concrete, it is necessary not only to select the appropriate type of fibre and its corresponding quantity (to achieve the desired properties of the concrete), but also to adequately handle the production technology. Fibres with high flexural stiffness and characteristic needle-shaped geometric shape are fundamentally different from other components of fresh concrete and are difficult to mix in fibre concrete production

Various fibres can be added to the concrete: steel, glass, basalt, polypropylene, carbon, organic and so on. Selected mechanical properties of fibres and cement matrix are shown in table 1. This wide range of fibres can be divided into two groups, depending what goal is being followed by adding fibres [1, 2].

If only the aim is to avoid the emergence and development of the number of shrinkages micro cracks at the initial stage of hydration, fibres of high fineness and relatively low Young modulus are used. However, it must be higher than the Young modulus of the solidified cement matrix. The cement matrix in this case is very uniformly reinforced with a large amount of fibres. The addition of such fibres does not substantially affect the physical and mechanical properties of the hardened concrete. The most widely used fibres of this type are some polypropylene and glass fibres.

In the latter case, the addition of fibre reinforcement to the concrete significantly affects some of its otherwise unfavourable properties in the hardened state. In general, increases the cross-tensile strength of the concrete, its toughness and impact resistance, ductility, resistance to temperature changes. Concrete has an increased energy absorption capacity and is more resistant to fatigue. The accompanying phenomenon of the presence of structural fibres in the concrete is the reduction of the volume changes caused by the shrinkage of the concrete. It is mainly a secondary reinforcement, not a main bearing reinforcement [3, 4].

Table 1. Mechanical properties of fibres and cement matrix [1, 2].

Fibre	Diameter (μm)	Density (kg/m^3)	Young modulus (GPa)	Tensile strength (MPa)	Elongation at break (%)
Steel	5 - 500	7 850	210	500 – 2 000	0.5 - 3.5
Glass	9 - 15	2 600	70 - 80	2 000 – 4 000	2 - 3.5
Carbon	7 - 10	1 900	230 - 240	2 600 – 4 000	1.4 - 1.8
Polypropylene	20 - 200	900	5 - 77	500 - 750	8
Flax	5 - 38	1 500	27.6	345 - 1035	2.7 - 3.2
Jute	10 - 25	1 300	26.5	393 - 773	1.5 - 1.8
Cotton	12 - 13	1 500	5.5 - 12.6	287 - 597	7 - 8
Aramid (Kevlar)	10	1 450	65 - 133	3 600	2.1 - 4
Cement matrix	-	2 500	10 - 45	3 - 10	0.02

Shrinkage is one of the negative properties of cement composites that arise due to volume changes due to physical and chemical processes occurring in the cement paste. These processes may result in cracks or other deformations. Surface cracks are most often caused by plastic shrinking by intense evaporation of water from the surface of fresh concrete. These occur most frequently on large areas of concrete (slabs, walls, sidewalks). Disruption of the surface entirety reduces the durability of the concrete structure. Negative effect is increased with increasing wind speed, reducing the relative humidity and with rising the temperature of environment, etc. [5 - 7].

Composition of cement composites significantly affects not only the drying shrinkage, as well as autogenous shrinkage. Shrinkage of cement composites increase with higher water to cement ratio, higher water and fine particles content, fineness of filler granularity and a greater proportion of binder sealant. Composition of concrete (composite) also has an impact on the type of shrinkage that occurs in concrete. Drying shrinkage is a substantial part of the regular concrete with higher water to cement ratio (above ca. 0.4). In the case of concrete with a low water to cement ratio increases the importance of autogenous shrinkage, which can form a dominant part of the total shrinkage [8]. In practical tests is a problem to differentiate these two types of shrinkage. Measured values generally represent the total shrinkage of concrete at a certain time interval.

Shrinkage of concrete can be a problem especially in the case of concrete with high water to cement ratio and finer filler such as self-levelling cement screeds. In these composites, the water to cement ratio is relatively high even with the use of quality super plasticizer agents. The basic technological measure to reduce the shrinkage is to treat the fresh concrete. In some cases, however, the correct treatment of concrete or the reduction of the water to cement ratio by use of the additives, does not necessary guarantee the prevention of cracking [9].

The future of fibre reinforced concrete is in their common application in structures together with the classic reinforced concrete. Not every type of fibre reinforced concrete is suitable for this purpose because different types of fibres affect the physical and mechanical properties of the resulting concrete. Only fibre-reinforced concrete can be considered, in which cracks are produced more evenly than in the case of a classically reinforced fibre-free element. Moreover, even after cracking, they exhibit a certain value of stagnant residual strength (so-called residual or averaged equivalent strength). At present, this is achieved by steel fibre reinforced concrete, fibre reinforced with structural Synthetic Fibre (SSF) or Alkali Resistant Glass Fibres (ARGF) [1].

Due to climate changes, volumetric changes, repeated shrinkage and cracks occur in cement-concrete pavements. Polymer or glass fibre may also be used to eliminate these changes in the

3. Results and discussion

3.1. Consistency and bulk density fresh cement mortars

The composition and the measured values of the bulk density of fresh cement mortars and of the consistency by flow table test are given in table 2.

Table 2. The composition and properties of fresh cement mortars.

Mixture	Fibres (%)		w/c ratio	Cement (g)	Sand (g)	Bulk density fresh mortar ($kg.m^{-3}$)	Flow table test (mm)
Matrix	none		0,65	450	1 350	2 007	154.0
ARGF1	1	ARGF				1 975	156.5
ARGF3	3					1 970	130.5
ARGF5	5					1 963	127.0
PF1	1	BeneSteel				1 969	153.5
PF3	3					1 965	152.0
PF5	5					1 958	135.0

The bulk density decreased with an increasing dose of fibres (Table 2). The type of fibres did not have a more significant effect. The effect of the fibre dose was also demonstrated by a consistency test. Decreasing flow diameter was caused by the increasing amount of fibres. The polymer fibres reduced the flow diameter less than the glass fibres.

3.2. Shrinkage

The course of shrinkage on hardened cement composites is shown in figure 2. In this figure can be seen, that the main shrinkage occurred within 28 days then the process was milder. The largest shrinkage was recorded on the non-fibre reference sample during the whole shrinkage monitoring period. In the first days of the measurement, the lowest shrinkage was observed using glass and polymer fibres at a dose of 5wt.%, the polymer fibres being better than glass fibres.

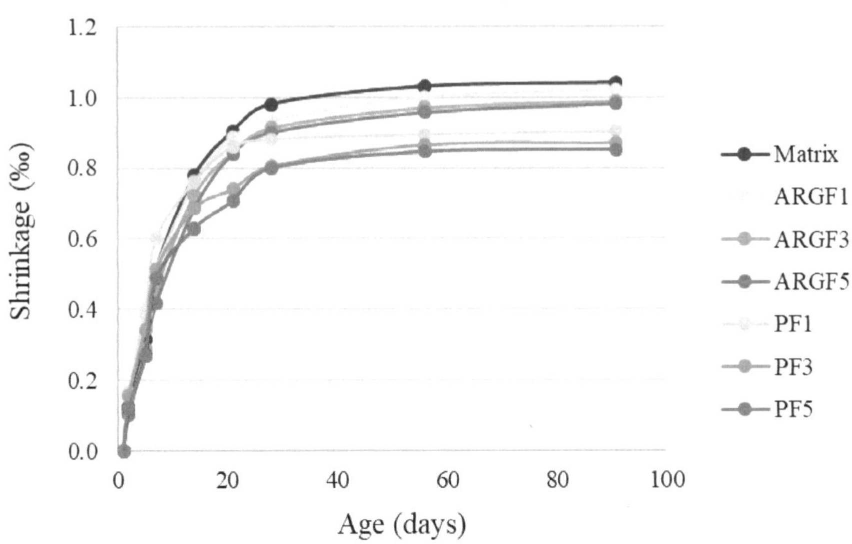

Figure 2. Influence of different types and dose of fibres on the shrinkage of cement mortars.

The change of weight loss of hardened cement composites with using a glass fibres did not significantly changed in the first days of measurement compared to a cement matrix sample. Using polymer fibres caused immediately to a higher loss of weight compared to a cement matrix sample on the first days of measurement. Over time, all samples were almost equal. Overall, samples using polymer fibres had a higher weight loss than samples with glass fibres.

3.3. Bulk density, compressive strength and flexural strength

The measured values of the bulk density of hardened cement mortars and of the compressive and flexural strength are given in table 3.

Table 3. The properties of hardened cement mortars.

Age	Mixture	Bulk density ($kg.m^{-3}$)	Compressive strength (MPa)	Flexural strength (MPa)
7	Matrix	2 160	31.0	5.4
	ARGF1	2 145	26.6	5.5
	ARGF3	2 100	25.4	5.8
	ARGF5	2 062	22.6	6.9
	PF1	2 096	22.8	5.1
	PF3	2 082	22.5	5.3
	PF5	2 047	22.7	5.3
28	Matrix	2 173	43.0	6.2
	ARGF1	2 140	36.3	6.6
	ARGF3	2 114	35.2	6.7
	ARGF5	2 066	34.6	7.4
	PF1	2 127	34.3	6.2
	PF3	2 086	32.8	6.2
	PF5	2 062	31.2	6.1
90	Matrix	2 190	43.3	6.6
	ARGF1	2 142	43.7	6.9
	ARGF3	2 125	43.2	6.9
	ARGF5	2 084	42.8	7.5
	PF1	2 156	45.4	6.8
	PF3	2 122	43.2	6.9
	PF5	2 095	42.2	6.9

The bulk density of hardened cement composites gradually increased with the time of sample storage. The growth is due to the continued hydration of the cement and the thickening of the structure resulting from the hydration products. The highest bulk density value was achieved on sample without the addition of fibres. The bulk densities of samples of the both types of admixtures decreased with increasing dose of fibres. This means, that the lowest values of the bulk densities reached samples with the addition of glass and polymer fibres at the dose of 5wt.%. The difference between glass and polymer fibres was not significant.

The fibre dose was showed on the bulk weight of hardened cement mortars. The highest value had the reference sample without a fibre dose throughout the measurement period. The bulk density has increased, which is related to the compaction of the structure by the hydration products. The bulk density dropped with an increasing dose of fibres, with more pronounced drops in glass fibres. By replacing a portion of the cement with fibres it results in less hydration products and thus a lower bulk density.

This is also the result of compression strength (Figure 3). From the beginning, the cement matrix sample reached higher strength values. After 90 days the results were reached, even the 1% dose reached the higher values in both cases of the fibres used. The most add-on of the fibre dose was shown in the flexural tensile measurement (Figure 4). The higher dose of glass fibres reached a higher increase in strength after 7 days than the other samples. It is more than 28% compared with the cement matrix. This trend was retained even after long-term measurements, but the difference was not so significant (decrease to 14%).

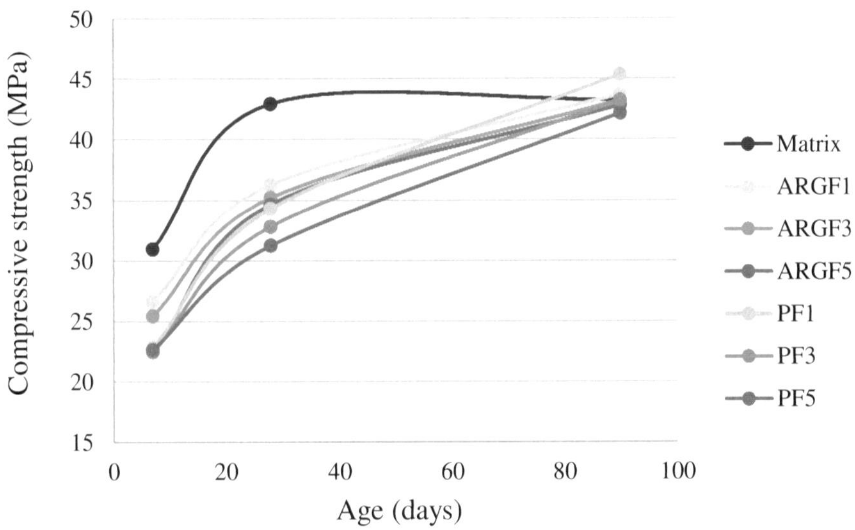

Figure 3. Influence of different types and dose of fibres on the compressive strength of cement mortars.

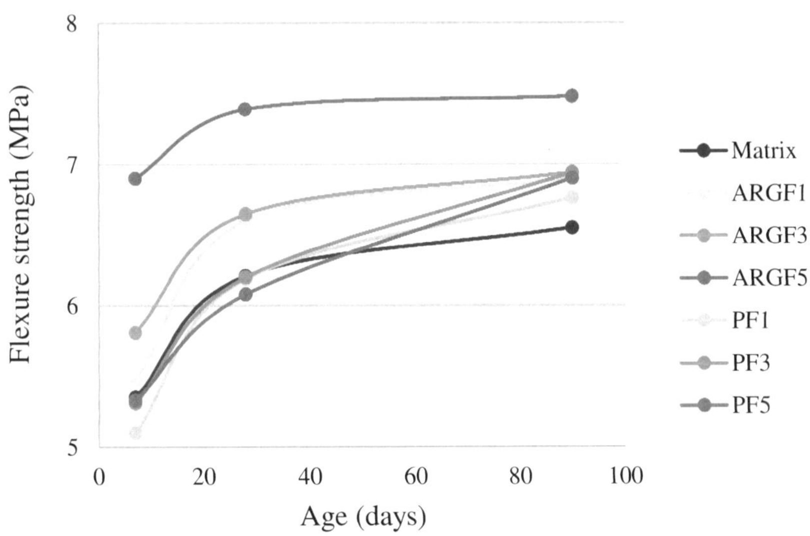

Figure 4. Influence of different types and dose of fibres on the flexural strength of cement mortars.

4. Conclusion

Based on the achieved results can be concluded that the addition of fibres has a positive influence on the shrinkage of hardened cement composites, especially on the first days of measurement. The increasing dose of fibres resulted in lower shrinkage. Polymer fibres had a more positive effect. Also, during the weight loss measuring, the polymer fibres had a greater impact on the weight loss than glass fibres. The higher dose of fibres led to the increasing weight change of mortars.

Addition of fibres resulted to in decreasing values of the bulk densities of fresh and hardened cement mortars. Also, the fibres thicken the consistency and thus reduce the flow diameter. The tensile strength at bending (flexural strength) was the most significant using glass fibre at dose 5wt.%. The results of these mechanical properties were better by a quarter than at the other samples. The fibre dose had on first a negative impact on strength measurements, but after a long-term measurement the differences were lost.

It can be concluded that by adding fibres, shrinkage of hardened cement composites is avoided, especially with the use of polymeric fibres. However, their addition is at the expense of worse consistency and lower compressive strength in the first weeks of hardening of cement composites.

Acknowledgements

This article was created with the support of the Ministry of Education, Science, Research and Sport of the Slovak Republic within the Research and Development Operational Program for the project "University Science Park of STU Bratislava", ITMS 26240220084, co-funded by the European Regional Development Fund and by the Science Grant Agency - project VEGA No. 1/0501/17.

References

[1] Bajza A and Rousekova I 2006 *The concrete technology* (Bratislava: Jaga Group) p 190
[2] Petrone G and Meruane V 2016 *Mechanical properties updating of a non-uniform natural fibre composite panel by means* Elsevier Science, pp 226-233
[3] Bentur A and Mindess S 2006 *Fibre Reinforced Cementitious Composites* London: CRC Press, p 624
[4] J Zhu et al 2018 IOP Conf. Ser.: Mater. Sci. Eng.**292** 012060
[5] Rahmani T, Kiani B, Bakhshi M and Shekarchizadeh M 2012 *Application of Different Fibres to reduce Plastic Shrinkage Cracking of Concrete* RILEM Bookseries 4, pp 635 – 642
[6] Malgorzata L 2015 *Assessment of mortar shrinkage in aspect of organic and inorganic modifiers use* Procedia Engineering 108, pp 309 – 315
[7] Junhua G et al 2019 IOP Conf. Ser.: Mater. Sci. Eng.**472** 012087
[8] Neville A. M. 1997 *Properties of Concrete* Harlow: Addison Wesley Longman Limited
[9] Matulova K and Uncik S 2013 *Effect of expansive cement and waste gypsum on shrinkage of concrete* Book of proceedings. Leuven, Belgium, pp 302-307
[10] Hodakova D, Zuzulova A, Capayova S and Schlosser T 2017 *Impact of Climate Characteristics on Cement Concrete Pavements*. Sofia: STEF 92 Technology, pp 467-472
[11] STN EN 196-1: 2016 *Methods of testing cement* Part 1: Determination of strength

IOP Conf. Series: Materials Science and Engineering **549** (2019) 012023 doi:10.1088/1757-899X/549/1/012023

Threshold of the chloride to hydroxide ions ratio in relation to steel reinforcement corrosion

P Rovnaníková[1] and D Vořechovská[1]

[1] Faculty of Civil Engineering, Brno University of Technology, Veveří 331/95, Brno 602 00, Czech Republic

E-mail: rovnanikova.p@fce.vutbr.cz

Abstract. A simple method for determining the corrosion dependence of steel reinforcement on the content of chloride and hydroxide ions is described. Steel test specimens were immersed for nine months in solutions with 0.03%, 0.3% and 3% NaCl with increasing concentrations of hydroxide ions, and pH of 8.5 to 13. The test specimens were cleaned of corrosion products and weighed at given time intervals of one month. The dependence of mass loss expressed as a percentage of the concentration of chloride and hydroxide ions was determined. The concentration threshold of the cCl^-/cOH^- ratio of 0.5 was evaluated from the results. Higher values of this ratio resulted in the formation of corrosion products on the surface of the test specimens.

1. Introduction

Chloride corrosion of steel reinforcement in concrete is a problem not only in coastal regions but also in regions where winter maintenance of roads with de-icing salts is carried out. Corrosion of steel reinforcement can be derived from the Pourbaix diagram, which maps out possible stable compounds on the Fe surface in aqueous solutions [1]. The pH values for passive stages of steel reinforcement can be derived. The relation between potential and pH varies in the presence of chlorides. Therefore, it is necessary to find a threshold concentration ratio between chloride and hydroxide ions. High concentration ratios are dangerous for steel reinforcement. Chloride ions can be present in free and bound form in concrete. Free chlorides are present in water-soluble compounds, and these ions participate in corrosion processes.

Numerous authors [2-7] have solved the free chloride concentration to hydroxide ion concentration threshold. Alonso et al. [8] summarised the results of the chloride and hydroxide ion ratio threshold of various authors. The critical value is between 0.12 and 6.0, depending on the environment in which the steel reinforcement is inserted.

The authors [8] based their results on electrochemical measurements of corrosion potential (E_{corr}), electrical (R_e) and polarisation (R_p) resistance. The corrosion current was calculated from the measurement results. They found that steel reinforcement depassivation occurs when the corrosion current is greater than 0.1 $\mu A/cm^2$.

The concentrations of chloride and hydroxide ions in the synthetic concrete pore solution are well defined. The concentration ratio thresholds were measured to be the lowest between 0.25 and 1.09 [2, 3, 9]. Higher threshold of concentration ratios of 6.0 were determined in cement mortars and concretes. Steel reinforcement corrosion in concrete depends on many factors, in particular the porous structure of the concrete, the diffusion coefficient for chlorides, the pH of the pore solution and the moisture of the concrete.

IOP Conf. Series: Materials Science and Engineering **549** (2019) 012023 doi:10.1088/1757-899X/549/1/012023

The article is focused on the monitoring of reinforcement corrosion in the form of steel test specimens in synthetic pore solutions with increasing pH and chloride ions and determining the critical concentration ratio of chloride and hydroxide ions via a simple method of steel mass loss.

2. Mechanism of steel reinforcement corrosion

Steel reinforcement in concrete is stored in a high concentration of hydroxide ions (pH ~ 12), which stabilise the phases on steel surface. Reducing the concentration of hydroxide ions below a certain threshold causes corrosion. Corrosion reactions are realised on the local electrodes:
anodic reaction

$$Fe \longrightarrow Fe^{2+} + 2\,e^-$$

cathodic reaction

$$H_2O + \tfrac{1}{2}\,O_2 + 2\,e^- \longrightarrow 2\,OH^-$$

When chlorides are present, they participate in the corrosion reactions. Their participation in the corrosion process can be described by the following chemical reactions [4]:

$$Fe^{2+} + 2\,Cl^- \longrightarrow FeCl_2$$

$$FeCl_2 + 2\,H_2O \longrightarrow Fe(OH)_2 + 2\,Cl^-$$

$$6\,FeCl_2 + O_2 + 6\,H_2O \longrightarrow 2\,Fe_3O_4 + 12\,H^+ + 12\,Cl^-$$

Chlorides enter and exit the reaction as Cl^-. The chloride anion can be labelled as a catalyst and can be reused.

3. Materials and methodology

The corrosion attack of steel in chloride solutions prepared with three concentrations of chlorides combined with OH^- ions in the pH range of 8.5 to 13 was monitored for nine months. The sodium chloride solutions contained 3, 0.3 and 0.03% of NaCl, which correspond to 0.529 mol/l, 0.0529 mol/l and 0.00529 mol/l of chloride anions, respectively. The corrosive solutions were prepared by dissolving NaCl in buffer solutions. To ensure pH stability, a buffer was prepared using boric acid and potassium chloride. This type of buffer maintained the pH stability of the corrosion solution for one month. The chloride content of the buffer was included in the chloride concentration in the corrosion solution. The pH of the corrosion solutions was measured every month. Although the pH was almost unchanged, fresh solutions were prepared each month.

The smooth steel specimens (specification of steel: ZTV d=10 S 235 JRG2 according to EN 10025) of 10 mm diameter and a length of 100 mm were used for the experiments. The test specimens were degreased with an organic solvent prior to immersion into the solutions and weighed. The solutions with the test specimens were placed into 500 ml polyethylene enclosed bottles (Fig. 1). The concentration ratios c(Cl)/c(OH) of the prepared solutions are presented in Table 1.

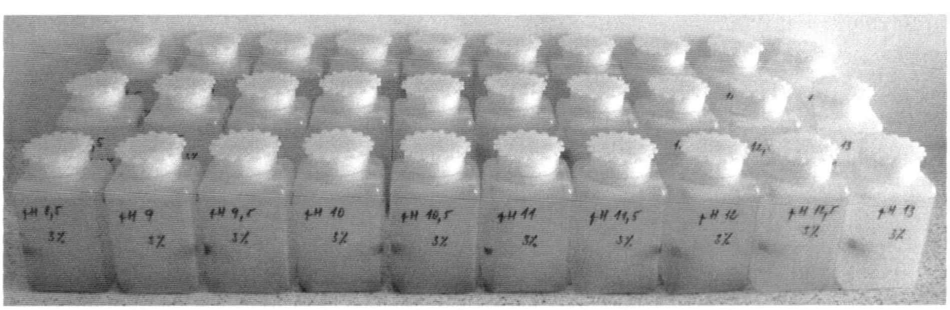

Figure 1. Steel test specimens in bottles with corrosion solutions.

The steel test specimens were removed from the solution at the selected time intervals (one month), cleaned of corrosion products, dried and weighed. Before removing the corrosion products, the samples were visually evaluated, and their documentation photo was taken. The mass loss was used for calculating the percentage changes of mass in time. From the results of corrosion losses, the threshold values of chloride and hydroxide ions concentration were found.

Table 1. The concentration ratios $c(Cl^-)/c(OH^-)$ in 3, 0.3 and 0.03% NaCl solutions.

pH	Cl^-/OH^- 3% sol. of NaCl	Cl^-/OH^- 0.3% sol. of NaCl	Cl^-/OH^- 0.03% sol. of NaCl
8.5	167356.0	16735.6	1673.6
9.0	52923.0	5292.3	529.3
9.5	16735.6	1673.6	167.4
10.0	5292.3	529.2	52.9
10.5	1673.6	167.4	16.7
11.0	529.2	52.9	5.3
11.5	167.4	16.7	1.7
12.0	52.9	5.3	0.5
12.5	16.7	1.67	0.2
13.0	5.3	0.5	0.1

4. Results and discussion

4.1. Visual assessment

Prior to removing the corrosion products, the corrosion rate was visually assessed. The rating was done on a three-point scale:

1. No corrosion (NC).
2. Pitting corrosion at several points (PC).
3. General corrosion on the whole surface (GC).

The results of the visual assessment after nine months are present in Table 2.

Table 2. Visual assessment of the steel specimens (nine months).

Assessment	pH range for 0.03% NaCl	pH range for 0.3% NaCl	pH range for 3% NaCl
NC	≥ 12	≥ 12.5	-
PC	9 - 11.5	9.5 - 12	10.5 - 13
GC	< 8.5	8.5 - 9	8.5 - 10

The image of the test specimens was documented with photos. The images of the steel specimens after nine months in the chosen corrosion environment are presented in Fig. 2, 3 and 4. There is significant surface corrosion on the whole specimens in solutions with higher concentration of chlorides. The test specimens immersed in 3% NaCl and pH 13 contained only a few small pitting places.

pH 8.5 – 0.03 % NaCl pH 13 – 0.03% NaCl

Figure 2. Image of steel specimens in 0.03% corrosion solution.

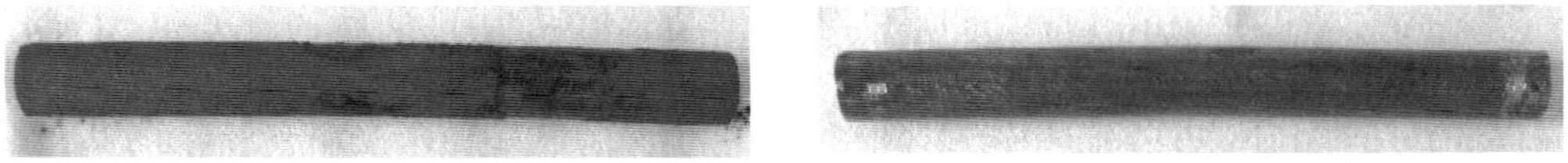

pH 8.5 – 0.3% NaCl pH 13 – 0.3% NaCl

Figure 3. Image of steel specimens in 0.3% corrosion solution.

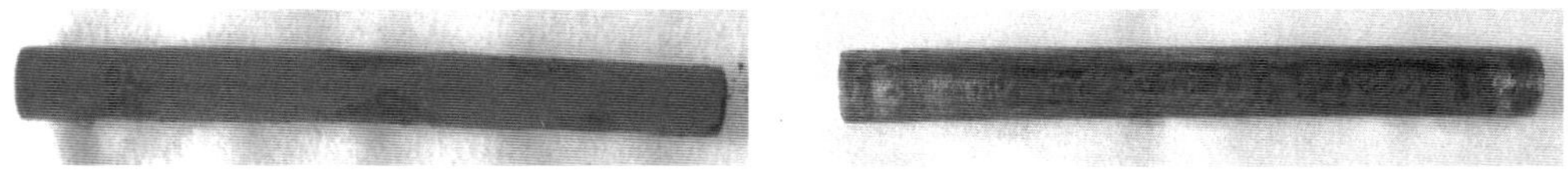

pH 8.5 – 3% NaCl pH 13 – 3% NaCl

Figure 4. Image of steel specimens in 3% corrosion solution.

4.2. Determination of threshold concentration ratio c(Cl)/c(OH)

The steel mass loss decreases with increasing pH value. The mass loss is significantly reduced in solutions with pH of 12.5: the steel did not corrode in the 0.03% NaCl solution. Results of mass loss vs. pH value are presented in Fig. 5, 6 and 7. Specification 1M to 9M is number of months.

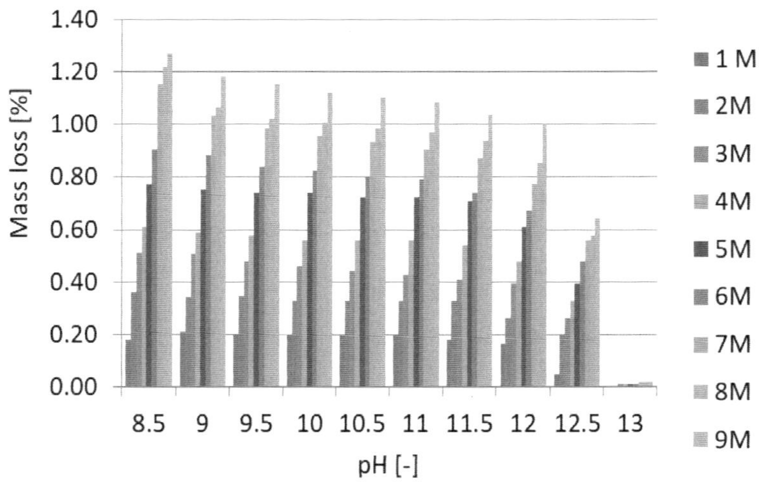

Figure 5. Mass loss vs. pH in 3% NaCl solution.

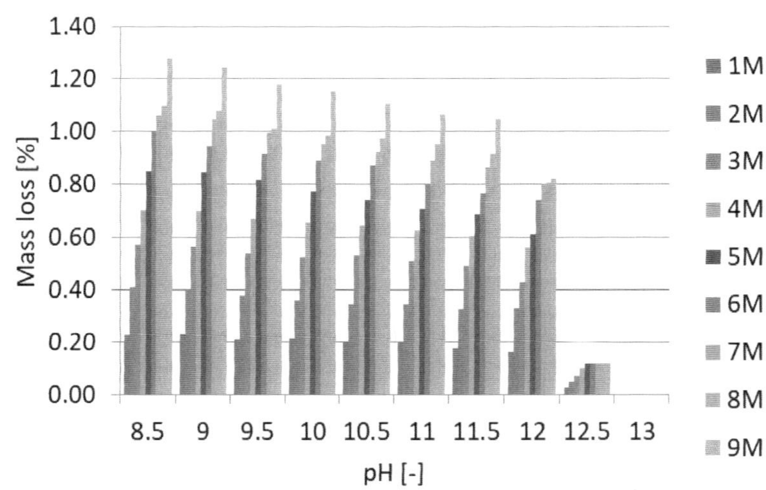

Figure 6. Mass loss vs. pH in 0.3% NaCl solution.

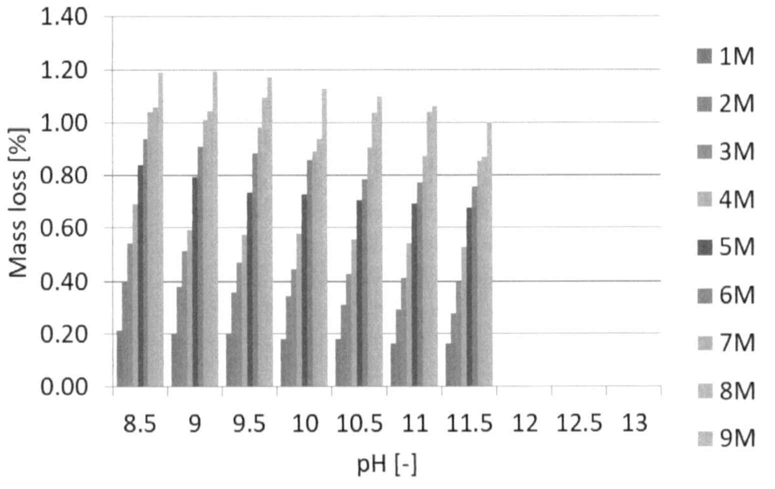

Figure 7. Mass loss vs. pH in 0.03% NaCl solution.

The mass loss results were correlated with the chloride and hydroxide ion ratios. The graphs in Fig. 8, 9 and 10 show the threshold level of the $c(Cl^-)/c(OH^-)$ concentration ratio. The NaCl solution (3% concentration) corresponds to a concentration ratio $c(Cl^-)/c(OH^-)$ of 5.3 for a pH value of 13. It is a high value of pH. The pitting corrosion was noted on the steel surface in our experiments.

A value of 0.5 was found for the threshold of the $c(Cl^-)/c(OH^-)$ concentration for the pH 13 and 0.3% NaCl. For the lowest NaCl concentration of 0.03%, a threshold of 0.5 for pH 12 was found.

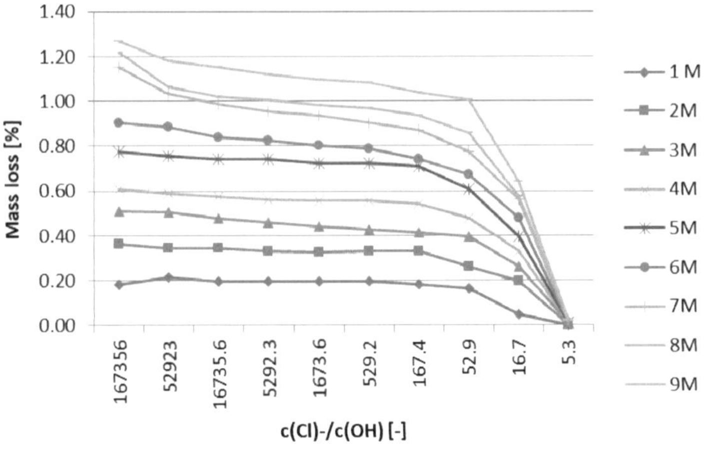

Figure 8. Mass loss vs. concentration ratio $c(Cl^-)/c(OH^-)$ in 3% NaCl solution.

5. Conclusion

This paper focused on the determination of the threshold value of the concentration ratio $c(Cl^-)/c(OH^-)$ via a simple method of mass loss. Corrosion experiments were performed in solutions of three concentrations of NaCl and increasing pH from 8.5 to 13.

The following conclusions can be drawn from the results:

- The concentration of hydroxide ions is the most important factor for steel reinforcement corrosion.
- Steel reinforcement corrosion is dependent on the $c(Cl^-)/c(OH^-)$ concentration ratio.
- The threshold value of the $c(Cl^-)/c(OH^-)$ ratio for steel reinforced corrosion is 0.5.

In conclusion, when assessing the corrosion of steel reinforcement in concrete, it is necessary to determine both the chloride ion content and the pH value.

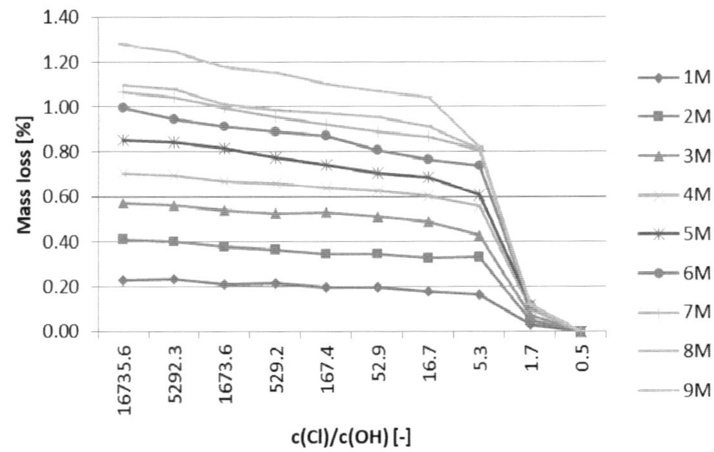

Figure 9. Mass loss vs. concentration ratio c(Cl⁻)/c(OH⁻) in 0.3% NaCl solution.

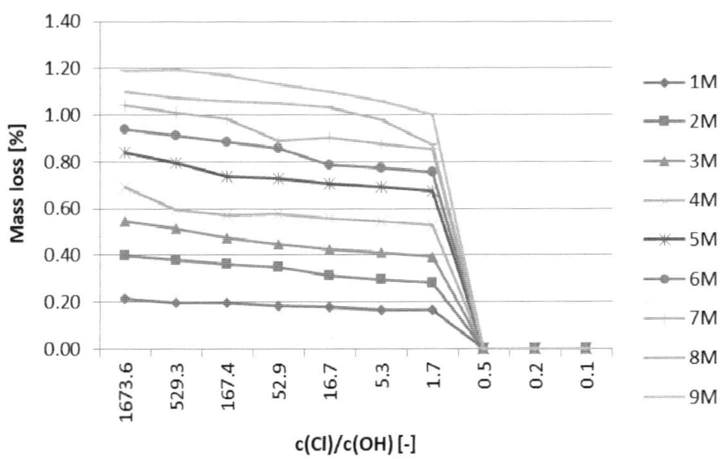

Figure 10. Mass loss vs. concentration ratio c(Cl⁻)/c(OH⁻) in 0.03% NaCl solution.

Acknowledgments
This outcome was achieved with the financial support of the Czech Science Foundation, project No. 18-07949S.

References

[1] Beverskog B and Puigdomenech I 1996 *Corr. Sci.* **38** 2121–2135
[2] Grimaldi G and Raharinaivo A 1999 *Bull. Lab. Pont. Chauss.* **223** 59–70
[3] Kalabisová E, Pražák M and Březina M 1998 *Conf. ACSCE 1998 (Malenovice)* 1–4
[4] Thangavel K and Rengaswamy N S 1998 *Cem. Constr. Comp.* **20** 283–292
[5] Ki Yong Ann, Ha-Won Song 2007 *Corr. Sci.* **49** 4113–4133
[6] Shi X, Xie N, Fortune K and Gong J 2012 *Constr. Build. Mat.* **30** 125–138
[7] Vořechovská D, Podroužek J, Chromá M, Rovnaníková P and Teplý B 2009 *Comput. Aided Civ. Infrastr. Eng.* **24** 446–458
[8] Alonso C, Andrade C, Castellote M and Castro P 2000 *Cem. Concr. Res.* **30** 1047–1055
[9] Goñi S and Andrade C 1990 *Cem. Concr. Res.* **20** 525–539

IOP Conf. Series: Materials Science and Engineering **549** (2019) 012024 doi:10.1088/1757-899X/549/1/012024

Innovation of University education by using available online LCA methods and their role in construction sector

M Ondova[1], A Estokova[1] and M Wolfova[1]

[1] Institute of Environmental Engineering, Faculty of Civil Engineering Technical University of Kosice, Vysokoskolska 4, Kosice 04200, Slovakia

E-mail: marcela.ondova@tuke.sk

Abstract. Currently, active education at universities is increasingly preferred. The introduction of innovative approaches to teaching is reflected in the increased number of practical lectures mainly. The use of new methods has also been applied to the subject of "LCA Building Materials", which is aimed at assessing the environmental impacts of building materials. Regarding the production of building materials and their use, the process of construction and operation of the building, as well as its disposal, the use of environmental assessment methods seems to be very desirable and perspective in regard to negative impacts on the environment and public health. This idea formed the basis for an active approach to teaching. During the lectures, theoretical knowledge is continuously supplemented by an illustrative example of calculating the environmental impacts of a selected family house. Through the available online LCA methods students evaluate their projects in parallel with the teacher, as if they were dealing with a real contract. Main goals of the project are to improve students' skills, to broaden their knowledge, to look at the issues comprehensively (construction industry, environmental protection, climate change etc.), and to apply knowledge to practice.

1. Introduction

One of the main aims of each society is to raise educational standards. High priority of universities is to support the students´ competitiveness and their ability to meet market requirements. New approach called "'Life Cycle Thinking" has become a central pillar in environmental policies and sustainable business decision-making. The environmental assessment ensures that the environmental impacts are considered before decisions are made.

Currently, one of the most used methods is Life Cycle Assessment (LCA). LCA method is very popular and perspective. Instead of separate fragments of environmental impacts LCA method considers the entire product's or service's life cycle. In the building industry LCA helps to reduce negative environmental impact of buildings, and to improve the future of public health.

LCA method has been developed according to ISO 14040 and ISO 14044 series standards, which describe the general principles and requirements to undertake LCA. Life Cycle Assessment is carried out in four distinct phases including goal and scope definition, life cycle inventory, life cycle impact assessment, and interpretation.

Nowadays, it can be used in various fields of building assessment [1]. Many authors monitor the environmental impacts of the production of building materials [2, 3]. González and Navarro [4] present that the building materials with low impacts can reduce CO_2 emissions up to 30%. Negative impacts can also be reduced by using renewable or recycled materials as potential secondary raw materials in building materials manufacturing [5 - 8]. Study of Turner and Collins proves, that alternative cements based on fly ash and ground granulated blast furnace slag can reduce CO_2 emissions by 13–22% [9]. The alternative material options have been studied by Caruso et al. [10]. Valencia and Gómez-Soberón [11] analyse suitable composition of interior partition walls with regard to their environmental impact. Environmental performance of the residential buildings was the topic of many researches. Evangelista et al. quantifies the environmental performance of residential buildings through the complete Life Cycle Assessment [12]. Carbon emissions and energy demand were monitored in his paper, but also in many other works [13, 14]. In general, the results of LCA research papers are different and have proven that optimal design does not exist. There for precise environmental analysis of each project is necessary [15]. Well-considered choice of materials for the construction of buildings should be based on scientific analysis [16].

Practical example of LCA study is given in this paper. It is focused on environmental impact of selected residential house regarding the material composition of each construction. LCA indicators: primary energy consumption (PEN_{RT}), global warming potential (GWP), acidification potential (AP), as well as environmental ($\Delta OI3$) and disposal (EI_{KON}) indicators were used to evaluate the environmental performance.

2. Material and methods

The exemplary study of the masonry family house (Figure 1) with common building materials was selected for determination of its environmental impacts.

Figure 1. Assessed family house.

The family house is a single-storey building, set in a flat ground and it is without basement. The family house consists of one housing unit with facilities and garage on the first floor. The roof of the building is hipped with a 15° slope with external rainwater drainage on the ground. Floor level (± 0.000) is distant approximately 300 mm from the landscape terrain. The main entrance to the building is from the north. The ground plan of the building is compact, polygonal. It is designed for 3-5 persons. Other technical specifications of assessed family house are presented in Table 1.

The foundation strips are 600 mm wide, made of C 12/15 concrete. Insulation against increased ground moisture is designed to penetrate the substrate, which is melted to the underlying concrete.

The external and internal load-bearing structures are the new brickwork structures made of POROTHERM bricks in the thickness of 300 and 380 mm. The internal partition structures are made of YTONG aerated concrete blocks in 150 mm thick cementitious adhesive.

The post columns are designed with monolithic reinforced concrete.

The building is insulated with contact insulation system with NOBASIL mineral wool thermal insulation (minimum thickness of 70 mm), in skirting part of extruded polystyrene XPS (minimum thickness of 50 mm).

Ceilings represent the supporting elements and beams - the reinforcing steel-concrete monolithic rim, non-bearing beams YTONG, above the door openings in the non-loadbearing partitions. Reinforced concrete parts are designed as monolithic, concrete C 16/20 - XC1. The ceiling structure above the first floor is designed as non-load-bearing and it is formed by a suspended plasterboard ceiling RIGIPS. In the ceiling construction thermal insulation is designed mineral wool (minimum thickness of 160 + 60 mm) sealed with a vapor barrier from below.

The basic load-bearing part of the roof is a standing stool, made up of beams, posts, middle binding crafter, on which wooden rafters are placed.

The roof slope is 15°. The roofing is designed with heavy concrete, type BRAMAC, including accessories.

Table 1. Technical specification of assessed family house.

Parameter	Family house
Number of rooms	5
Useful area [m^2]	112.1
Built-up area [m^2]	175.25
Living area [m^2]	82.3
Garage area [m^2]	27.75
Terrace area [m^2]	31.92
Accessories [m^2]	29.8
Total built-up volume [m^3]	701.0

Assessment was carried out in four LCA phases including goal and scope definition, life cycle inventory, life cycle impact assessment, and interpretation. The functional unit established in all assessed scenarios was 1 m^2 of the gross floor area. The study period of 100 years was considered for the wall constructions, and 25 years for both exterior and interior plasters. According to EN 15804 standard the evaluation was performed within the „cradle to gate" boundaries using the eco2soft.
Baubook eco2soft is a tool that allows the calculation of ecological figures for buildings, including HVAC (Heating, Ventilation, and Air Conditioning). Based on the existing component calculator, it displays global warming potential (GWP), acidification potential (AP) and primary energy content (PEN_{RT}) of all building elements. It also features the calculation of environmental indicator ΔOI3, as required for the energy pass of municipal buildings. In addition, the disposal indicator (EI) of a building that is the area-weighted mean value of the disposal indices of the constructions, can be also calculated.

Inventory analysis focused on all environmental inputs and outputs associated with a selected product was processed through the Ecoinvent database.

3. Results and discussion

All calculated environmental indicators mentioned above are given in Table 2.

Table 2. Resulting environmental parameters (OI3, PEN_{RT}, GWP, AP and EI) of assessed family house per functional unit [m^2] and study period (100 years).

Environmental indicator	Family house
OI3	592
PEN_{RT} [MJ]	8104
GWP [kg CO_2 equ.]	539
AP [kg SO_2 equ.]	1.74
EI	0.67

The results presented in Table 2 show that the environmental profile of the assessed family house is slightly above the common average. However, in order to achieve better environmental impact values, it is necessary to use alternative built-in materials or structural elements.

To find more optimal solution for a construction by using variations of original project (usually two more variants) is common practice in LCA method. By using the parameters obtained in the evaluation of the original variant, it is possible to propose individual improvements of each structure.

In this case it is a comparison of two external walls prepared using different thermal insulation systems (XPS, mineral wool).

Figure 2 shows the results of assessed external wall insulated with XPS. Indicator OI3 is on the scale in class "B", EI_{KON} reached value 0.78 points/m^2, and the values of the other environmental indicators are as follow: primary energy (PEN_{RT}) 1531 MJ/m^2; GWP 93.9 kg CO_2/m^2, and AP 0.293 kg SO_2/m^2.

This information can be used as input data for other calculated variations in order to reduce negative environmental impact of each unit and the whole family house, ultimately.

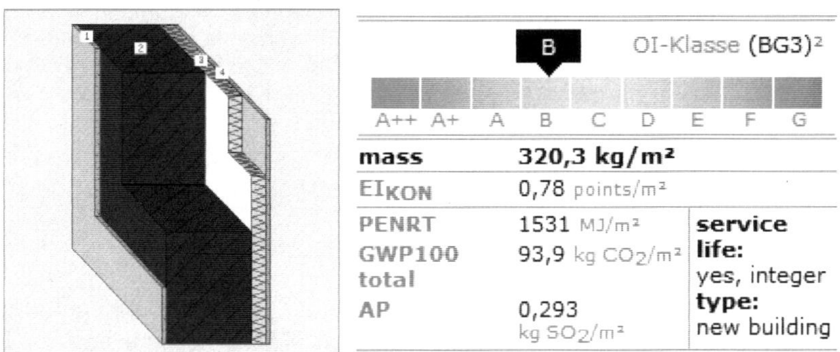

Figure 2. Results of assessed external wall insulated with XPS
(1- CR lime cement finish plaster: d = 0.20cm, λ = 1.050 W/m.K.; 2 - POROTHERM 30 N+F: d = 30.00cm, λ = 0.205 W/m.K.; 3 – XPS: d = 5.00cm, λ = 0.036 W/m.K.; 4 - silicate plaster: d = 1.50cm, λ = 0.800 W/m.K.).

For comparison, Figure 3 shows alternative external wall insulated with mineral wool.

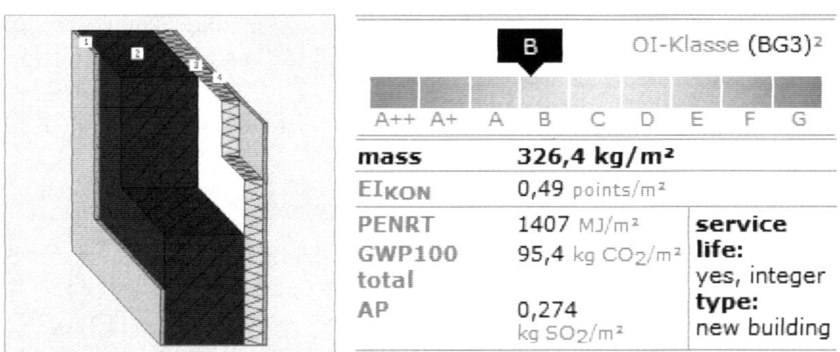

Figure 3. Results of assessed external wall insulated with mineral wool
(1- CR lime cement finish plaster: d = 0.20cm, λ = 1.050 W/m.K.; 2 - POROTHERM 30 N+F: d = 30.00cm, λ = 0.205 W/m.K.; 3 – mineral wool: d = 7.00cm, λ = 0.044 W/m.K.; 4 - silicate plaster: d = 1.50cm, λ = 0.800 W/m.K.).

Calculated values are significantly reduced in all evaluated indicators. The biggest benefit was marked at EI and PEN_{RT} indicator. Regarding recyclability and energy intensity the use of mineral wool is preferred.

Based on these results it is necessary to realize that increased attention should be paid to the environment while choosing the building materials. Slight change in the material composition of the structure can reduce the environmental burden while ensuring the required functional and thermal insulation parameters. The comprehensiveness of LCA evaluation can help in the decision-making process regarding the suitable selection of materials. Another preferred option is the selection of materials based on the EPD (Environmental Product Declarations) certificate, which guarantees a reduced negative impact on the environment.

4. Conclusion

The process of buildings construction consists of a few complex phases (design, construction, operation, maintenance of buildings). Usually, it requires enormous amount of raw materials, and energy and water resources. Moreover, it produces large quantities of waste that causes environmental pollution. Sustainable construction (green buildings) is the only answer for healthier environment. It requires more efficient use of resource in construction, reconstruction, operation and maintenance. Life Cycle Assessment and subsequent smart choice of building materials appear to be a starting point for building sustainability.

The achievement of a "cleaner" future with preservation of today's comfort is the main goal for coming generations. Therefore, the innovation of university education from passive to active is necessary. Primary is the introducing to practical subjects with modern online access to education. Subsequently, responsible and active preparation of students at universities is highly positively perceived from the perspective of their future job as well as very desirable by employers. In addition, the LCA experience is a very important parameter for the future of construction in connection with our environment.

Acknowledgments

This research has been carried out within the Slovak Cultural and Education Grant Agency (contract No. 073TUKE-4/2018) and Scientific Grant Agency of the Ministry of Education, Science, Research and Sport of the Slovak Republic and the Slovak Academy of Sciences (grant number 1/0648/17).

References

[1] Buyle M *et al* 2013 *Renewable and Sustainable Energy Reviews* **26**, 379-388
[2] Abeysundara Y, Babel S 2010 *Env Progress & Sust Energy* **29** (3), 370-381
[3] Muñoz I *et al* 2018 *Construction and Building Materials* **168**, 178-186
[4] Gonzalez M, Navarro J 2006 *Build Environ* **41**, 902–909
[5] Milutienė E *et al* 2012 *Clean Techn Environ Policy* **14** (6), 1075-1084
[6] Junak J, Stevulova N 2013 *Visnik Nacional'nogo universitetu L'vivska politechnika: teorija i praktika budivnictva* **756**, 63-68
[7] Vaclavik V *et al* 2016 *Solid State Phenomena* **244**, 77-87
[8] A Brenek *et al* 2015 *Advanced Structured Materials* **70**, 177-188
[9] Turner L K, Collins F G 2013 *Constr Build Mater* **43**, 125–130
[10] Caruso M C *et al* 2018 *IOP Conf. Ser.: Mater. Sci. Eng.* **442** 012010
[11] Valencia E Y, Gómez M J – Soberón 2019 *LCA for the Study and Investigation of Buildings*
[12] Evangelista P P A *et al* 2018 *Construction and Building Materials* **169**, 748–761
[13] Velagapudi S *et al* 2014 *Env Progress & Sust Energy* **33** (4), 1366-1372
[14] Li D Z *et al* 2013 *Build Environ* **59**, 448-455
[15] Benedetto L De, Klemes J 2008 *Chem Eng T* **13**, 343-350
[16] Ondova M, Estokova A 2016 *Environ. Prog. Sustain. Energy* **35**, 1113–1120

IOP Conf. Series: Materials Science and Engineering **549** (2019) 012025 doi:10.1088/1757-899X/549/1/012025

Presence of magnesium-alumina spinel in forsterite ceramics and its influence during sintering and on resulting properties of fired body

M Nguyen[1] **and R Sokolář**[1]

[1] Brno University of Technology, Faculty of Civil Engineering, Institute of Technology of Building Materials and Components, Veveří 331/95, 602 00 Brno, Czech Republic

E-mail: nguyen.m@fce.vutbr.cz

Abstract. The main objective of this paper is to describe the creation and behaviour of magnesium-alumina spinel (MA-spinel) phase from raw materials mixtures in forsterite refractory and its influence on resulting physical properties and mineralogical composition of fired body. Forsterite ceramics is mostly used for its high refractoriness as a refractory material in kilns and furnaces and also in electrotechnical engineering for its high coefficient of linear thermal expansion for ceramic-metal joints. Currently, forsterite is not produced in the Czech Republic. The primary objective is to synthesize forsterite ceramics via solid-state reaction utilizing various raw materials used in industry with stoichiometric ratio of forsterite. The resulting properties of fired body will be determined using X-ray diffraction analysis for mineralogical composition, thermal dilatometric analysis and vacuum water absorption method for physical properties of fired body and refractoriness of pyrometric cones will be conducted.

1. Introduction

Forsterite is a magnesium silicate mineral with chemical formula $2MgO \cdot SiO_2$. It is magnesium-rich end-member of the olivine solid solution series. Naturally occurring olivines are fairly abundant mineral in the earth's crust but they contain about 10-15% of fayalite (iron silicate) with chemical formula $2FeO \cdot SiO_2$. While the melting point of forsterite is relatively high (1890 °C), the melting point of fayalite is relatively low (1205 °C) for refractory applications. The amount of fayalite up to 10% in olivine has not very significant impact on refractoriness. [1-2]

Forsterite ceramics primary use in the industry is for refractory bricks [3] due to the high refractoriness of forsterite (up to 1850 °C) and refractoriness under load above 1600 °C. [4] This is significant for utilization of forsterite as a lining of the sintering zone of rotary kilns for cement production and metallurgical furnaces, because forsterite practically does not react with clinker or iron at high temperatures. [1] Another significant attribute of forsterite is its coefficient of linear thermal expansion which has linear increase similar to coefficient of metals with increasing temperatures. [5] Due to this attribute, forsterite is also used in electrical engineering for joining the ceramics with metal. [4] Forsterite may even be suitable for biomedicine as a bone implant biomaterial. [6]

Addition of aluminium oxide into the raw material mixture results in a creation of MA-spinel ($MgO \cdot Al_2O_3$) in addition to forsterite. MA-spinel in forsterite refractory improves its mechanical properties and sintering while retaining its refractory properties. Inexpensive source of aluminium oxide is fly ash which is secondary energetic product in power plants combusting coal. [7-10]

2. Materials and methods

2.1. Raw materials

Raw materials used for synthesis of forsterite were raw magnesite (magnesium carbonate), talc, quartz sand, olivine, calcined caustic magnesite (CCM), fly ash from power plant Mělník and kaolin Zettlitz Ia. Pharmaceutical quality raw magnesite (purity 98%, d_{50} = 2 μm) and talc (purity 97%, d_{50} = 5 μm) were acquired from the company Fichema Ltd., micro-milled quartz sand ST9 (purity 99.6%, d_{50} = 6 μm) was acquired from the company Sklopísek Střeleč Inc., finely ground olivine was acquired from the company Olivin, Norway. CCM 85 (min. 85% of MgO) was acquired from the magnesite plants in Jelšava, Slovakia. Fly ash as a secondary energetic product of coal firing in power plant Mělník and kaolin Zettlitz Ia was used for better binding strength of a green body.

Five different raw material mixtures were designed with a stoichiometric ratio of forsterite. First mixture (REF) was intended as a reference value for other mixtures and was composed of pure pharmaceutical quality raw materials – magnesite and talc to form pure forsterite during synthesis. Second mixture (OLI-CCM) was composed of CCM 85 and olivine. Third mixture (FA-CCM) was composed of fly ash from power plant Mělník and CCM 85. Fourth mixture (OLI-QS-CCM) was composed of olivine, quartz sand ST9 and CCM 85. Fifth mixture (MAG-FA) was composed of raw magnesite and fly ash from power plant Mělník. All mixtures contained 10 wt% of kaolin Zettlitz Ia for improved binding strength of the green body and improved sintering during firing. Weight composition of raw materials with stoichiometric ratio of forsterite for all individual mixtures is presented in table 1. Chemical composition of used raw materials is presented in table 2.

Table 1. Composition of individual mixtures from raw materials.

Raw materials	REF (wt%)	OLI-CCM (wt%)	FA-CCM (wt%)	OLI-QS-CCM (wt%)	MAG-FA (wt%)
Magnesite	53.8	-	-	-	62.2
Talc	36.2	-	-	-	-
CCM 85	-	42.2	50.0	49.9	-
Quartz sand ST9	-	-	-	15.6	-
Olivine	-	47.8	-	24.5	-
Fly ash	-	-	40.0	-	27.8
Kaolin Zettlitz Ia	10.0	10.0	10.0	10.0	10.0

Table 2. Chemical composition of used raw materials.

Raw materials	MgO (%)	SiO_2 (%)	CaO (%)	Fe_2O_3 (%)	Al_2O_3 (%)	K_2O+Na_2O (%)	LOI[a] (%)
Magnesite	44.2	0.1	0.4	0.0	0.0	0.0	52.0
Talc	31.5	59.1	1.0	0.7	1.0	0.2	6.5
CCM 85	85.0	0.5	5.2	7.3	0.8	0.2	1.0
Quartz sand ST9	0.0	99.6	0.1	0.1	0.2	0.1	0.0
Olivine	24.1	64.7	0.7	8.8	1.0	0.5	1.0
Fly ash	1.4	57.3	2.2	5.1	29.3	1.7	1.2
Kaolin Zettlitz Ia	0.5	46.8	0.7	0.9	36.6	1.2	13.2

[a] Loss on ignition

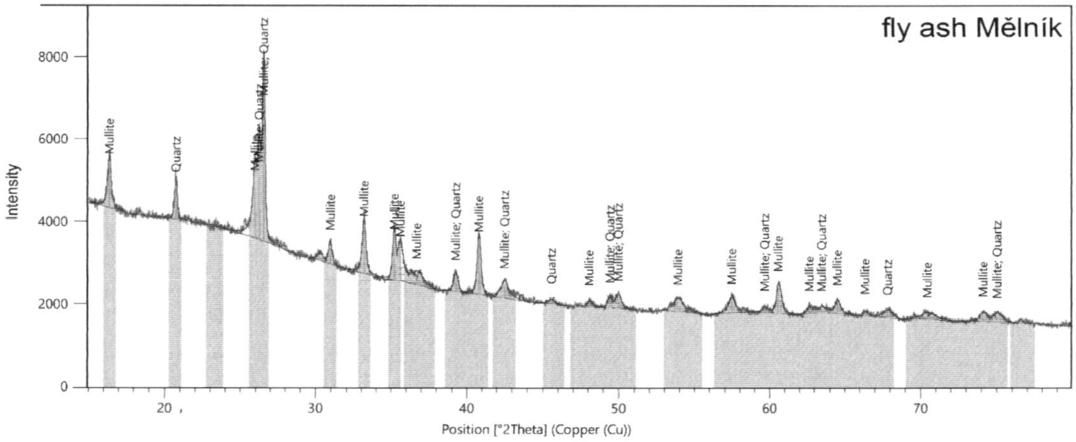

Figure 1. X-ray diffraction analysis of raw fly ash from power plant Mělník.

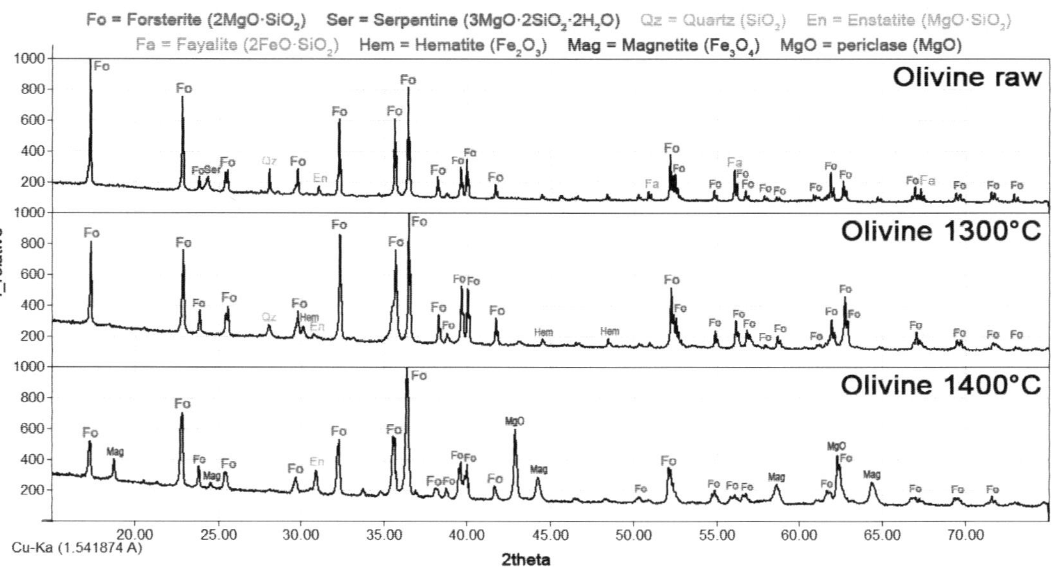

Figure 2. X-ray diffraction analysis of raw and heat treated (1300 °C and 1400 °C) olivine.

It is apparent from figure 1 that major crystal phases in raw fly ash from power plant Mělník are mullite and quartz. The curved background of diffraction curve also indicates the presence of amorphous glass phase and iron oxide. Figure 2 contains XRD analysis of raw and heat-treated olivine. Crystal phases present in raw olivine are forsterite, fayalite, enstatite, quartz and serpentine. Crystal phases present in heat treated olivine at 1300 °C are forsterite, enstatite, quartz and hematite. And in heat treated olivine at 1400 °C are forsterite, periclase, magnetite and enstatite.

2.2. Preparation of test samples

All mixtures were subjected to homogenization in a mechanical rotary homogenizer for 24 hours to ensure homogeneity of all mixtures. Homogenized mixtures were then mixed with water to gain an optimal viscosity of plastic paste for plastic moulding of test samples. Dimensions of test samples were approximately 20 × 25 × 100 (mm^3). Pyrometric cones were also prepared from the same plastic paste as test samples according to the official standard EN 993-13:1995. Green bodies and pyrometric cones were then dried in a laboratory dryer at 105°C. After drying, all test samples and pyrometric cones were fired in an electrical laboratory furnace with air atmosphere and controlled environment.

Maximum firing temperature was set to 1300 °C with heating rate of 4 °C per minute and soaking time of 2 hours at maximum temperature.

2.3. Experimental procedures

Several laboratory tests were performed on test samples and pyrometric cones after firing. Apparent porosity, water absorption and bulk density were determined utilizing the vacuum water absorption method described in the official standard EN 993-1:1995. Change in dimension (length) of test samples during firing was determined according to the official standard EN 993-10:1997. Refractoriness of pyrometric cones was determined according to the official standard EN 993-12:1997 using an electrical laboratory furnace with an observation opening which enables the use of a digital camera to record the real time behaviour of pyrometric cones during increasing temperature up to 1800°C, considering optical properties. Thermal dilatometric analysis was done in accordance with the official standard EN 993-19.

Mineralogical composition of fired test samples was identified using X-ray diffraction analysis (PANALYTICAL Empyrean XRD) with Cu cathode as a radiation source, $K\alpha 2/K\alpha 1$ ratio of 0.5, accelerating voltage 45 kV, beam current 40 mA, diffraction angle 2θ in the range from 15° to 70° with a step scan of 0,01°. Acquired XRD patterns were then evaluated using a PC software Match! version 3.8 and Crystallography Open Database (COD). Semi-quantitative analysis of all fired samples was obtained by means of the reference-intensity ratio (RIR) method using the same software. Final chemical composition of all mixtures after firing is presented in table 3.

Table 3. Final chemical composition of test samples from all mixtures after firing.

Mixture	MgO (%)	SiO_2 (%)	CaO (%)	Fe_2O_3 (%)	Al_2O_3 (%)	K_2O+Na_2O (%)
REF	53.6	39.1	0.9	0.54	5.6	0.29
OLI-CCM	48.8	36.4	2.7	7.53	4.2	0.37
FA-CCM	44.8	28.9	3.6	6.17	15.6	0.91
OLI-QS-CCM	49.8	37.0	2.9	6.03	4.0	0.26
MAG-FA	44.8	32.3	1.4	2.65	17.9	1.04

3. Results and discussion

The presence and influence of MA-spinel ($MgO{\cdot}Al_2O_3$ or $MgAl_2O_4$) in forsterite ceramics were studied to understand the decomposition and crystallization of various minerals from raw material mixtures via solid-state reaction and the impact on resulting properties of forsterite ceramics. Firing shrinkage is apparent from table 4. Relatively high firing shrinkage of mixtures REF and MAG-FA is due to decomposition of magnesite (magnesium carbonate) into magnesium oxide and carbon dioxide. Released carbon dioxide also lead to high porosity and water absorption of these mixtures and created very porous structure of those test samples.

Opposite phenomenon occurred in the mixture FA-CCM which also had relatively high firing shrinkage, however apparent porosity and water absorption of these test samples were the lowest of all samples. This phenomenon occurred due to the creation of eutectic melt in the mixture in the presence of flux oxides (Fe_2O_3, Na_2O and K_2O). Flux oxides created optimal conditions for creation of amorphous glassy phase and enhanced sintering. Firing shrinkage of mixtures OLI-CCM and OLI-QS-CCM was relatively low (1.5% and 2.9%) even though the amount of iron oxide was similar to the mixture FA-CCM. This phenomenon occurred because significant amount of iron oxide in these mixtures is from olivine, bound in the form of fayalite $2FeO{\cdot}SiO_2$ (confirmed by XRD analysis in figure 2) which conducts differently with increasing temperature. R. Michael et al. [11] also confirmed this behaviour of olivine in high temperatures.

Table 4. Test results performed on fired test samples from all mixtures.

Test sample	Firing temperature (°C)	Firing shrinkage (%)	Apparent porosity (%)	Water absorption (%)	Bulk density (kg·m^{-3})
REF	1300	26.9	56.8	41.6	1340
OLI-CCM	1300	1.5	40.3	20.0	2010
FA-CCM	1300	13.3	24.4	10.6	2310
OLI-QS-CCM	1300	2.9	39.4	19.8	1990
MAG-FA	1300	36.5	40.4	23.7	1680

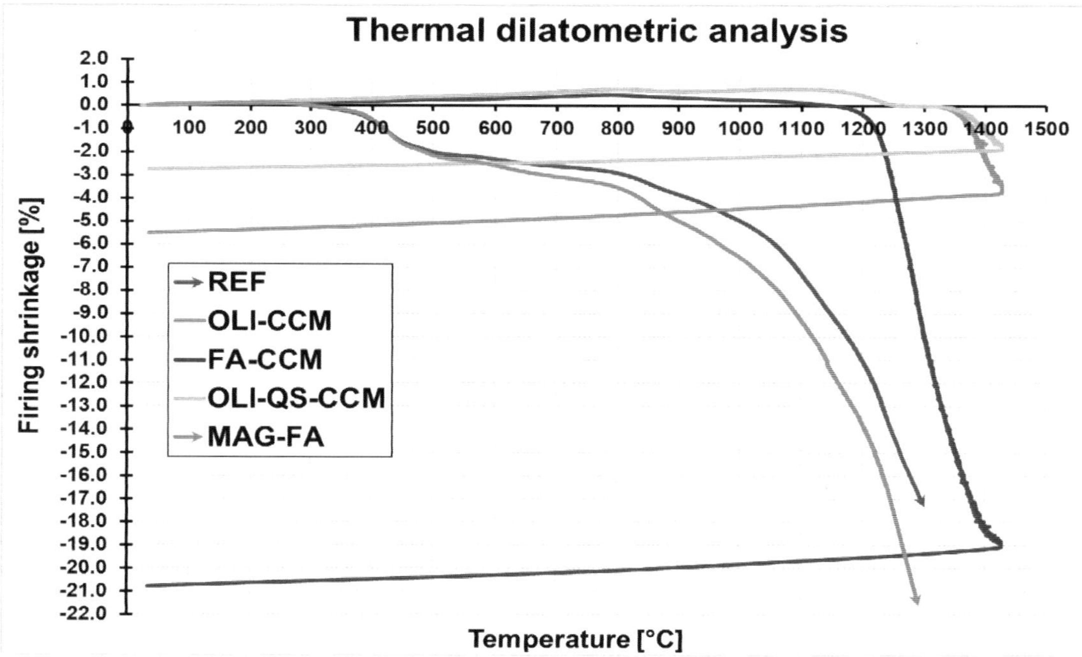

Figure 3. Thermal dilatometric analysis of all designed mixtures.

Figure 3 contains thermodilatometric curves of all mixtures. Mixtures REF and MAG-FA contained magnesite which started decomposing around 400 °C. Shrinkage of these test samples was too high that it exceeded the measuring range of the testing equipment. Sintering of the mixture FA-CCM occurred around 1200 °C. Sintering of the mixtures OLI-CCM and OLI-QS-CCM started around 1350 °C.

Figure 4 contains the X-ray diffraction analysis patterns of all test samples from all mixtures. Major crystal phase in all mixtures is forsterite. Minor crystal phases are periclase (MgO) in mixtures OLI-CCM, FA-CCM, OLI-QS-CCM and MA-spinel ($MgO·Al_2O_3$) in all mixtures except reference (REF) mixture. REF mixture is composed of pure forsterite with no traces of minor crystal phases. Mixture FA-CCM contains forsterite as a major crystal phase with periclase and MA-spinel as minor crystal phases. These minor crystal phases formed from mullite ($3Al_2O_3·2SiO_2$) which is present in fly ash (figure 1) and decomposes at temperatures above 1200 °C due to the presence of flux oxides (Fe_2O_3, Na_2O, K_2O) and with magnesium oxide forms MA-spinel and also amorphous glass phase which is presented as curved background of XRD patterns. No traces of mullite were found in the XRD patterns after firing test samples with fly ash as a raw material. Due to the lack of sufficient amount of silicon dioxide, unreacted magnesium oxide (periclase) is present in this sample.

Similar phenomenon also occurred in mixture MAG-FA. Mullite from fly ash decomposed and formed MA-spinel with magnesium oxide from magnesite and the remaining magnesium oxide reacted with quartz (figure 1) from fly ash into forsterite. Amorphous glass phase was also created. Mixture OLI-QS-CCM also contained minor crystal phases of periclase and MA-spinel. MA-spinel was created from aluminium oxide contained in kaolinite. Due to the presence of flux oxides, amorphous glass phase was also formed and therefore there was a surplus of magnesium oxide which remained unreacted.

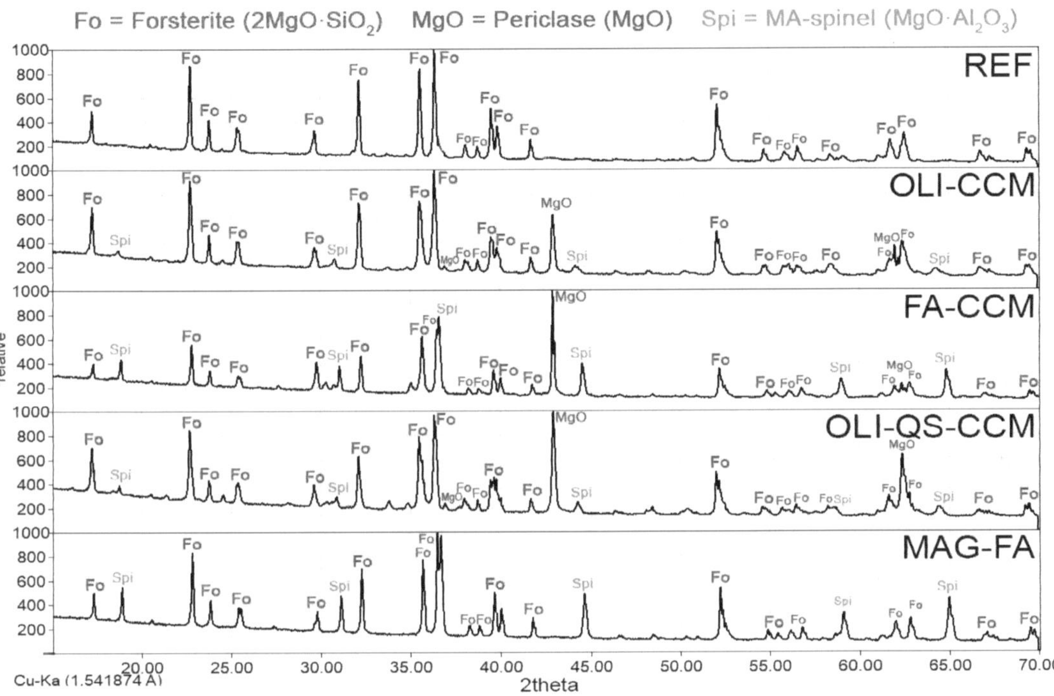

Figure 4. X-ray diffraction analysis of test samples from all mixtures fired at 1300°C.

Table 5 contains the semi-quantitative analysis by means of the reference-intensity ratio (RIR) method confirmed the results of X-ray diffraction analysis. Major crystal phase (73-100%) in all mixtures is forsterite and minor crystal phases are periclase (10-16%) and MA-spinel (5-20%). Quantities of these minor phases are in relation to the chemical composition of raw materials mixtures and consecutive chemical reactions between oxides during solid-state reactions.

Table 5. Semi-quantitative analysis of all mixtures by means of the RIR method.

Mineral quantity	REF (%)	OLI-CCM (%)	FA-CCM (%)	OLI-QS-CCM (%)	MAG-FA (%)
Forsterite (2MgO·SiO$_2$)	100.0	83.3	72.5	78.6	80.4
Periclase (MgO)	-	11.1	9.7	16.3	-
MA-spinel (MgO·Al$_2$O$_3$)	-	5.6	17.8	5.1	19.6

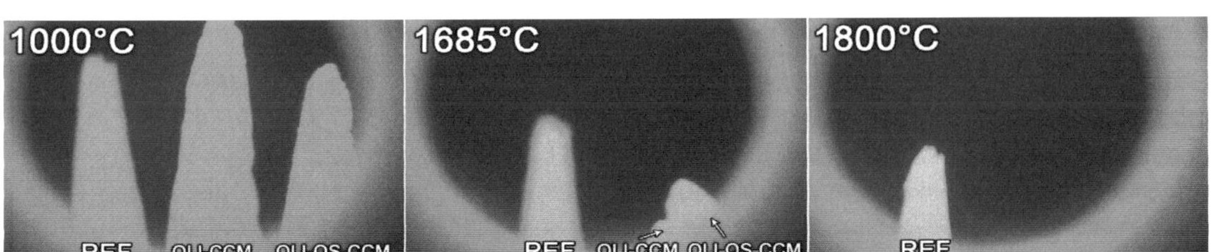

Figure 5. Refractoriness test of first set of pyrometric cones.

Figure 5 represents the first set of pyrometric cones for refractoriness test and was comprised of mixtures REF, OLI-CCM and OLI-QS-CCM. Both pyrometric cones from mixtures OLI-CCM and OLI-QS-CCM bent at 1685 °C meanwhile pyrometric cones from REF mixture endured up to the maximum achievable temperature of laboratory furnace which is 1800 °C. Refractoriness of pure forsterite is 1850 °C. [1] Lower refractoriness of mixtures OLI-CCM and OLI-QS-CCM was probably because of creation of eutectic melt due to the presence of impurities CaO, Fe_2O_3 and alkalis (Na_2O, K_2O). These two mixtures have one of the highest amount of CaO, Fe_2O_3 of all mixtures.

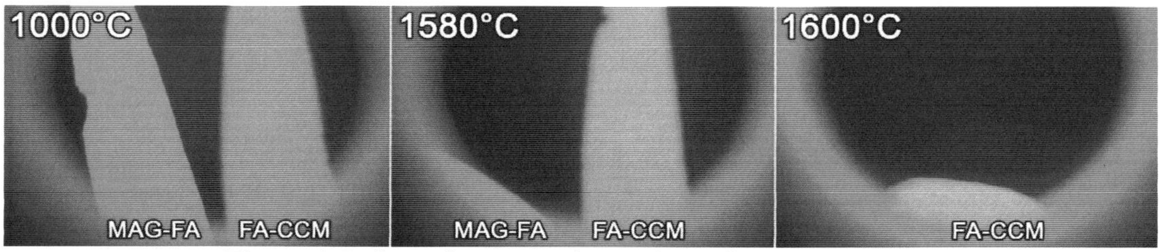

Figure 6. Refractoriness test of second set of pyrometric cones.

Figure 6 represents the second set of pyrometric cones for refractoriness test and was comprised of mixtures MAG-FA and FA-CCM. Pyrometric cone from mixture MAG-FA bent at 1580°C meanwhile pyrometric cone from mixture FA-CCM bent at 1600°C. Lower refractoriness of these mixtures was probably because of creation of eutectic melt due to the presence of impurities in the mixtures – CaO, Fe_2O_3 and alkalis Na_2O and K_2O. Mixtures MAG-FA and FA-CCM had higher amount of alkalis, so the refractoriness was even lower than that of mixtures OLI-CCM and OLI-QS-CCM.

4. Conclusion

Forsterite ceramics was successfully synthesized from all designed mixtures at 1300 °C. X-ray diffraction analysis confirmed presence of forsterite in all mixtures and periclase and/or MA-spinel in some mixtures. RIR method confirmed results of XRD and revealed that forsterite was major crystal phase in all mixtures with amounts from 72.5% to 100%. Minor crystal phases present in test samples were unreacted magnesium oxide (periclase) and MA-spinel ($MgO \cdot Al_2O_3$).

Unreacted magnesium oxide was present in mixtures which contained caustic calcined magnesite (CCM) as a raw material, namely mixtures OLI-CCM, OLI-QS-CCM and FA-CCM. In mixtures which contained magnesite as a raw material (mixtures REF and MAG-FA), periclase was not found. This was probably because all magnesium oxide from magnesite reacted to form either forsterite or MA-spinel. The MA-spinel was formed due to the recrystallization of mullite (present in fly ash) and subsequent reaction with magnesium oxide in mixtures with fly ash (FA-CCM and MAG-FA) in larger quantity (18-20%) or from kaolin in mixtures OLI-CCM and OLI-QS-CCM in smaller quantity (5-6%). All mixtures also contained certain amount of amorphous glass phase and iron oxide which can be distinguished by curved background in XRD patterns.

Test results of vacuum water absorption method revealed that test samples from mixtures which contained magnesite as a raw material (REF and MAG-FA) had high firing shrinkage, apparent porosity and water absorption due to the decomposition of magnesite which has high loss on ignition. These samples were very porous with relatively low bulk density, which is suitable for example for thermal insulation purposes.

Sample from mixture FA-CCM also had relatively high firing shrinkage, however its apparent porosity and water absorption were the lowest. This was caused due to the presence of fly ash in the mixture which has impurities that enabled sintering and densification of the sample. Samples which contained olivine as a raw material had relatively low firing shrinkage while their apparent porosity, water absorption and also bulk density were relatively high. Due to the presence of flux oxides, samples were densified and solid-state reaction to form forsterite occurred, while their dimensions were almost unchanged.

Refractoriness of all mixtures was conducted on pyrometric cones from these mixtures. The lowest refractoriness had mixture MAG-FA at 1580°C, followed by mixture FA-CCM at 1600 °C, followed by mixtures OLI-CCM and OLI-QS-CCM at 1685 °C. Highest refractoriness had mixture REF which consisted of pure forsterite and its value was over 1800 °C which is maximum achievable temperature of laboratory furnace. Impurities such as CaO and flux oxides Fe_2O_3, Na_2O and K_2O had negative impact on refractoriness in the rest of the mixtures. These mixtures contained 1.4-3.6% of CaO, 2.65-7.53% of Fe_2O_3 and 0.26-1.04% of alkalis Na_2O and K_2O.

Utilization of secondary energetic product – fly ash can be very beneficial for synthesis of forsterite-based ceramics. Fly ash as a raw material is a source of silicon dioxide SiO_2 in the form of quartz for forsterite synthesis and also source of aluminium oxide Al_2O_3 in the form of mullite which forms small quantity of MA-spinel in forsterite ceramics which improves both mechanical properties and also enhances formation of amorphous glass phase which lowers apparent porosity and water absorption while increases bulk density of test samples at firing temperature of 1300 °C. The only disadvantage appears to be lower refractoriness caused by impurities which can be improved by lowering the amount of fly ash in the mixture.

Acknowledgements
This paper was supported under the Internal Grant Agency of Brno University of Technology, Faculty of Civil Engineering, specific junior research grant number FAST-J-19-5815 with project name: *Influence of accessory oxides in raw materials on properties and synthesis of forsterite ceramics.*

References

[1] Budnikov P P 1960 *Technologie keramiky a žárovzdorného zboží.* 1st ed. (Prague : State publishing of technical literature)
[2] Kingery W 1960 *Introduction to ceramics* (New York: Wiley) pp 261-284
[3] Zhao F, Zhang L, Ren Z, Gao J, Chen X, Liu X and Ge T 2019 A novel and green preparation of porous forsterite ceramics with excellent thermal isolation properties *Ceramics International* **45**(3) pp 2953-2961
[4] Hanykýř V and Kutzendörfer J 2000 *Technologie keramiky.* 1st ed. Hradec Králové : Vega ISBN 80-900-8606-3
[5] Bouhifd M, Andrault D, Fiquet G and Richet P 1996 *Thermal expansion of forsterite up to the melting point. Geophysical research letters* **23**(10) pp 1143–46
[6] Ni S, Chou L and Chang J 2007 Preparation and characterization of forsterite (Mg_2SiO_4) bioceramics *Ceramics International* **33** pp 83-88 DOI: 10.1016/j.ceramint.2005.07.021
[7] Mustava E, Khalil N and Gamal A 2002 Sintering and microstructure of spinel–forsterite bodies *Ceramics International* **28**(6) pp 663-667 DOI: 10.1016/S0272-8842(02)00025-1
[8] Furlani E, Aneggi E and Maschio S 2014 Sintering Behaviour of Waste Olivine and Olivine/Alumina Blends *Materials* **7**(6) pp 4773-4788 DOI: 10.3390/ma7064773
[9] Petric N, Martinac V, Tkalcec E, Ivankovic H and Petric B 1989 Thermodynamic analysis of results obtained by examination of the forsterite and spinel formation reactions in the process of magnesium oxide sintering *Ind. Eng. Chem. Res.* **28**(3) pp298-302 DOI: 10.1021/ie00087a008
[10] Tavangarian F and Emadi R 2011 Synthesis and characterization of spinel–forsterite nanocomposites *Ceramics International* **37**(7) pp 2543-2548 DOI: 10.1016/j.ceramint.2011.03.056
[11] Michael R, Ammar M R, Simon P, de Bilbao E and Poirier J 2014 Behaviour of Olivine Refractories at High Temperature: Agglomeration in a Fluidized Bed Reactor *Refractories Worldforum* **6**(1) pp 95-98

IOP Conf. Series: Materials Science and Engineering **549** (2019) 012026 doi:10.1088/1757-899X/549/1/012026

Crystallization process in the preparation of the tetracalcium ferroaluminate

S Ravaszová[1] and K Dvořák[1]

[1] Brno University of Technology, Faculty of Civil engineering, Veveří 331/95, 602 00 Brno, Czech Republic

E-mail: ravaszova.s@fce.vutbr.cz

Abstract. The study of the origin, stability and properties of various calcium silicate polymorphs is important for a better understanding of the processes involved in the production of Portland clinker. At the same time, the synthetic pure clinker phases can serve in further investigations, such as hydration or the effects of plasticisers. The aim of this article is to prepare the pure phase of tetracalcium ferroaluminate and to observe the crystallization process based on the calculation of the size of the crystallite using the simplest method, the Scherrer's equation. Based on the results of the X-Ray analysis, the temperature of 1350 °C with a soaking time of 3 hours is sufficient to produce the tetracalcium ferroaluminate pure phase. This firing mode is also characterized by a very good development of the crystalline structure.

1. Introduction

Portland cement is a complex product made from very simple and abundant materials: limestone and clay (or shale). Very precise proportions of these two basic materials have to be mixed with some additions to create a raw meal with a precise chemical composition. Usually, Portland cement raw meal is heated up to a temperature of about 1450 °C. The chemical composition of the raw feed is adjusted to obtain the formation of four minerals that react with water to create bonds: tricalcium silicate C_3S, dicalcium silicate C_2S, tricalcium aluminate C_3A and tetracalcium ferroaluminate C_4AF. The first two minerals constitute the silicate phase of Portland cement, and the tricalcium aluminate and tetracalcium ferroaluminate its aluminous phase e also called the interstitial phase, because in the burning zone this part of the clinker reaches a more or less viscous state that bonds together the two silicate phases [1,2,3]. Alite content is by far the most common in currently produced Portland clinker. Except for brownmillerite, all major clinker minerals show strong polymorphism.

The composition of the raw material and the setting of firing parameters are the most influential factors that determine the final phase composition of clinker [4,5].

Due to previous unsuccessful attempts to synthesize the pure C_4AF phase, the work deals with optimization the manufacturing process, verifying purity and monitoring the crystallization process in the C_4AF production process. The previous failure occurred in attempts to synthesize C_4AF using the Wesselsky-Jensen method [6] mainly due to the selected firing mode.

Therefore, nine different firing modes have been tried to find the best possible firing method. At the same time, much more emphasis was placed on precision in the early stage of the preparation of the raw material mixture when the homogenization time was extended and the medium used in the process of milling the raw material mixture in the planetary mill PULVERISETTE 6 was changed.

Pressing of the pellets, were replaced by the spontaneously formed pellets after drying from wet suspension with 10 mm in diameter, which were directly dosed into the platinum crucibles for solid state reaction [7,8].

2. Materials and methods

The Wesselsky-Jensen method was used to prepare tetracalcium ferroaluminate. This method indicated the molar ratio of used raw materials. Calcium carbonate $CaCO_3$ (p.a. 99,8%), aluminum oxide Al_2O_3 (p.a. 99,0%) and iron oxide Fe_2O_3 (p.a. 99,8%) were used to prepare a tetracalcium ferroaluminate. The raw material mixture was homogenized with the water in a Pulverisette 6 planetary mill for 10 minutes at 400 rpm. Fritsch planetary ball mill Pulverisette 6 in a bowl with the volume of 0.5 dm^3 was used. The bowl was filled by 25 steel grinding balls with 20 mm in diameter. The comminution of the material takes place primarily through the high-energy impact of the grinding balls. The grinding bowl, containing the grinding balls and the material to be ground, rotates around its own axis on the main disk whilst rotating rapidly in the opposite direction. Centrifugal force causes the 3ground sample material and the grinding balls to separate from the inner wall of the grinding bowl. The grinding balls then cross the bowl at high speed and further grind the sample material by impact against the opposite wall. The homogenized raw material mixture was dried in a BINDER C170 laboratory dryer for 24 hours.

The spontaneously formed pellets with 10 mm in diameter were directly dosed to platinum crucibles.

Superkhantal High temperature furnace Clasic 2017S was used for solid state reaction. Solid state reacton was carried out at 3 temperatures: 1350, 1400 and 1450 °C with 3 soaking times: 1, 3 and 5 hours. At the end of each firing mode, the platinum crucibles were removed from the furnace and immediately cooled down with air flow. The samples were adjusted by milling to X-Ray analysis. At first, the samples were milled in a RS 200 vibratory disc mill for 30 seconds at 900 rpm. The Vibratory Disc Mill RS 200 comminutes by impact and friction. The grinding set is firmly attached to the vibration plate with a quick-action lever. The plate with the grinding set is subjected to circular horizontal vibrations. The centrifugal force acting on the grinding rings in the dish results in extreme pressure. Impact and frictional forces acting on the sample produced analytical fineness in 1-3 minutes. The circular vibrations are produced by a frequency controlled 1.5 kW 3-phase motor.

The material was further milled with isopropyl alcohol in McCrone Micronising Mill to complete sample preparation for X-Ray analysis.

Phase composition control was performed by X-ray analysis on Panalytical Empyrean XRD with Cu-Kα as the radiation source, λ=1,540598 for Kα1, accelerating voltage 45 kV, beam current 40 Ma, diffraction angle 2θ in the range from 5° to 80° with a step scan of 0.01°.

Input values were obtained to determine the effect of the firing temperature and the soaking time on the evolution of C_4AF crystallites. The crystallite size was evaluated on selected diffractive lines at appropriately selected crystallographic planes that had a major crystallographic direction. Two distinct crystallographic planes were selected. They are determined by the Miller Index "hkl": 200 and 411, which were shown on the diffraction pattern at angles of 2θ: 50.4 and 33.7°. Calculation of the size of crystallites was based on the measurement of FWHM (full width of half maximum). The measurement of FWHM was carried out using the HIGHSCORE PLUS program. The FWHM values were obtained in the program HIGHSCORE PLUS for selected diffraction lines. From the list of peaks and their FWHM, the data for the lines that responded to the Kα2 radiation were deleted.

The instrument extension was set to the LaB_6 standard (lanthanum hexaborite). LaB_6 is considered to be a fully crystalline material in which the size of crystallites is theoretically equal to infinity. Diffraction lines of LaB_6 were determined in the same positions as the selected diffraction lines in the same way in the HIGHSCORE PLUS program. The size of the crystallite was evaluated by the modified Scherrer equation [9]:

$$L = \frac{K.\lambda}{\cos\theta} \cdot \frac{1}{\beta} = \frac{K.\lambda}{\cos\theta} \cdot \frac{1}{\sqrt{B^2 - b^2}} \quad (1)$$

for the $\lambda = 1.540598$ for the Cu cathode and the Scherrer constant $K = 0.94$, crystallite size values were obtained without device extension.

The last step was scanning on the scanning electron microscope TESCAN MIRA3 XMU. Three samples were selected, which were prepared using a colloidal graphite solution in isopropyl alcohol and a rotary pump pumped by a carbon sputter.

3. Results

Figure 1 shows brownmillerite samples removed from the laboratory furnace and cooled down by air flow for individual temperatures and soaking times.

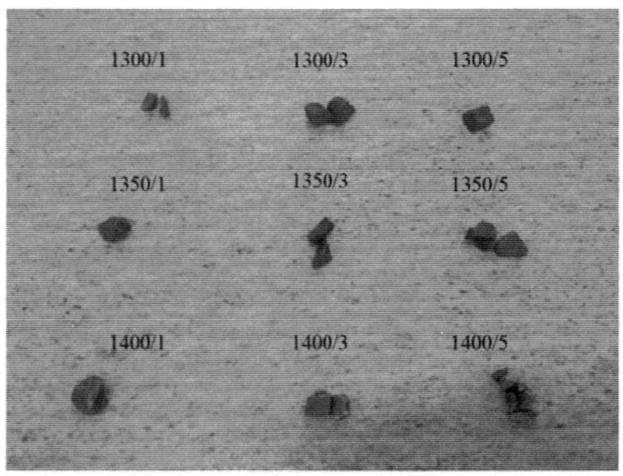

Figure 1. Brownmillerite samples removed from the laboratory furnace.

C_4AF was sintered at temperatures of 1300, 1350 and 1400 °C with soaking times of 3 hours. There are colour differences from the picture between samples. Samples sintered at 1300 °C are brown to rust in colour, while samples sintered at higher temperatures are gray to black. For samples sintered at temperature of 1400 °C a greater amount of gloss can be observed. The gloss of the sample is more intense with longer soaking time.

The quality control of the mineralogical composition of prepared tetracalcium ferroaluminates were performed by X-Ray analysis, the scans of which are shown in Figure 2.

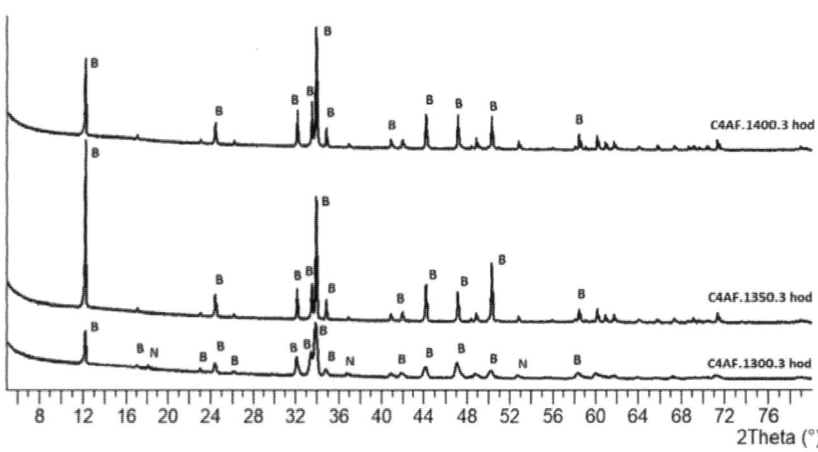

Figure 2. X-Ray diffraction patterns of C_4AF (B-Brownmillerit, N-$Ca_{12}Al_{14}O_{32}$).

At a firing temperature of 1300 °C, impurities in the form of a mineral from the group $Ca_{12}Al_{14}O_{32}$ are identified on the diffraction pattern. A temperature of 1350 °C is already sufficient to fully crystallize C_4AF. Therefore, the temperature of 1400 °C is no longer necessary in terms of material quality.

For the clarity of the resulting crystallite size values, 3D graphs were constructed to express the dependence of crystallite size on firing temperature and soaking time, which are shown in Figure 3.

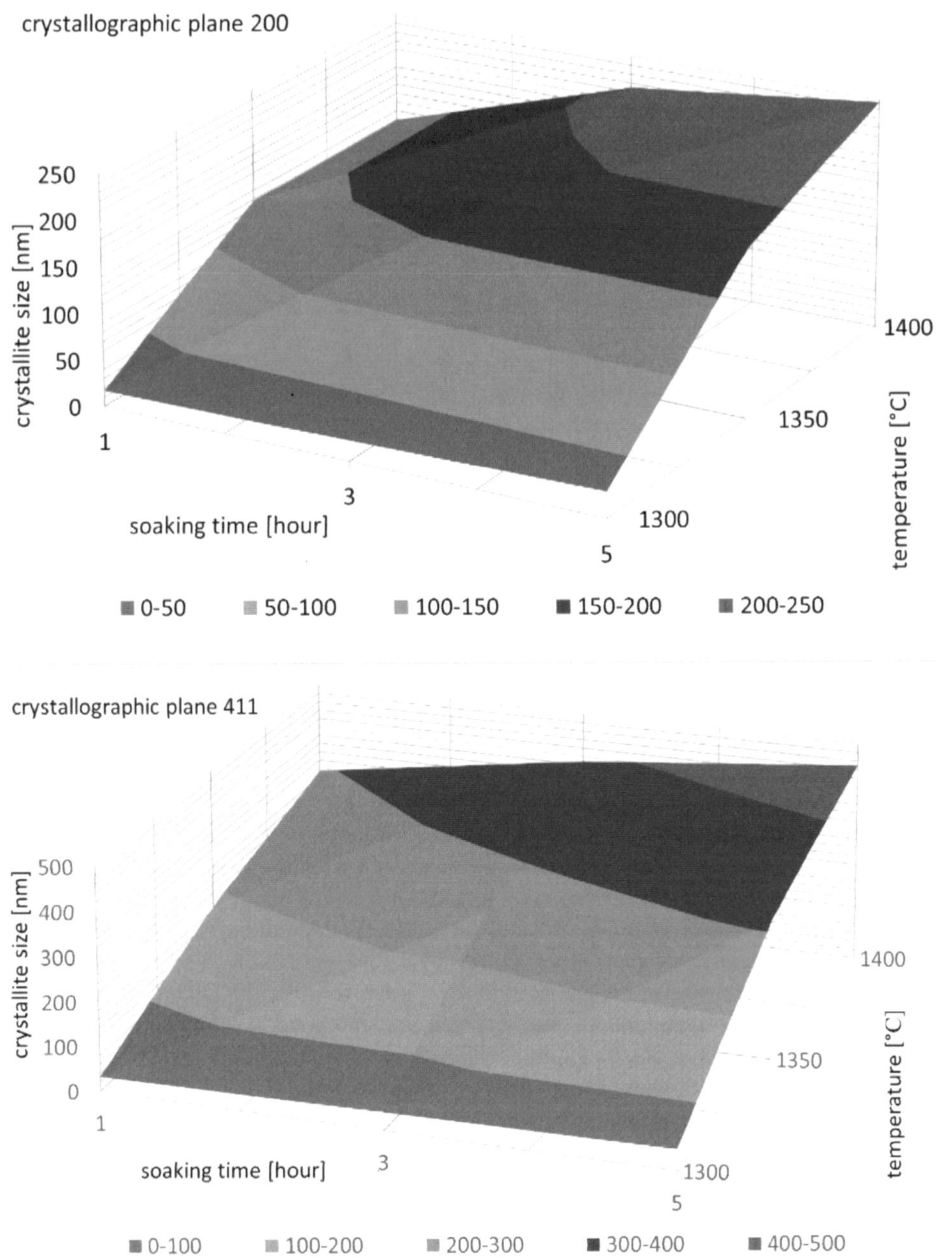

Figure 3. Graphical representation of the influence of the temperature and the soaking time on the size of C_4AF crystallites in crystallographic planes 200 and 411.

The trend of changing the size of the crystallites is very similar in both selected crystallographic planes. The most significant changes in crystallinity occur at a temperature of 1350 °C with soaking time of 3 hours.

The grain shape was evaluated by optical microscopy. The Figure 4 shows C_4AF images at 3 different temperatures with soaking of 3 hours.

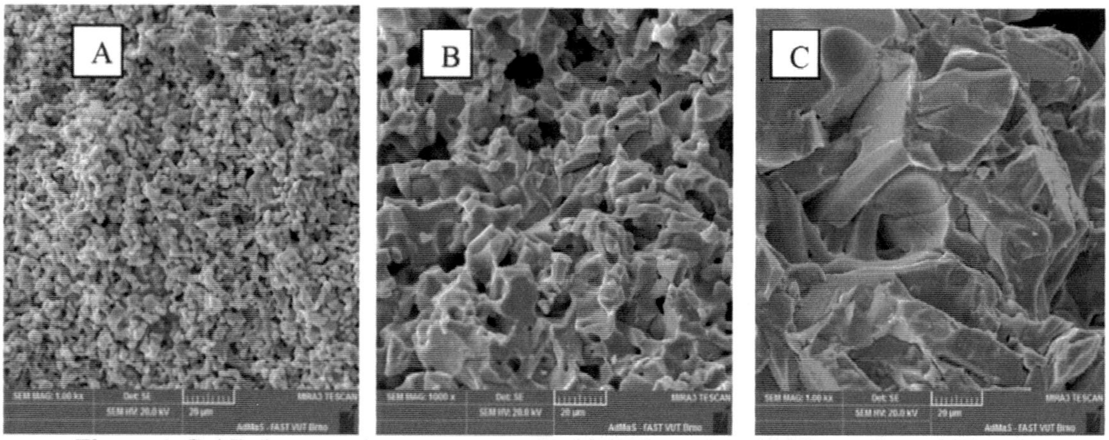

Figure 4. C_4AF electron microscope scans (A-1300/3hour, B-1350/3hour, C-1400/3hour).

4. Discussion

The article deals with the preparation and monitoring of the influence of temperature and soaking time on the C_4AF crystallization process.

Based on the results of the X-Ray analysis, it can been state that the temperature of 1300 °C even with 5 hours of soaking time was insufficient to fully crystallize the material, because the residues with a different mineralogical composition are identified on the diffraction patterns. At a temperature of 1350 °C with soaking time of 3 hours, the pure C_4AF phase has already been prepared and thus a temperature of 1400 °C was no longer necessary in terms of firing mode.

The simplest Scherrer-method was used to evaluate the effect of the firing process on the crystallization of the tetracalcium ferroaluminate. The trend in the development of C_4AF crystallites can be considered very similar in both selected crystallographic planes. In general, the higher the firing temperature and the soaking time, the higher the crystallite size. The most significant changes in the size of crystallites occur at a temperature of 1350 °C with 3 hours in both crystallographic planes. This temperature and soak result in a continuous increase in crystallite size to 180 nm in the crystallographic plane 200 and to 256 nm in the crystallographic plane 411. With increasing soaking time, there is only a slight increase in crystallite size to 183 nm in plane 200 and 313 nm in the 411 plane. Consideration should also be given to the fact that values above 200 nm are the limits of the applicability of the method and the device extension. Electron microscopy images also confirmed the trend in C_4AF crystal structure development. At a firing temperature of 1300 °C, the crystalline structure is relatively fine with small crystals, while at 1350 °C it reaches the top and large C_4AF crystals are observed in the scan. Conversely, at 1400 °C, the surface appears to be sintered.

5. Conclusion

Nine different firing modes have been tried to obtain the most advantageous or optimal C_4AF firing process. After evaluating the results of X-Ray analysis and scanning electron microscopy, firing at a temperature of 1350 °C (with a heating rate of 8 °C/minute) with soaking time of 3 hours seems to be the most suitable firing process to obtain a pure material. Samples synthesized by this firing process were shown as the desired pure C_4AF phase in the X-ray analysis. Firing at higher temperatures and longer soaking times was therefore unnesessary. At lower temperature, impurities in the form of $Ca_{12}Al_{14}O_{32}$ occurred in the mineral. The firing at 1350 °C with 3 hours of soak also showed a very good development of the crystal structure, where the magnitude of the crystallites reached values of about 180 nm, which was also confirmed by electron microscope images.

Acknowledgements
This work was financially supported by project No. FAST-J-19-5795: "Synthesis of Portland Clinker Phases" and project No. 17-24954S "The conditions of Thermodynamic Stability and Transformation of Aft Phases".

References
[1] Taylor HFW 1997 *Cement chemistry* (London: Thomas Telford Ltd.)
[2] Hewlett P 2004 *Lea's Chemistry of Cement and Concrete 4* (Oxford: Elsevier Butterworth-Heinemann)
[3] Duda HW 1988 *Cement Data-book* (Wiesbaden: Berlin: Bauverl)
[4] Maki I and Goto K 1982 *Cement Concrete Research* **12** 301-308
[5] Jeffery J W 1952 *Acta Crystallographica* **5** 26-35
[6] Wesselsky A and Jensen O M 2009 *Cement Concrete Research* **39** 973-980
[7] Ravaszová S and Dvořák K 2018 *IOP Conf. Series: Materials Science and Engineering* **385** 012044
[8] Ravaszová S 2019 *21st International Conference of doctoral Students* (Brno) 473-478
[9] Patterson A L 1939 Phys. Rev. **56**

Usability of waste perlite in the technology of production of autoclaved aerated concrete

J Hustavova[1], P Sebestova[1], L Meszarosova[1], V Cerny[1] and R Drochytka[1]

[1] Brno University of Technology, Faculty of Civil Engineering, Veveří 95, 602 00 Brno, Czech Republic

E-mail: meszarosova.l@fce.vutbr.cz

Abstract. The building industry constantly seeks ecological solutions and accents utilization of industrial wastes. For this reason, even less examined alternatives are considered. This is true also for manufacture of autoclaved aerated concrete. This paper addresses usability of waste perlite in the technology of production of autoclaved aerated concrete. Waste perlite has been examined to great extent in particular in the area of cement based composite materials, where its positive influence is known. This leads to the idea of finding out if waste perlite could contribute positively to the technology of autoclaved aerated concrete. In this basic research, four replacement amounts of quartz sand with waste perlite were tested: 10%, 30%, 50% and 100%. Replacement of 30% appears to be most promising. At this amount, the compressive strength increased and at the same time, volume weight was reduced by 40%.

1. Introduction

Autoclaved aerated concrete is ranked among calcium-silicate composites, or so-called aerated silicates. Porous structure gives autoclaved aerated concrete its characteristic properties like low volume weight and low thermal conductivity. This material is also inflammable; therefore, it has a wide range of application as a fireproof layer. Basic raw materials for manufacture of autoclaved aerated concrete are quartz sand, lime, cement, gypsum and aluminium powder [1][2]. Quartz sand, which has highest proportion of volume of all input materials, is gained from natural resources. However, these resources become gradually exhausted, quality of sand decreases and its price rises. Not only because of these reasons but also to contribute to our environment, there are efforts of replacing primary raw materials at least partially. Ash materials have been commonly used for the technology of manufacture of autoclaved aerated concrete [3][4]. Ash materials are a source of amorphous silicic oxide, which is more reactive than its natural crystalline form. Secondary raw materials usually contain amorphous silicic oxide, which is advantageous for technology of manufacture of autoclaved aerated concrete. Various materials have been examined in the efforts of finding alternative replacement of silica sand, for example ground packing glass or slag from waste incineration [5]. These researches are only starting, but they are very promising. These alternative raw materials include wastes from production of expanded perlite (WEP).

Perlite is defined as a volcanic glass and therefore it can be found in amorphous form. Its microstructure binds approx. 5% of water. This water leaves in the form of steam during rapid heating at about 1000 °C. During this relative swift reaction, highly porous structure of perlite is formed (expanded perlite). Thus, the volume of perlite is about 20 times multiplied and its specific surface increases, too. Perlite contains mainly SiO_2 and Al_2O_3, further oxides are detected in lower amounts: Na_2O, K_2O, MgO, Fe_2O_3 and CaO. [6][7][8][9][10]. Because of amorphous form of SiO_2 and Al_2O_3, perlite is considered to be pozzolanic material and it seems to be a promising secondary raw material [10][11]. WEP is used in the area of cement based composites and thermo-insulating mortars or various grouting materials. Researchers Pachta et al. [8] examined influence of addition of waste perlite in grouting mixes based on lime binder. It was found that addition of waste perlite considerably increased compressive strength of hydraulic grouting mixes with no negative impact on rheology. Waste perlite was also examining in production of alkali activated slag composites (AACSs). However, WEP was in this case treated by grinding unlike Patcha et al., where WEP was used without any treatment. When used as addition for AASCs, it was found, that increasing amount of addition causes reduction of compressive strength even though no significant changes were observed in the microstructure. Kotwica et al. [10] examined the influence of ground waste perlite as pozzolanic material that could be used as a partial replacement of cement binder. It was found that ground WEP in the amount of 35% of cement binder increased overall strength of cement by 50%. In the past years it was proved that addition of waste perlite in cement composites has positive effect on microstructure and compressive strength but mainly on reduction of hydration heat [7].

As regards use of waste perlite in the technology of autoclaved aerated concrete, it is almost unexplored area. Rózycka and coll. [9] researched influence of WEP on properties of AAC, in particular its thermal conductivity. They state that as the amount of replacement with WEP in AAC increases, volume weight is reduced as well as compressive strength. On the other hand, the addition of WEP had a positive impact on the formation of microstructure or development of 1.1 nm tobermorite. They state that this could be caused by morphology of SiO_2 or by high fineness of WEP. It is known that when the specific surface of siliceous component is higher, the reactions are faster, hence tobermorite 1.1nm is formed easily [9]. However, this is not always true. Hartmann [11] defines in his paper the ideal limit of grain size between 0.08 mm and 0.03 mm. In case the grains are smaller than 0.03 mm, no crystallization of 1.1nm tobermorite occurs.

Tobermorite is a calcium hydro-silicate mineral (CSH), which is formed in microstructure of calcium silicate composites during hydrothermal treatment, which is between 150 °C and 220 °C and pressure of vapor 0.8 – 1.6 MPa and length of hardening 3 – 10 hours. These parameters vary based on the type of product. During this hydrothermal treatment, crystalline or amorphous SiO_2 and calcic components are dissolved in water. Thus, CSH phase is formed which subsequently becomes the binding phase and binds individual grains of non-reacted silica, which is usually in excessive amount so that some part of it can serve as filler. Thus, mechanical strength of the product is achieved. [2][3][11][12][13].

2. Raw materials and methods

Basic raw material for manufacture of autoclaved aerated concrete is quartz sand, lime, cement, gypsum and aluminum powder. Basic requirement for unslaked lime is high purity and gradual increase of temperature when hydrating. Reactivity of lime for autoclaved aerated concrete is expressed in time, when the lime paste reaches the temperature of 60 °C. Lime used in the research reached the temperature of 60 °C in 6 minutes and contains only 5% of impurities, hence it fulfills the requirement of high purity and gradual reactivity. Used cement CEM I 52.5 N assures sufficient manipulation strength before autoclaving. Energetic gypsum contains over 90% of dihydrate of calcic sulphate, which supports formation of thixotropic structure of fresh mix and supports formation of new crystalline calcium hydro-silicates (CSH) Aluminum powder is the main agent for formation of the macrostructure. It reacts with calcium hydroxide and gaseous hydrogen is released, forming porous structure of autoclaved aerated concrete. Aluminum powder is selected with respect to the standard CSN 42 0825 Aluminum powders and pastes for autoclaved aerated concrete. Silica sand contains over 92% of crystalline silica oxide and low amount of other oxides.

2.1. Waste perlite as filler

Because of high content of amorphous silica oxide, waste perlite was tested in the research as a replacement of silica sand. Volume weight of WEP was determined to be 300 kg·m^{-3}. This low value is considered advantageous and overall volume weight of resulting sample is expected. Determined water absorbing capacity of WEP was 260%. Chemical composition of waste perlite is in Table 1.

Table 1. Chemical composition of waste perlite.

Oxide	CaO	MgO	K_2O	Al_2O_3	P_2O	Fe_2O_3	MnO	Na_2O	SiO_2	TiO_2
[%]	1.26	0.228	4.56	13.1	<0.023	2.02	0.057	2.14	**74.0**	0.157

2.2. Mix design and preparation

The mix-design corresponds with the mix-design used in real technology of production of autoclaved aerated concrete. To achieve high volume weight, higher amount of aluminum powder is used. WEP was used as a substitution of silica sand by 10% 30% 50% and 100%. Dosing of WEP was by volume. Water-cement ratio was constant for all mix-designs. Reference mix-design is stated in Table 2.

Table 2. Raw materials.

Raw materials	REF [%]
Lime	9
Cement	14
Gypsum	3
Aluminum powder	0.137
Quartz	75

Preparation consists of several stages. First, fresh mix is prepared. Lime and cement are thoroughly mixed in homogenizer. Aluminum powder is mixed with detergent and water. Sand, energo-gypsum and waste perlite are mixed in a mixer and together with water. Then, lime is added in the mixer and after mixing, suspension of aluminum powder. Time of mixing is 2 minutes from adding reactive components in water. Prepared fresh mix was poured into warmed up forms with dimension 100x100x100 mm and then the forms were placed in a dryer at 40 °C so that aeration could take place. At this stage, it is important to ensure conditions remain unchanged. After aeration and setting of the fresh mix, forms are removed, and samples placed in the autoclave, where final hardening of the mix takes place. Development of temperature is graphically presented in Figure 1.

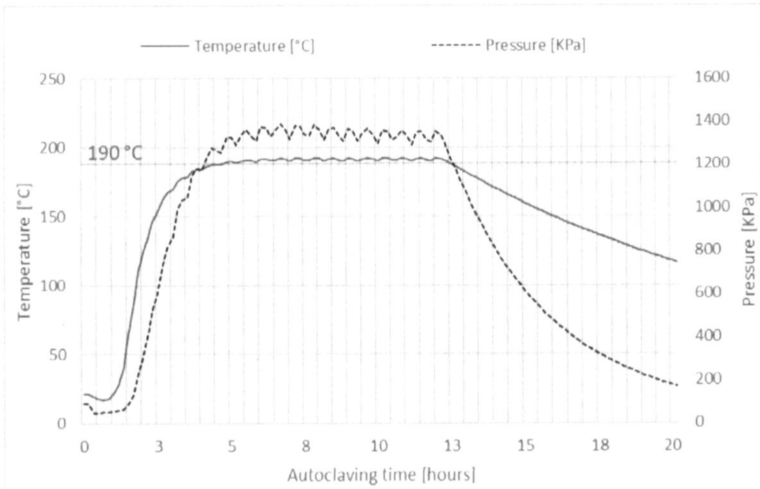

Figure 1. Autoclaving process.

2.3. Physical-mechanical properties

Compressive strength was determined on dried samples at (105±5) °C. The test was carried out in accordance with the standard CSN EN 679 Determination of strength of compressive strength of autoclaved concrete. Test specimen is placed in the press in the direction perpendicular to the growth of the material and is loaded at constant rate (0.1±0.05) MPa/s. Size of tested specimens was 100×100×100 mm. Volume weight was determined in accordance with the standard CSN EN 678 Determination of volume weight of dry autoclaved aerated concrete. Macrostructure was determined with digital optical microscope Keyence VHX-950F.

2.4. Preparation of samples for examination of microstructure

2.4.1. X-ray diffraction

Out of all samples, cores were prepared, which were ground to max. 0.1 mm by vibration grinder for the study of microstructure during the first stage of preparation of samples. Ground samples were then ground in isopropanol suspension with addition of comparing standard in the MC Microne grinder to the size <0.02 mm. The suspension was then dried and the powder was applied on a carrier of the Empyrean Panalitical apparatus to carry out quantitative XRD analysis.

2.4.2. Scanning electron microscope

To study microstructure with scanning electron microscope, representative fragments with size 5x5x5 mm were selected. The examination was done with microscope TESCAN MIRA 3 XMU.

3. Results

3.1. Compressive strength, volume weight and slump

Compression strength, bulk density and flow test of samples with WEP replacement are displayed in Figure 2. Based on determination of compressive strength of samples, dependence of compressive strength on the amount of WEP substituting silica sand was noted. The dependence is not linear. Compressive strength decreases more from 30% replacement with WEP. On the other side, samples with 10% and 30% of replaced amount showed compressive strength about 18% higher than reference samples. Probably, formation of CSH or 1.1 nm tobermorite was supported because of high specific surface of WEP and content of amorphous SiO_2.

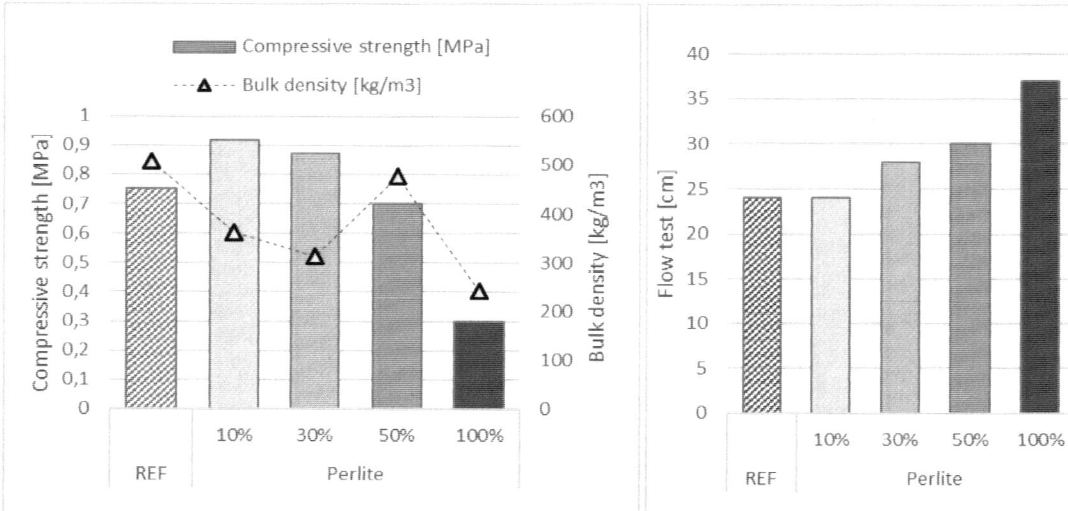

Figure 2. The charts of compression strength, bulk density and flow test of samples with WEP replacement.

When the substitution with waste perlite is used, the volume weight compared to the reference sample significantly drops. As assumed, the lowest determined volume weight was that of 100% replacement of silica sand with WEP – the value decreased by more than 50%. Volume weight of samples with 10% and 30% replacement decreased by nearly 40%. Volume weight of sample with 50% replacement increased because of collapse of the porous structure during aeration. Reduction of volume weight is caused by very low volume weight of WEP.

3.2. Porous structure

Porous structure of samples is clearly visible in the Figure 3. The first image shows porous structure of reference sample and detailed view of a selected pore. The diameter of the pores is approximately 2 mm. After 10% replacement, slight increase of pore size and strengthening of inter-pore walls compared to reference sample, can be observed. This aspect causes increase of compressive strength compared to the reference sample. This trend continues in the sample with 30% replacement. Increased pore size and strengthening of inter-pore walls compared to reference and 10% replacement sample can be observed. During aeration of fresh mix of sample with 50% replacement, the structure collapsed. For this reason, the sample shows irregular porous structure with prevailing non-aerated matrix. Sample with 100% replacement shows extremely large irregular pores, which were created probably because of high slump value. The mix was not sufficiently viscous and individual pores agglomerated.

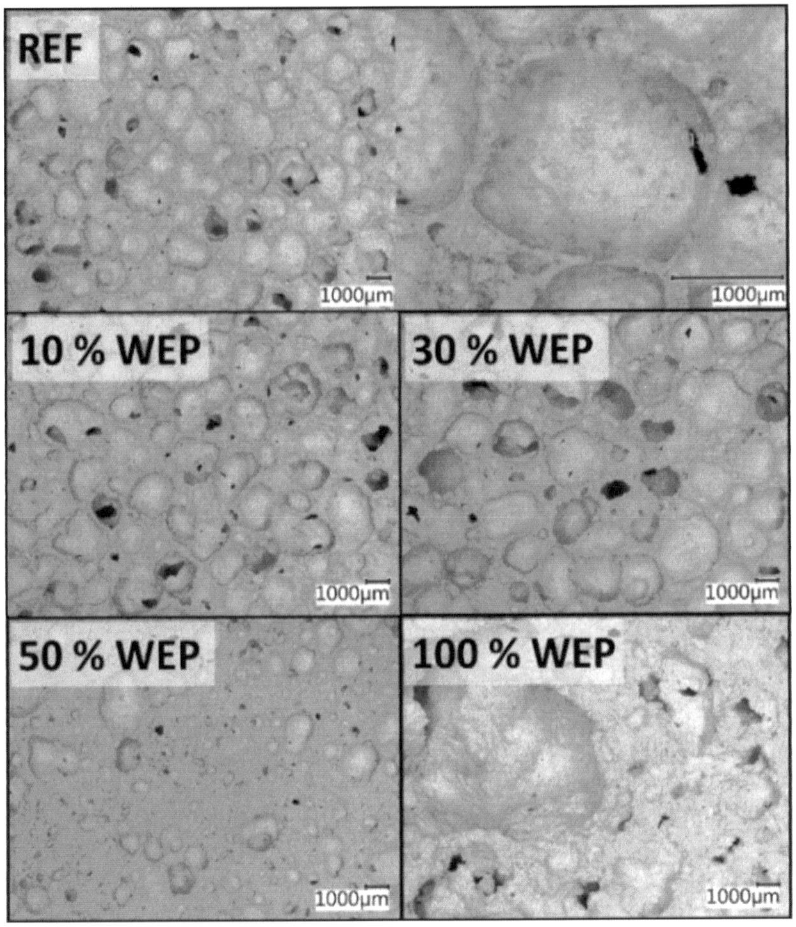

Figure 3. The structure of samples with WEP replacement.

3.3. Microstructure

According to X-ray diffraction analysis, it was found that samples with 10%, 30%, 50% substitution and reference sample contained the same basic minerals, quartz, portlandite and tobermorite. Crystalline SiO_2 entered the sample with quartz sand, and therefore its intensity and amount decreased with increase of sand replacement with WEP. As for tobermorite, the dependence of intensity and quantity on substitution has not been confirmed here. The intensity of the tobermorite diffraction line is almost unchanged at 10% and 30% replacement of the silica sand with WEP. The same was seen in the SEM images, where there was no difference between 10% and 30% substitution and therefore no change in the morphology of tobermorite. Conversely, with 50% replacement, the most tobermorite was detected.

According to SEM images (Figure 4), it is evident that the morphology of tobermorite was different with this replacement. In the reference sample, tobermorite was found in the form of long slats, in the sample with 10% and 30% substitution, tobermorite was found in the form of thinner and shorter slats, and in the sample with 50% substitution, tobermorite was detected in the morphology of the low leafs. This is probably due to the higher proportion of amorphous SiO_2 from WEP. It has been suggested that with 100% substitution of silica sand with WEP, 1.1 nm tobermorite should be detected in the highest amount with respect to the reactive form of SiO_2 in WEP. However, the results showed that 1.1 nm tobermorite is not present. This is also likely to cause very low compressive strength of the samples.

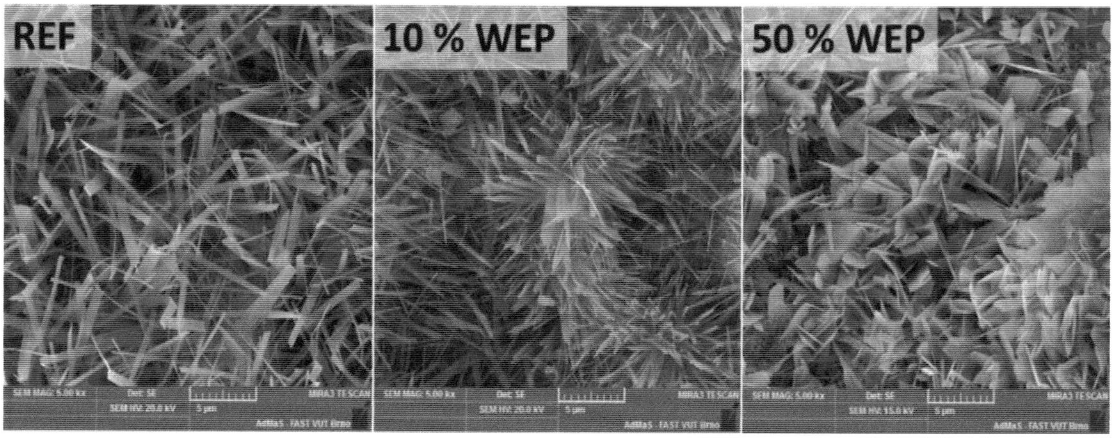

Figure 4. SEM images of samples with WEP replacement.

4. Conclusion

The research focused on possibilities of using WEP in the technology of autoclaved aerated concrete. The aim was achieving lower volume weight of autoclaved aerated concrete by using WEP. Volume weight was considerably reduced as early as with 10% replacement of sand with waste perlite. In spite of reduced volume weight, compressive strength of samples was higher up to the amount of 30% replacement. This trend can be explained by increase of thickness of walls between pores. At 50% replacement amount, thixotropic structure was not formed and after aeration, the porous structure collapsed.

Sample with 100% replacement showed very irregular porous structure with extremely large pores. However, the volume weight of the sample was lower by 50% in spite of unsuitable structure. For this reason, it is not always necessary to create porous structure and hence using expansion to decrease the volume weight. For further research of 100% replacement of silica sand with waste perlite it is necessary to adjust water cement ratio and amount of aluminum powder.

Acknowledgments

The paper was supported by the project FAST-S-19-5779 with name "Problems of effective use of secondary raw materials in the area of silicate and non-silicate based composites" and with financial support of the project GA17-14198S with name "Kinetics of silicon micro-structure creation in dependence on hydrothermal conditions and type of used materials".

References

[1] Drochytka R 1999 Pórobeton (Aerated autoclaved concrete) VUTIUM (Brno) ISBN 80-214-1476-6

[2] Chen Y, Chang J, Lai Y and Chou M M 2017 A comprehensive study on the production of autoclaved aerated concrete: Effects of silica-lime-cement composition and autoclaving conditions *Construction and Building Materials* **153** 622-629

[3] Kurama H, Topçu İ B and Karakurt C 2009 Properties of the autoclaved aerated concrete produced from coal bottom ash *Journal of Materials Processing Technology* **209** 767-773

[4] Zou J, Guo Ch, Jiang Y, Wei C and Li F 2016 Structure, morphology and mechanism research on synthesizing xonotlite fiber from acid-extracting residues of coal fly ash and carbide slag *Materials Chemistry and Physics* **172** 121-128

[5] Černý V, Drochytka R, and Šebestová P 2018 Options for the implementation of new secondary raw materials in autoclaved aerated concrete *6th International Conference on Autoclaved Aerated Concrete*. ce/papers. 431-437. ISSN: 2509-7075.

[6] Pachta V, Papadopoulos F and Stefanidou 2019 Development and testing of grouts based on

perlite by-products and lime *Construction and Building Materials* **207** 338-344

[7] Long W, Tan X, Xiao B, Han N and Xing F 2019 Effective use of ground waste expanded perlite as green supplementary cementitious material in eco-friendly alkali activated slag composites *Journal of Cleaner Production* **213** 406-414

[8] El M, Nehme A and Nehme S 2017 Utilization of industrial waste perlite powder in self-compacting concrete *Journal of Cleaner Production* **156** 507-517

[9] Różycka A and Pichór W 2016 Effect of perlite waste addition on the properties of autoclaved aerated concrete Construction and Building Materials **120** 65-71

[10] Kotwica Ł, Pichór W, Kapeluszana E and Różycka A 2017 Utilization of waste expanded perlite as new effective supplementary cementitious material *Journal of Cleaner Production* **140**, 1344-1352

[11] Harmann A 2004 Untersuchungen zum Kristallisationsverhalten und zur Morphologie von 11 Å Tobermorit in Abhängigkeit von der Reaktivität der Kieselsäurequelle und dem Ionenbestand der Hydrothermallösung *Fachbereich Geowissenschaften und Geographie der Universität Hannover (Dissertation thesis*)

[12] Kikuma J, Tsunashima M, Ishikawa T, Matsuno S, Ogawa A, Matsui K and Sato M 2011 Effects of quartz particle size and water-to-solid ratio on hydrothermal synthesis of tobermorite studied by in-situ time-resolved X-ray diffraction *Journal of Solid State Chemistry* **184**, 2066-2074

[13] Mitsuda T 1982 . Influence of starting materials on the hydrothermal reaction in the CaO-SiO_2-H_2O system *Journal of the Japanese Association of Mineralogists* 317-329.

IOP Conf. Series: Materials Science and Engineering **549** (2019) 012028 doi:10.1088/1757-899X/549/1/012028

Bulk density decreasing of lightweight silicate based composite

L Meszarosova[1] and R Drochytka[1]

[1] Brno University of Technology, Faculty of Civil Engineering, Institute of Technology of Building Materials and Components, Veveří 95, 602 00 Brno, Czech Republic

E-mail: drochytka.r@fce.vutbr.cz

Abstract. The newly developed silicate composite combines advantages of two types of bulk density decreasing methods. Direct bulk density decreasing method is based on introduction of pores in the structure of binder and indirect bulk density decreasing is achieved by means of using fine grained porous aggregate. This combination makes it possible to achieve maximal decrease of bulk density of the mix. Material of this type shows very low bulk density, approx. 300 kg·m^{-3} and appropriate application would be for non-structural, filling material or insulating material. Achieved results will serve as base for further modification of mix-design of aerated concrete utilizing secondary raw materials.

1. Introduction

The main target of the research was verification of possibilities of preparation of material based on aerated concrete without the autoclaving process. Lightened materials based on silicates (aerated concrete, perlite concrete and foam concrete (light weight cellular concrete) present a very good material base for filling materials in all respects. Their production is not costly. Production processes can be modified so that secondary raw materials can be used.

Classic autoclaved aerated concrete is prepared as a mix of cement, lime and gypsum used as binding elements, fine silicates, aerating powder and water. Because of chemical reaction, the mix expands and volume of the mix doubles and creates highly porous structure. On average, 80% of volume of the material is created by pores, 50% of air pores and 30% of micro-pores. Microstructure of autoclaved aerated concrete is created mainly by tobermorite. [1] [2]

Macroscopic pores are formed thanks to expansion of the material and caused by aeration; micro-pores are formed in the walls between macroscopic pores. Addition of fly ash enhances porous structure and inter-facial transition zones between pores and matrix. N. Narayaan a K. Ramamurthy in their work describe changes of strength of aerated concrete depending on the way of curing, in particular impact on compressive strength and shrinkage. X-ray diffraction of K. Ramamurthy shows that the main reaction products of aerated concrete are those of group of tobermorite and calcium silicate hydrates (C-S-H). The reaction goes as follows: C-S-H rich in calcium → C-S-H → 11,3Å tobermorite. The product of the reaction is a mix of crystalline, semi-crystalline and nearly amorphous phase of tobermorite. Crystalline structure of tobermorite is variable and it is characterized by proportion of tobermorite to C-S-H. [3][4]

R. Demirboğa and R. Gül described in their paper from 2003 reduction of bulk density and heat conductivity of samples by addition of silica fumes and fly ash. Thus they followed K. Lu-Shu, who described and experimentally formulated dependence between bulk density and thermal conductivity: thermal conductivity reduces with decreasing bulk density. [5]

Y. Xia et al. dedicates one of his research to comparison of properties of autoclaved and non-autoclaved aerated concrete. This non-autoclaved material has a substantial advantage in facilitating the manufacturing process and reducing the cost of products. When Y. Xia and studied the microstructure of non-autoclaved concrete, he identified tiny AFt crystals and amorphous C–S–H gels. [6]

Many authors describe using porous aggregate in theirs research. [7][8][9] Expanded perlite has excellent thermo- and sound-insulating properties, it is volumetrically stabile, has low bulk density, is resistant to humidity, frost (-200 °C) and high temperatures (900 °C) and has excellent sorption properties. It is naturally occurring aluminosilicate volcanic stone that expands up to 20 times its volume when quickly heated to the softening temperature (900-1200 °C). [10] It is used as a filler for manufacture of thermo-insulating boards, backfills, filler for perlite plastering, mortars, lightened concretes, heat-resistant concrete, perlite mattresses, insulation of tanks for liquidized gases, filler in paints, filtration, in agriculture for lightening soil and regulation of soil humidity. Added into concrete, it enhances its properties, lightens it, reduces thermal conductivity, acoustic conductivity and increases resistance to shrinkage. [11][12][13]

2. Materials and methods

Following input materials were used: expanded perlite EP 100, unslaked lime and cement CEM II 32.5 R, aluminum powder and fly ash for experimental partial replacement of binder.

Binders based on silicates like cement and unslaked lime have relatively high bulk density, which increases bulk density of resulting mix. To achieve density reduction, light-weight filler was used (expanded perlite) in combination with aluminum powder for direct decrease of bulk density.

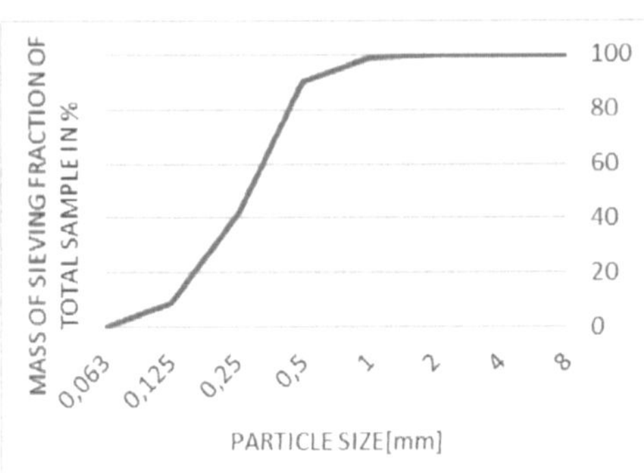

Figure 1. Particle size distribution of expanded perlite.

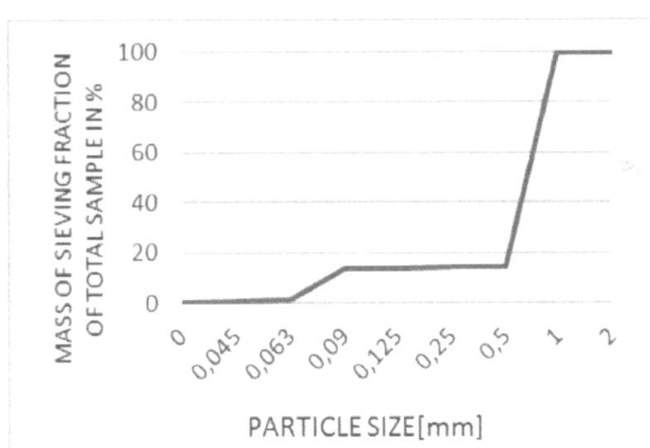

Figure 2. Particle size distribution of used fly ash.

As lightweight aggregate the expanded perlite was used, its properties are in table1.

Table 1. Properties of expanded perlite.

Bulk density	150	$kg\cdot m^{-3}$
Specific thermal capacity	0.85	$kJ\cdot kg^{-1}\cdot K^{-1}$
Heat conductivity coefficient	$0.16\text{-}3.92\cdot 10^{-7}$	$m^2\ s^{-1}$
Specific thermal conductivity	0.05	$W\cdot m^{-1}\cdot K^{-1}$
Maximal temperatures for use	900	°C

The fly ash was used as replacement of part of binder. Its chemical composition is stated in Table 2.

Table 2. Chemical composition of fly ash.

SiO_2	TiO_2	Al_2O_3	Fe_2O_3	FeO	MnO	MgO	CaO	Na_2O	K_2O	anne-aling loss
43.2	0.7	19.5	3.3	1.1	<1.0	2	17.2	<1.0	1.1	5.31

Used aluminum powder is fine powder made of aluminum with purity of 99.7%, 89% of particles have dimension below 45 μm, particles are spherical. It is a combustible solid, which releases combustible gases when in contact with water.

Physically-mechanical properties were determined on test specimens in the shape of a block with dimension 40×40×160 mm. Volumetric weight, compressive strength and flexural strength were determined. Mineralogical composition and microstructure by means of X-ray diffraction analysis was carried out (on the apparatus PANalytical Empyrean), electron microscopy (TESCAN MIRA 3 XMU) and optical microscopy (LEICA DM4000 M LED).

3. Discussion of results

Reference mix for this type of material was made of cement binder, filled with single fraction perlite EP100 in the amount 43% by weight. It seems as disadvantageous to measure expanded perlite by weight, but it is caused by low bulk density of expanded perlite. This proportion was determined on the basis of previous experiments, where it was found, that the mix is sufficiently workable, contains sufficient proportion of cement matrix which can be lightened with air pores. First, the reference samples with bulk density around 500 $kg \cdot m^{-3}$ were examined. Then, the influence of addition of various air entrainers was tested.

For example, calcium carbide at 0.1% by amount of cement had no influence on change of bulk density. At 5%, it destructed the whole structure. Moreover, at higher amount of calcium carbide, lot of acetylene is released, which blows up the mix, it is combustible and has a very unpleasant smell.

Another examined mix was perlite concrete with aerating agents (urea and sodium hypochlorite 5% by weight). The reaction should release gaseous nitrogen which should aerate the mix. However, neither this reaction reduced bulk density. On the surface of the material, undesired crystals were formed.

After these unsuccessful experiments, addition of Al powder was examined in subsequent stages. In the first stage of designing the material, the optimal amount and proportions of individual components of binder were experimentally determined so that lowest possible bulk density and highest strengths could be achieved (at least 0.1 MPa). The composition of mixtures is in the Table 3.

Table 3. The mixture composition.

Mixture	Perlite EP100	Cement CEM II/B-M(S-L)32.5R	Lime CARMEUSE CL 90	Al-powder	Water/binder ratio
1a	25	50	25	0.8	2
1b	25	40	35	0.8	2
1c	25	25	50	0.8	2
2a	35	25	40	0.7	3
2b	35	35	30	0.7	3
2c	35	45	20	0.7	3
3a	50	15	35	0.5	6
3b	50	25	25	0.5	6
3c	50	35	15	0.5	6

IOP Conf. Series: Materials Science and Engineering **549** (2019) 012028 doi:10.1088/1757-899X/549/1/012028

3.1. *Physical-mechanical properties*

As the results imply, decreasing proportion of binder reduces not only bulk density but also strengths. The most significant drop of strength was observed in mix with 50% of binder and proportion of cement in binding component 50% by weight.

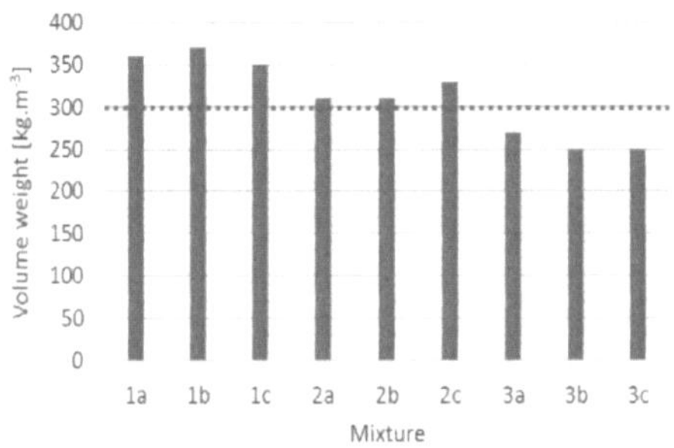

Figure 3. Bulk density, the maximal required value is 300 kg·m^{-3}.

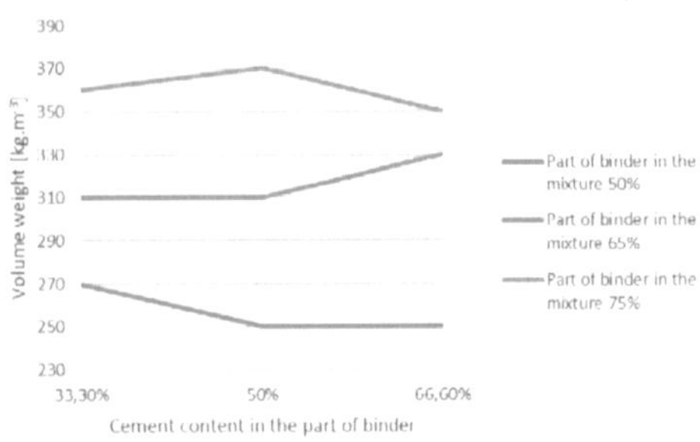

Figure 4. Bulk density, the maximal required value is 300 kg·m^{-3}.

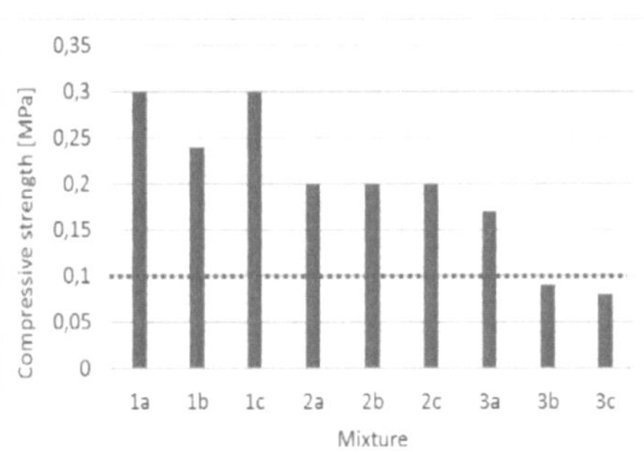

Figure 5. Compressive strength, the minimal required value is 0.1 MPa.

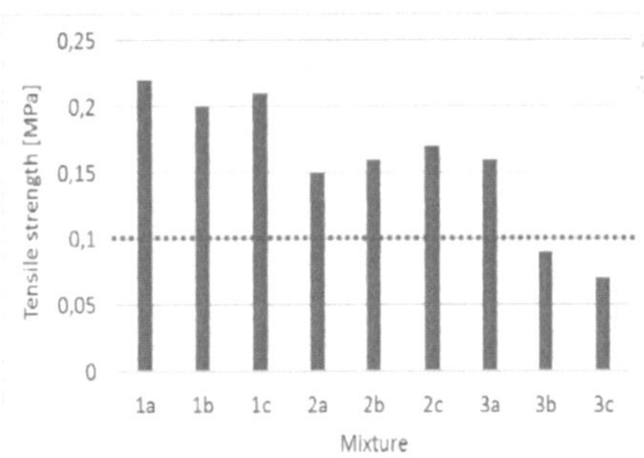

Figure 6. Flexural strength, minimal required value is 0.1 MPa.

There was an effort to replace part of the binder content with fly ash. But in this case it was not beneficial. The 10 % replacement of binder caused increase of bulk density of 20%, which means 320 kg·m^{-3} replacement of 20 % caused increase of 25 % (it means 340 kg·m^{-3}), and that caused, that the limit was exceeded. The bulk density should have the value under 300 kg·m^{-3}.

For this reason, mix with 75% of binders (without the replacement of binder with fly ash) was selected for further experiments. This material has bulk density 270 kg·m^{-3}, compressive strength 0.17 MPa and flexural strength 0.16 MPa.

3.2. *Microstructural analysis*

The Figure 5 shows that aluminum powder made the structure of the sample sufficiently porous and shape of individual pores is relatively regular, and their distribution is even in whole cross section.
The size of pores in samples was from 230 µm to 630 µm. Figures 7-9 shows grains of filler (expanded perlite EP100) coated with cement-lime matrix.

Figure 7. Macroscopic structure of the sample.

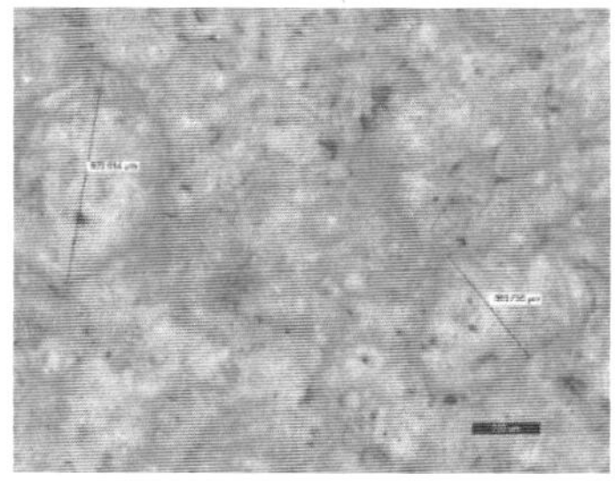

Figure 8. Porous structure of the sample determined with optical microscope.

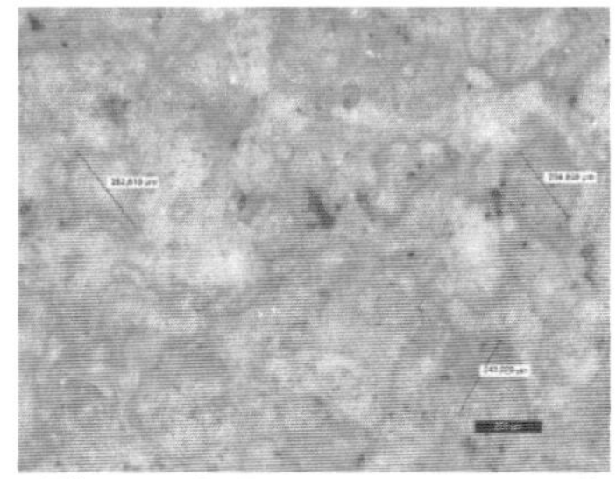

Figure 9. Porous structure of the sample determined with optical microscope.

The image with 5000× magnification shows silicate composite exposed only to laboratory conditions, grains of portlandite, calcite and amorphous C–S–H gels in hydrated cement matrix are visible. This same concluded Xia, Y. in his work [6], too.

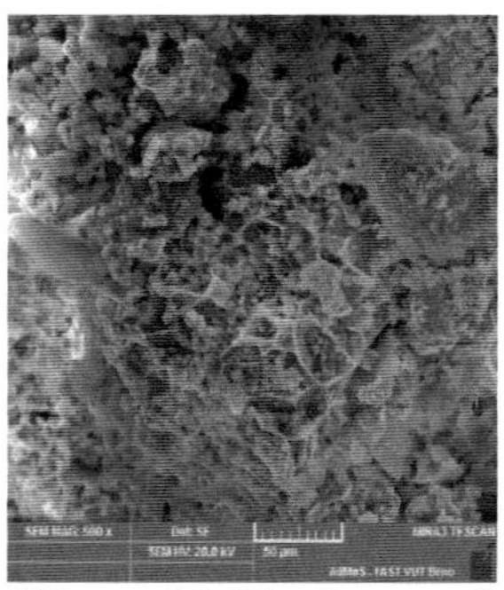

Figure 10. The image of the cement matrix, at 1 month age, at 500× magnification.

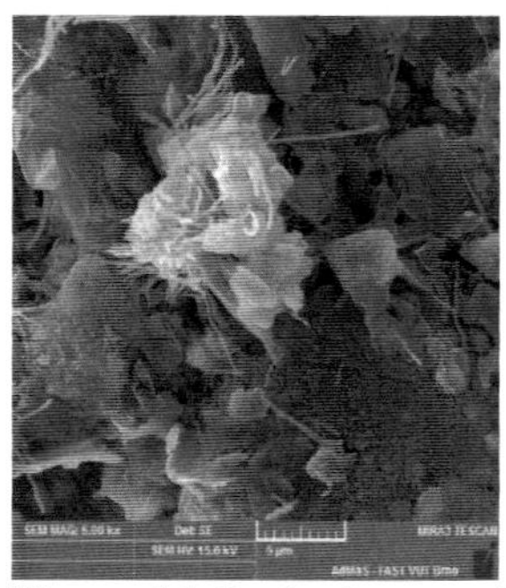

Figure 11. In the image of sample at 5000× magnification, crystals of portlandite in hydrated cement matrix are apparent, age 1 month.

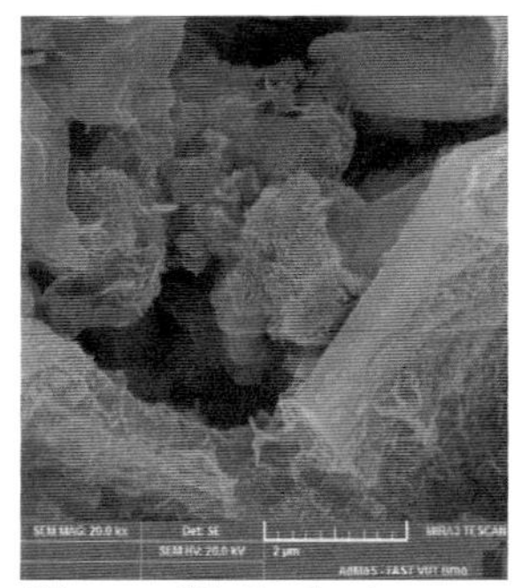

Figure 12. In the image at 20 000× magnification, hydrated cement matrix is visible, age 1 month.

Figure 13. The image of the cement matrix, at 1 year age, at 500× magnification.

Figure 14. In the image of sample at 5000× magnification, crystals of portlandite in hydrated cement matrix are apparent, age 1 year.

Figure 15. In the image at 20 000× magnification, hydrated cement matrix is visible, age 1 year.

X-ray diffraction analysis and differential thermal analysis implies that samples contain β-silica, calcite and portlandite, which was confirmed also by microscopic analyses.

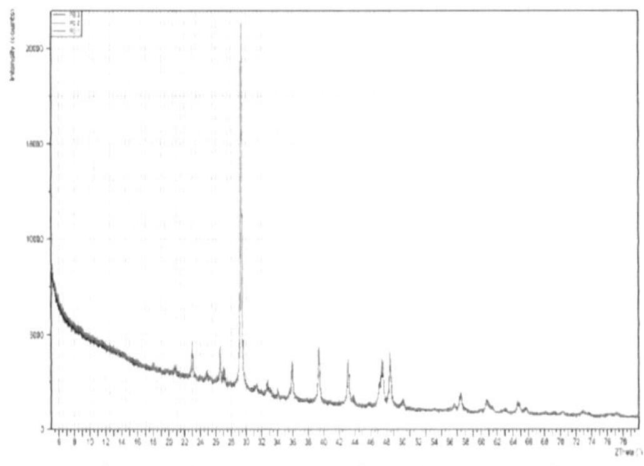

Figure 16. X-ray diffraction analysis.

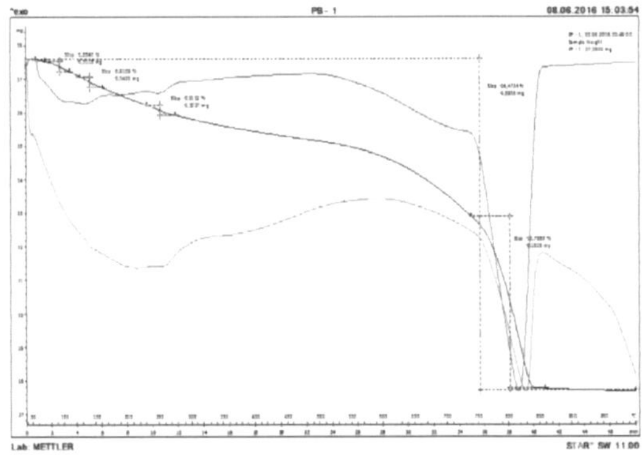

Figure 17. Differential thermal analysis.

4. Conclusion

The properties of non-autoclaved aerated concrete, created by mix of unslaked lime, cement CEM II/B-M (S-L) 32.5 R, fine grained perlite EP 100 and aluminum powder, were experimentally determined.

The results imply that it is advantageous to use material with direct and indirect lightening of the structure. It is possible to achieve bulk density lower than 300 kg·m^{-3}. Without the process of autoclaving, it is possible to achieve compressive strength about 0.1 MPa. This material could be used as material with several ways of use, for example as filling material with low bulk density or insulating material, its preparation could be done in-situ. The achieved results will serve as base for further modifications of mix-design of autoclaved aerated concrete with secondary raw materials, in the first stage with waste perlite.

Acknowledgments

The paper was supported by the project FAST-S-19-5779 with name "Problems of effective use of secondary raw materials in the area of silicate and non-silicate based composites"

References

[1] Kus H., Carlsson T. (2003) Microstructural investigations of naturally and artificially weathered autoclaved aerated concrete, Cement and Concrete Research **33**, pp 1423-1432

[2] S. Aroni, G.J. De Groot, M. J. Robinson, G. Svanholm, F.H. Wittman (Eds.) (1993)Autoclaved Aerated Concrete: Properties, Testing and Design, RILEM Recomendet Practice, E&FN Spon, London

[3] Ramamurthy. K., Narayanan N. (2000) Microstructural investigations on aerated concrete, Cement and Concrete Research **30**, pp 457-464

[4] Narayanan N., Ramamurthy. K. (2000) Structure and properties of aerated concrete: a review, Cement and Concrete Composites **22**, pp 321-329

[5] Demirboğa R., Gül R. (2003) Thermal conductivity and compressive strength of expanded perlite aggregate concrete with mineral admixtures. Energy and Buildings **35**, pp 1155-1159

[6] Xia, Y., Yan, Y., & Hu, Z. (2013). Utilization of circulating fluidized bed fly ash in preparing non-autoclaved aerated concrete production. Construction and Building Materials, **47**, pp 1461–1467. https://doi.org/10.1016/J.CONBUILDMAT.2013.06.033

[7] Melichar, T. Dufka, A., Bydzovsky, J. (2017). Changes in the composite structure and parameters after an exposure to a synergic action of various extreme conditions. Materiali in Technologije, **51**(2), pp 243-249.

[8] Chung, S.-Y., Abd Elrahman, M., Kim, J.-S., Han, T.-S., Stephan, D., & Sikora, P. (2019). Comparison of lightweight aggregate and foamed concrete with the same density level using image-based characterizations. Construction and Building Materials, **211**, 988–999.

[9] Videla C., López M. (2002) Effect of lightweight aggregate intrinsic Strength on lightweight

concrete compressive strength and modulus of elasticity, Materiales de Construcción, Vol **52**, No 265

[10] Papa E., Medri V., Natali Murri A., Laghi L., De Aloysio G., Bandini S., & Landi E. (2018). Characterization of alkali bonded expanded perlite. Construction and Building Materials, **191**, pp 1139–1147.

[11] Casini, M. (2019). Insulation Materials for the Building Sector: A Review and Comparative Analysis. Reference Module in Materials Science and Materials Engineering.

[12] Liu W. V., Apel D. B. Bindiganavile (2014) Thermal properties of lightweight dry-mix shotcrete containing expanded perlite aggregate. Cement & Concrete Composites **53**, pp 44-51

[13] Brouk J.J. Perlite aggregate: its properties and uses. J Am Concr Inst 1949; **46**(11): pp 185-90

IOP Conf. Series: Materials Science and Engineering **549** (2019) 012029 doi:10.1088/1757-899X/549/1/012029

Temperature and chemical resistance of repair composite and microstructure monitoring depending on different type of binder and filler

J Hodul[1], L Mészárosová[1], T Žlebek[1] and R Drochytka[1]

[1] Brno University of Technology, Faculty of Civil Engineering, Veveří 331/95, Brno 602 00, Czech Republic

E-mail: hodul.j@fce.vutbr.cz

Abstract. Currently, we are carrying out several repairs to concrete structures that are permanently exposed to chemically aggressive environments and the effects of temperature and cyclic mechanical stress. It is therefore necessary to develop a highly resistant patching composite that would return the damaged objects to their original state or improve resistance to external attack. In this paper, results of the research, in which monitoring of the temperature was performed and chemical resistance of the polymer composites of repair, specifically with the epoxy matrix and a different type of filler, is presented. As a filler, we use secondary raw materials that are currently produced largely as a by-product of industrial production in the Czech Republic and abroad. It has been found that the type of filler used has not an impact on temperature resistance. but in chemical resistance plays a considerable role also used filler. The microstructure was also monitored before and after chemical stress.

1. Introduction

Newly developed mass is used for the rehabilitation of existing structures experiencing mechanical stress in chemically aggressive environments with higher temperatures. Epoxy resins (ER) are very useful in this environment because they are known to resist famously demanding conditions.

ERs fall into the category of polymers – thermosets that comprise molecules of two or more oxiranes. Their use is divided into several areas, notably electrical, fire retardant, paint (paints, varnishes, adhesives, etc.), and composite [1] ERs can be cured with various curing agents, especially those based on diols, diacids, anhydrides and diamines [1].

In the use of repair materials in demanding industrial and energy plants, chemical resistance is an important parameter. In this environment, aggressive liquids often appear, such as those with a strongly basic and strongly acidic character. The effect of various chemicals on the polymers may lead to a change in their physical properties. There may, for example, be swelling, colour change, chemical degradation of the polymer chains, bursting, crosslinking, or other possible disturbances. In a design environment, organic or inorganic acids are occurred, while organic acid (HCOOH, CH_3COOH, and CH_3CH_2COOH) are relatively weak acids and are therefore much less dissociated. Acting mainly as a solvent, and their effect leads to the formation of blisters on the surface and separation of the segments of the macromolecular chains [2]. Kalenda and Kalendová demonstrated that increased chemical resistance of polymer compositions can be achieved by introducing fillers that are capable of reacting with diffusing acid. Inert fillers or pigments (TiO_2, graphite, carbon black, chromium oxide), present in the ER matrix, increase the diffusion of aggressive (corrosive) media because the penetration at the interface of the pigment/binder extends along the pigment particles [2].

For characterization of the polymer materials heat resistance, glass transition temperature (T_g). This temperature characterizes an upper temperature limit at which the polymer is still useful, was also showed to be very close to between the onset of the so-called diffusion control regime during the isothermal cure of thermoset and its glass transition, the influence of T_g on the chemorheology of a thermoset, and the correlation between internal stresses and T_g. Many authors have dealt with these dependences, e.g. Hale et al. [3], who created a model equation expressing the relationship between T_g and other specific parameters. He found out that increase of T_g is caused by a decreasing in chain-end concentration, as well as formation of effective cross-links and further decrease in the configurational entropy.

2. Identification of materials

As a part of the research, we used two types of polymer resins. The developed patching composites labelled RC1 had a binder component epoxy comprising a resin, a filler such as a reference quartz sand Dorsilit (REF), waste packing transparent glass (WGT), and waste from the production of insulating mineral wool panels (WMTI). The development of repair composites began with a performance of 30%; the maximum tested performance was 85%, and after it exceeded 50%, the amount of filler was increased successively by 5%, because it is from this filling that the properties of the polymer mass are markedly changed. The amount of filler was expressed in weight percentage on the total weight of the fresh mixture.

Patching composites labelled RC2 as a binder component having a different type of epoxide, consisting of an ER (alkoxymethyl) oxirane (C12-C14 alkyl), solvent naphtha (petroleum), and light aromatic. Polyamine-based hardener contained benzyl alcohol, polymer with benzenamine, hydrogenated formaldehyde, 2,4,6-tris (dimethylaminomethyl) phenol, 4,4'-methylenebis (cyclohexylamine), followed by filler components that used quartz sand (Dorsilit), transparent waste glass (WGT), pulverised fly ash contaminated due to denitrification of the flue gas (PFA) and waste from mineral thermal insulation production without organic compound (WMTI). For composites of repair, RC2 was selected with an initial quantity of filler of 40%, making it suitable as a patching material with low viscosity in compositions such as RC1 with planned utilisation in the form of grout.

Table 1. Input materials used as fillers – Properties and requirements for pre-treatment.

Marking materials	Requirements for pre-treatment	Density [kg/m^3]
REF	Need not be modified	2662
WGT	Grinding and sorting to fractions	2534
PFA	Need not be modified	2390
WMTI	Need not be modified	2620

3. Tests performed

3.1. Chemical resistance

Any changes to the properties were affected by the type and concentration of the chemicals, in addition to the time of exposure to the chemicals on the surface of the polymers. According to the formulas (25%RC1+75% REF, 25%RC1+75% WGT, 40%RC1+60% PFA, 30%RC2+70% REF, 30%RC2+70% WGT, 35%RC2+65% WMTI), set samples were prepared different polymer repair composite. Application was done through suitable means in a thick layer on the laboratory specimen slide, which had been cleaned and dried with acetone. Samples were allowed to polymerise for 7 days on a clean substrate at 20 ± 2 °C and then immersed in a sealable glass tube with the appropriate chemicals. Checking of the samples were carried out for 7 to 30 days after immersion in aggressive liquid solutions. The evaluation results were mainly visual, and the reaction mass according to chemicals classified by the evaluation system are indicated in Tab.1. Second, some specimens were examined using the optical digital microscope VHX-950F Keyence.

Table 2. Evaluation system for accelerating chemical resistance tests designed with respect to the expected behaviours of the tested samples in the aggressive media.

*******	No changes
******	Colour changes
*****	Swelling and colour changes
****	Peeling off from the glass
***	Peeling off from the glass and swelling included colour changes
**	Peeling off from the glass and softening
*	Complete disruption

Based on the expected use of the repair materials developed in industrial and energy plants, solutions were selected that are used in these plants to a great extent. Acetic acid (CH_3COOH), phosphoric acid (H_2SO_4), sodium hydroxide (NaOH), sodium chloride (NaCl), hydrogen peroxide (H_2O_2), and distilled water (H_2O), followed by petroleum products and gasoline were chosen.

3.2. Temperature resistance

Determination of heat resistance of polymers is usually carried out by setting the glass transition temperature (T_g) through differential scanning calorimetry (DSC) or dynamic mechanical analysis (DMA). Samples for T_g were tested using DMA 2980 (TA Instruments). This is an analysis in which through periodic sinusoidal deformation of the sample clamped in the jaws of the system, the force required for this deformation is measured. Based on the course of the deformation and force (deformation subsequently converted to deformation stress) and their phase shift is determined by a dynamic modulus of elasticity and its components: elastic (Storage) and the Loss (loss) modulus. Measurements were performed in the temperature range of 40-130 °C at a heating rate of 5 °C / min, amplitude of 10 mm, and a frequency of 1 Hz. The dimensions of the samples were 35 (17.6 active) × × 3.3 × 12.0 mm, and from each material, two samples were obtained.

3.3. Coefficient of linear thermal expansion

Coefficient of thermal expansion (α) was determined for the chosen repair and grout materials. The samples' dimensions were 20 × 20 × 100 mm, and measurement was carried out in the temperature range of 20-60 °C for approximately 18 hours. Coefficient of thermal expansion (α) is an important parameter, particularly for polymers used in engineering applications. Low value of α is often desirable for achieving dimensional stability.

4. Results and discussion

4.1. Chemical resistance

Evaluation of the chemical resistance of tested repair composites was conducted after 30 days of exposure to various corrosive media. It was found that the selected repair materials were resistant to strong inorganic acids and hydroxides (see Fig. 1 and Fig. 2). However, the materials were not resistant to acetic acid (CH_3COOH) at a concentration of 10%. After immersion in distilled water and saturated sodium chloride (NaCl), even after 30 days, there was no change. Gasoline, which is a primarily aliphatic hydrocarbon mixture obtained by fractional distillation of crude oil with added aromatic hydrocarbons and additives also had no effect on the degradation of the epoxy repair mass.

Table 3. Evaluation of chemical resistance of tested repair composites (RC1, RC2) after 30 days of immersion in different liquid aggressive solutions.

Aggressive medium	Concentration	RC1 75% REF	RC1 75% WGT	RC1 60% PFA	RC2 70% REF	RC2 70% WGT	RC2 65% WMTI
H_2SO_4	40%	*******	*******	*******	*******	*******	******
NaOH	40%	*******	*******	*******	*******	*******	*******
CH_3COOH	10%	*	*	*	***	***	**
Gasoline	-	*******	*******	*******	*******	*******	*******
NaCl	10%	*******	*******	*******	*******	*******	*******
H_2O_2	30%	*****	*****	**	*****	*******	******
Distilled water	-	*******	*******	*******	*******	*******	*******

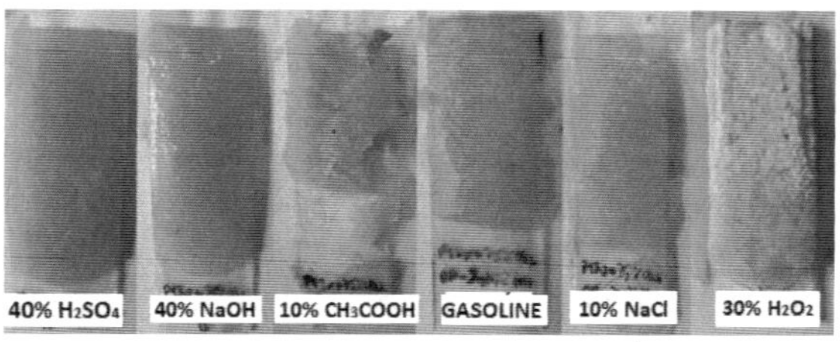

Fig. 1. Samples RC1 75% WGT after 30 days immersion in aggressive liquid solutions.

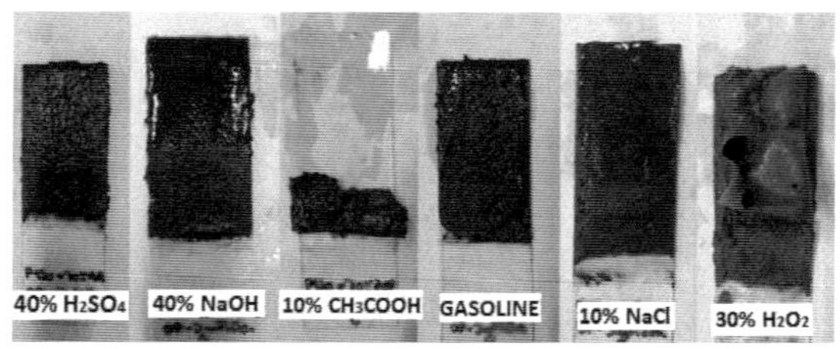

Fig. 2. Samples RC1 60% PFA after 30 days immersion in aggressive liquid solutions.

For the selected composite (RC1 65% OIDM) showed, indicating somewhat lower chemical resistance than for other formulations (see Fig. 3 and Tab. 3). This difference may be due to a different filler – particularly the particle size and shape and its chemical composition.

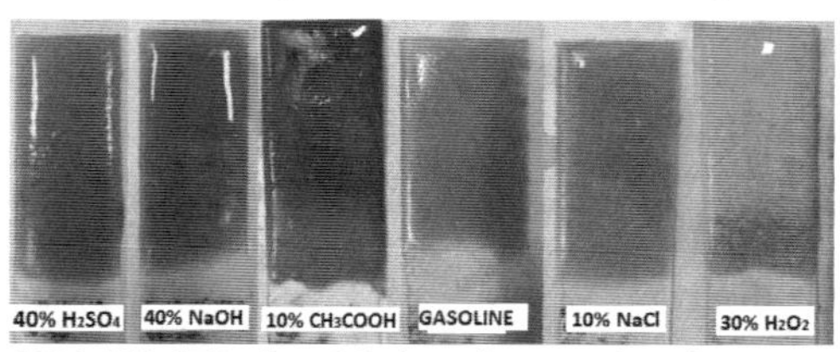

Fig. 3. Samples RC2 65% WMTI after 30 days immersion in aggressive liquid solutions.

The surface of some samples was further investigated by optical microscopy after chemical stress to observe the influence of aggressive media on infringement material. Based on the images from the optical microscope, the action of the sulfuric acid did not disturb the surface (Fig. 4, Fig. 5). On the samples that were exposed to acetic acid, considerable deterioration of the surface and cracking were found (Fig. 6).

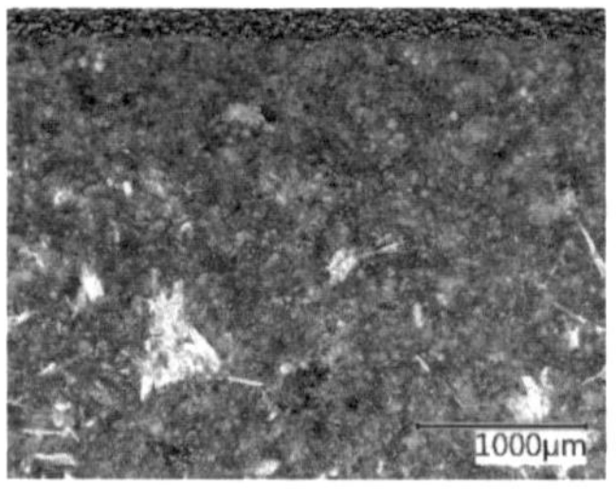

Fig. 4. Surface of sample RC1 60% PFA after immersion in 40% H_2SO_4 (magn. 32×).

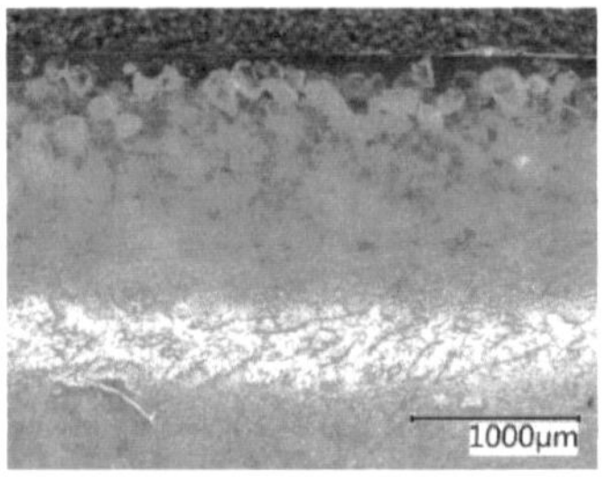

Fig. 5. Surface of sample RC1 75% WGT after immersion in 40% H_2SO_4 (magn. 32×).

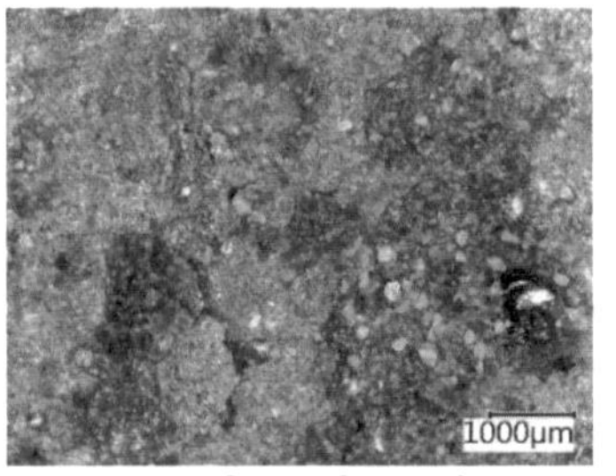

Fig. 6. Surface of sample RC2 75% WGT after immersion in 40% H_2SO_4 (magn. 16×).

4.2. Temperature resistance

Examination of the heat resistance by repair composites was conducted by determining glass transition temperature (T_g). T_g is a very important parameter for epoxy thermosets and epoxy matrix composites because it determines the conditions for the use of these types of materials. In most applications, epoxy resins are used at a temperature below T_g (the glassy state). Different T_g values can be justified through different ways of preparing materials [4]. Results of the determination of T_g of DMA are shown in Fig. 7 and Fig. 8; the abscissa indicates temperature in [°C], and the y-axis shows the values the of elastic modulus (left) and loss modulus (right). The highest T_g was calculated from the value of the dissipation factor (tan δ), which, upon reaching the T_g, was at maximum. Graphical comparison set T_g can be observed in Fig. 9. Based on the results, it is evident that T_g does not significantly affect the filler used, but rather the binder type (ER) and the principle of its curing. All test materials were a T_g higher than 70 °C, ensuring safe use even at higher temperatures in severe environments.

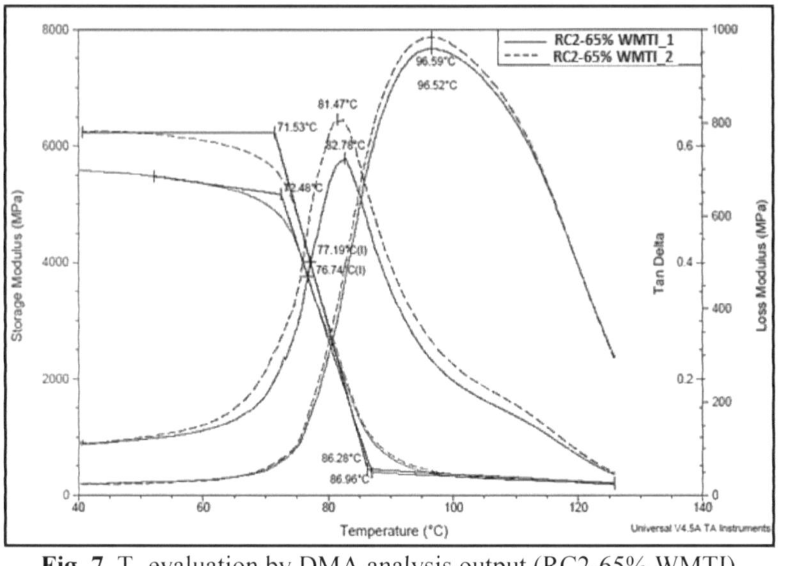

Fig. 7. T_g evaluation by DMA analysis output (RC2-65% WMTI).

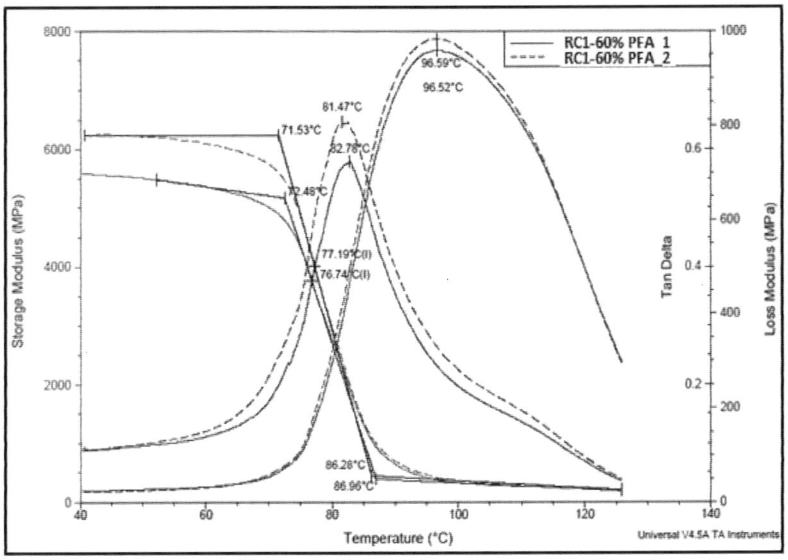

Fig. 8. T_g evaluation by DMA analysis output (RC1-60% PFA).

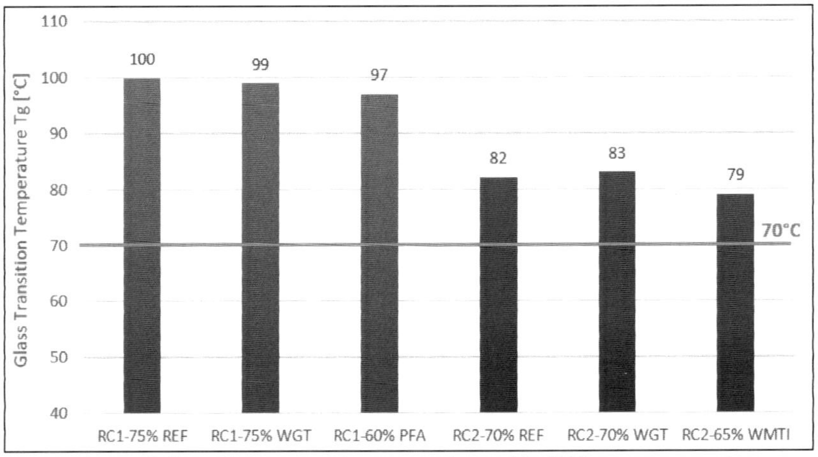

Fig. 9. T_g comparison of the polymer repair composites with different type and amount of fillers.

4.3. Coefficient of linear thermal expansion

The dilatometric curve of the selected repair composite RC1, which is primarily intended for use in the remediation of thermally and chemically stressed structures, is shown in Fig. 10. The results of the coefficient of thermal expansion (α) are given in Tab. 4

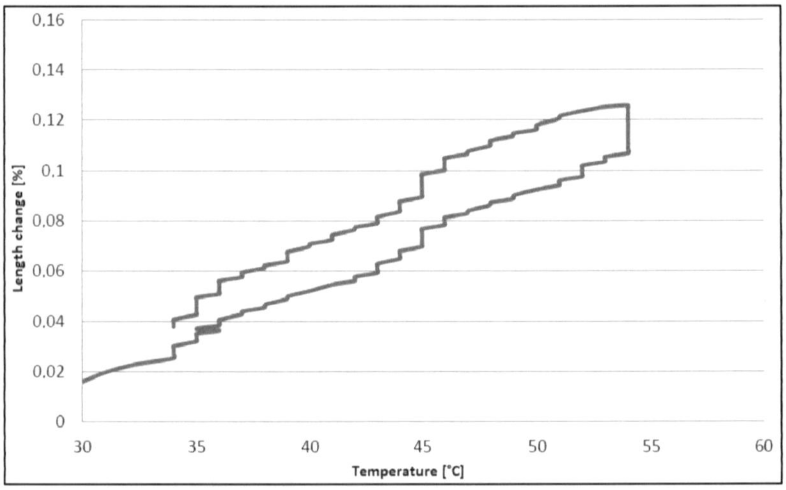

Fig. 10. Dilatometric curve of the sample RC1-60% PFA.

Table 4. Values of the coefficient of linear thermal expansion (α) of different polymer repair composites.

Formulation	ΔT [K]	L_0 [mm]	α [K^{-1}]
RC1-75% REF	30	99.1965	$40.1 \cdot 10^{-6}$
RC1-75% WGT	29	99.0962	$55.3 \cdot 10^{-6}$
RC1-60% PFA	24	99.2088	$52.8 \cdot 10^{-6}$
RC2-70% REF	27	99.1794	$55.5 \cdot 10^{-6}$
RC2-70% WGT	26	99.1856	$55.3 \cdot 10^{-6}$
RC2-65% WMTI	33	99.1565	$40.9 \cdot 10^{-6}$

The determined values of the coefficient of thermal expansion (α) confirmed the generally known evidence that polymer-based materials exhibit higher values of α compared with silicate materials. In the case of repair materials for the repair of concrete structures, the same or similar coefficients of thermal expansion (α) that show a concrete bed are required [5]. For conventional concrete, $\alpha = 14 \cdot 10^{-6}$ K^{-1} [6]. The developed materials did not fulfil this requirement completely because of the achieved values of α between $40 \cdot 10^{-6}$-$55 \cdot 10^{-6}$ K^{-1} (Tab. 4). In case of the rehabilitated part of the structure loading by cyclic temperature variation, small cracks could occur in the tested repair materials due to the higher stress at the connection point of the remediation material to the substrate. However, because of the high cohesion of repair materials to the substrate and a lower modulus of elasticity, repair composites can be developed that are capable of effectively resisting even alternating temperature changes. This assumption still needs to be verified in practice.

5. Conclusion

Based on the determined values of thermal resistance by means of DMA analysis, and the results of the T_g, it can be concluded that the type of filler used has no effect on the temperature resistance of repair composites developed using epoxy resin (ER), which is planned mostly used for the repair of concrete structures. Heat resistance is mainly affected by the type of binder used. The results showed that even the coefficient of thermal expansion is influenced primarily by the type of binder used, in this case the epoxy resin, and not by the type and amount of filler.

Regarding the chemical resistance, the worst degradation of the test fuel was caused by acetic acid (10% CH_3COOH). Therefore, repair composites should not be used in environments where an aggressive medium is present. It was also found that these composites are resistant to strong solutions of inorganic acids and alkalis.

Acknowledgements
This work was financially supported by the junior specific research project No. FAST-J-19-5784 "Research of special polymer coating using hazardous waste and secondary raw materials" and by the project of Ministry of Industry and Trade of the Czech Republic No. FV20303 "Progressive polymer materials for chemically high aggressive environments using secondary raw and hazardous wastes".

References

[1] Wisanrakkit G, Gillham J K and Enns, J B 1990 The Glass Transition Temperature (t_{g},) as a Parameter for Monitoring the Cure of an Amine/Epoxy System at Constant Heating Rates, *Journal of Applied Polymer Science* **41** pp 1895-1912
[2] Kalenda P and Kalendová A 1994 Improved Chemical Resistance of Epoxy Resin-based Coating Compositions *Dyes and Pigments* **27**(4) pp 305-312
[3] Hale A., Macosko C W, Bair H E., Laboratories T B and Hill M 1991 Glass Transition Temperature Function of Conversion in Thermosetting Polymers *Macromolecules* **24**(9) pp 2610-2621
[4] Zhou J and Lucas J P 1999 Hygrothermal effects of epoxy resin. Part II: variations of glass transition temperature, *Polymer* **40**(20) pp 5513-5522
[5] Drochytka R, Dohnálek J, Bydžovský J, Pumpr V, Dufka Á, Dohnálek P 2012 Technical conditions for rehabilitation of concrete structures III Brno: *Association for the rehabilitation of concrete structures* ISBN 978-80-260-2210-7
[6] Kratina J 2013 Linear thermal expansion of materials. *E constructor* [online]

IOP Conf. Series: Materials Science and Engineering **549** (2019) 012030 doi:10.1088/1757-899X/549/1/012030

Polymer based grout with specially treated hazardous waste

T Žlebek[1], J Hodul[1] and R Drochytka[1]

[1] Brno University of Technology, Faculty of Civil Engineering, Veveří 331/95, Brno 602 00, Czech Republic

E-mail: zlebek.t@fce.vutbr.cz

Abstract. The present research verified the possibility of using specially treated hazardous waste (solidification product) derived from a solidification agent (fly ash from fluidized bed combustion (FA)) and hazardous waste (end-product) as a filler in polymeric grout. Hazardous waste was pre-treated by solidification technology, while within the performed research, a total of two solidification paths were verified. One of them was a dry homogenisation (TOS1) and the other technology was wet granulation of the input materials (TOS2). The use of dry homogenization solidification technology, which is economically less demanding, seems to be more advantageous. Using such specially treated filler was verified by some selected tests and the results indicated grout with a relatively high compressive strength of about 50 MPa and a flexural strength of about 20 MPa. The developed polymeric grout also showed a very good adhesion to a cast basalt, which was tested by the pull-off test and subsequently observed in detail by optical microscope.

1. Introduction

To produce high-quality jointing compounds in demanding industrial applications, the most commonly used binder is polymer-based, which is distinguished by very good adhesion to most materials [1]. Some of the best binders are polymeric binders based on epoxy resin (EP), which exhibit excellent adhesion to most materials and, after curing, excellent physical and mechanical properties, minimal shrinkage and high chemical resistance [2,3]. Thus, this current research selected polymeric composition based on epoxy resin (ER) as a binder. Using the ER as a binder for a patching material was also discussed by Hodul et al. [4], wherein they dealt with a possibility of using fly ash contaminated by flue gas denitrification process as filler to polymer repair materials. Hazardous waste (HW) can be used as a filler in polymer compositions. However, it needs to be pre-treated appropriately. The most common type of treatment of HW is solidification and stabilisation (S/S) [5]. It is the mixing of HW with solidification agents that bind dangerous substances into a solid matrix to prevent the release of contaminated components into the environment. The most frequently used reagents are lime, cement or fly ash. This type of treatment has been also studied by Hodul et al. [6] in their experiment, in which the neutralization sludge solidification engaged. Lopéz et al. [7] who tried to stabilise the HW containing large amounts of mercury. The aim is to maximise the use of all waste in order to minimise their very expensive treating and landfill, which contributes to environmental protection. The aim of this work was to verify the possibility of using solidification product formed from selected secondary raw materials such as a solidifying agent and hazardous waste as a filler in polymeric grout. This material is primarily intended for grouting elements in basalt sewer sewers and certain industrial plants where it will be exposed to a heavy chemical aggressive environment. In connection with the continuous production of hazardous waste (HW) in industry and in the effort to avoid landfilling these materials, we need to find a suitable use of such hazardous waste. HW was treated with the appropriate solidification technology in order to produce suitable fillers useful in polymeric grout. As part of the research, a total of two solidification technologies were verified. One of these was the dry homogenisation of all input materials (solidification technology TOS1) and a second one was wetted granulation (solidification technology TOS2). Solidification TOS2 was performed by granulation with wet water spray mixture made up of hazardous waste, fly ash from fluidized bed combustion (FA) and cement on a granulating plate. Once cured and dried, the pellets were then ground and a very fine filler in the form solidification product was created. Solidification by the way of dry homogenisation (TOS1) was carried out by intimate mixing of hazardous waste and FA in a proper homogeniser for 24 hours. The possibility of using such a specially treated filler was verified by the selected tests, which was the determination of the compressive and three-point flexural strength, the determination of adhesion to the basalt tile, determination of tensile strength and elongation at break, using an optical microscope that monitored distribution of the filler in an epoxy matrix, and also the connection interface with basalt grout tile. Due to the possibility of some aggressive components releasing hazardous waste from this filler, it cannot be used to grout elements in hygienically clean operations.

2. Identification of the materials

2.1. The filler material prepared by dry homogenisation (TOS1)
The first solidification technology was dry homogenisation of the individual components using 5% of hazardous waste resulting from the incineration of municipal solid waste (end-product). Furthermore, fly ash form fluidized bed combustion (FA) was selected as the solidification agent. The mixture was allowed to homogenise for 48 hours and then applied as the filler to the epoxy grout.

2.2. The filler material prepared by the wet granulation (TOS2)
The second solidification technology was represented by wet granulation of input. 5% amount of hazardous waste resulting from the incineration of municipal waste (end-product) was used again and as the solidification agent were used FA and cement. All materials were irrigated on the granulation plate with water to form a pellet-like small bowls. These pellets were subsequently dried and ground to a very fine fraction and the obtained filler was used in the grout.

2.3. Binder component – epoxy resin
A two-component, solvent-free and chemically resistant epoxy resin was used as a binder. The mixing ratio of the two components A: B was 1.8: 1.

Component A (epoxy resin) contained the following compounds: epoxy resin (alkoxymethyl) oxirane (C12 -C14 alkyl), formaldehyde, reaction products with 1-chloro-2,3-epoxypropane and phenol.

Component B (hardener side) comprised: 4,4-methylenebis (cyclohexylamine) -phenyl-methanol, formaldehyde polymer with benzenamine, hydrogenated.

The epoxy resin was chosen due to its excellent physical and mechanical properties, high chemical resistance and minimal shrinkage [8].

IOP Conf. Series: Materials Science and Engineering **549** (2019) 012030 doi:10.1088/1757-899X/549/1/012030

3. Examinations carried out

3.1. Determination of compressive strength and three-point bending strength

These tests were performed according to standard EN 13892-2 Methods of test for screed materials - Part 2: Determination of flexural and compressive strength. 20%, 30% and 40% filling of the epoxy grout with the filler from dry homogenisation (TOS1) and the filler from wet granulation (TOS2) was tested. The determination of flexural strength was carried out on 20 × 20 × 100 mm test beams (Figure 1). Initially, the flexural strength test was performed, and the compressive strength test was performed on the beam fractions (Figure 2).

Figure 1. Determination of flexural strength.

Figure 2. Determination of compressive strength.

3.2. Determination of cohesion with cast basalt

This test was performed according to standard DIN EN 1542 Products and systems for the protection and repair of concrete structures - Test methods - Determination of consistency by tear test. On the surface of the grout were glued metal targets with a 50 mm diameter (Figure 3) which, after curing of the epoxy adhesive, were trimmed to the octagonal shape up to the base basalt paving. Subsequently, using the pull-off tester DYNA Z16 Proceq (Figure 4), the pull-off force and place of breakage was determined (Figure 5), and the cohesion with cast basalt was found.

Figure 3. Metal targets glued for the determination of cohesion to cast basalt.

Figure 4. Performing the pull-off bond strength test.

Figure 5. Place of failure after pull-off test in the contact zone grout/cast basalt surface.

3.3. Monitoring of the contact zone between the grout and cast basalt paving

Detail of grout connection to the cast basalt tiles surface was monitored using a digital microscope WHX-950F. This microscope is a digital CMOS image sensor with virtual pixels 1600 (H) × 1200 (V) frame rate of 50 F / allowing magnification to 200x. The contact zone was observed at a magnification of 64× and 6×.

3.4. Determination of tensile properties

This test was performed according to standards EN ISO 527-1 Determination of tensile properties - Part 1: General principles and EN ISO 527-2 Determination of tensile properties - Part 2: Test conditions for moulding and extrusion plastics. The test was performed according to the standards mentioned in special test specimens in the shape of dog bone (Figure 6) on a test press (Figure 7). Loading was carried out at rate of 5 mm/min and the initial gauge length of extensometer was 25 mm. Tensile properties were calculated from the measured values, as was the highest tensile strength (tensile strength at the break) and respective elongation at the break.

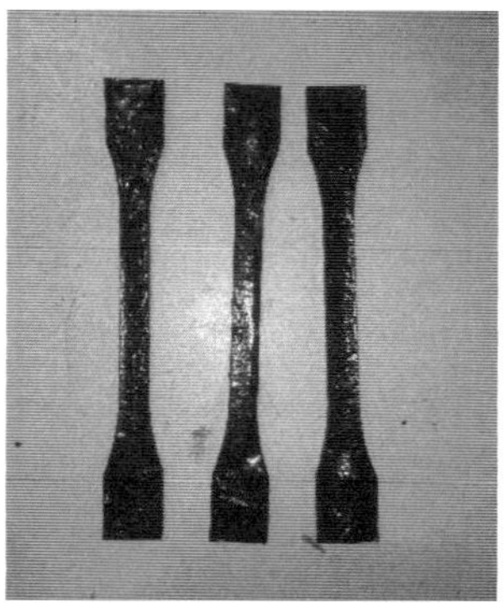

Figure 6. Specimens in the shape of dog bone for determination of tensile properties.

Figure 7. Loading test specimen of the polymer grout by a tensile force in the test device.

4. Results and discussion

4.1. Determination of compressive strength and three-point flexural strength

The results of the compressive strength (Figure 8) and three-point flexural strength (Figure 9) indicate that with an increasing percentage of filling, the resulting values decreased, and therefore for further testing , the 30% filling which guarantees relatively good properties and a relatively high percentage of the solidification product. Flexural strength of mass with 30% of the filling was about 25 MPa and compressive strength was about 50 MPa. When compared to reference materials, it is seen that almost the same characteristics of jointing grout material had 20% filler TOS2. At higher percentage of filling the strength was slightly smaller than the reference value, but the differences were not significantly large. As it can be seen, the type of solidification of HW had a significant effect on the final strength grout. The biggest difference is apparent in samples with 20% of filling, when with sample containing filler TOS2 was an increased compressive strength, probably due to better homogeneity of the particles of filler into the polymeric matrix. It can be also affected by the specific surface area and shape of the filler particles. This fact can be observed in both the compressive strength and flexural strength.

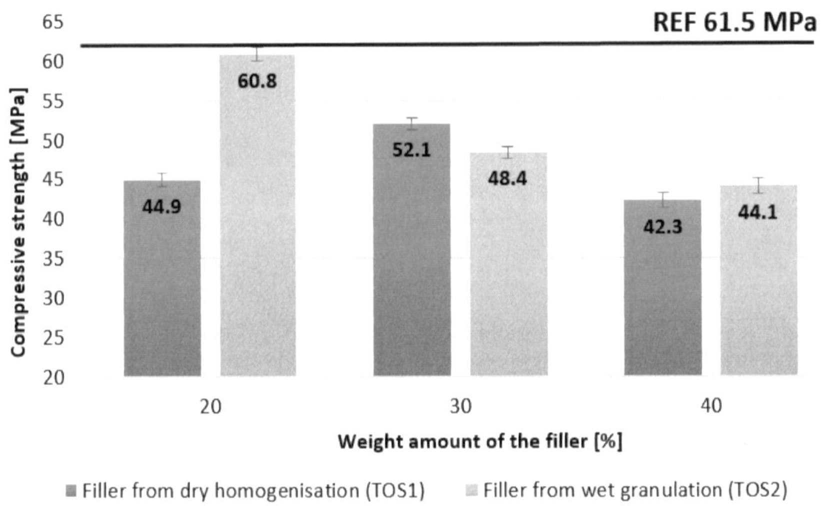

Figure 8. Results of the developed epoxy grout compressive strength.

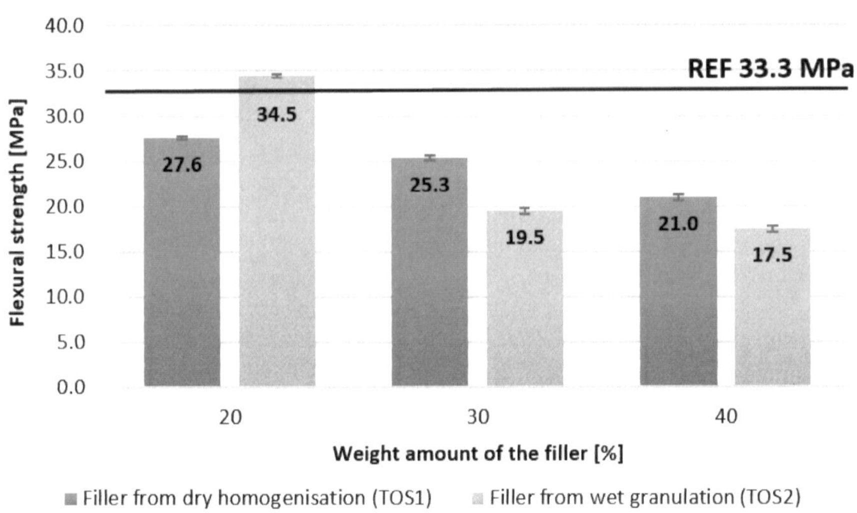

Figure 9. Results of the developed epoxy grout three-point flexural strength.

4.2. Determination of cohesion with cast basalt

In determining the cohesion of polymer grout with cast basalt tiles, it was found that the cohesion of the two developed grout jointing materials is approximately 3.5 MPa (see Fig. 10), which is very satisfactory value for the purpose of intended use. Reference material containing approximately 30% wt. of silica flour showed almost the same cohesion with cast basalt (3.55 MPa) as developed grout containing solidification product. The violation has occurred in all the samples on the surface of the pavement of basalt - see Figure 5 Epoxy resins generally have a high cohesion with almost all materials. It was demonstrated here that the cast basalt with a smooth surface is not an exception. It was successfully shown that the developed grout can be safely used for grouting cast basalt elements.

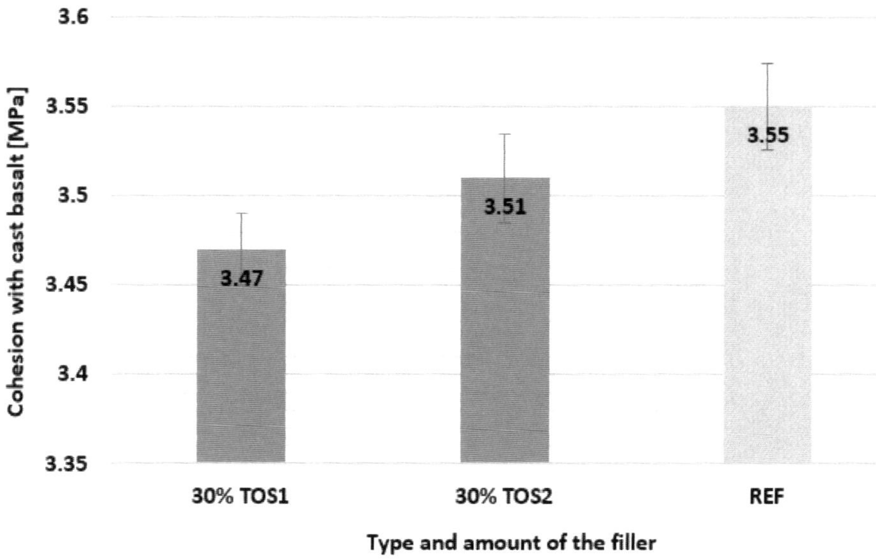

Figure 10. Results of the cohesion with cast basalt tiles

4.3. Monitoring of the contact zone between the grout and cast basalt

To find the detail of grout and cast basalt connection was monitored using a digital microscope contact zone between these materials. In 6x magnification (Figure 11), air voids in the grout were noticed likely due to mixing. However, in the contact zone, cast basalt tiles had no imperfections or free space, indicating a high cohesion between the materials. For more detailed observation of the connection between the developed epoxy grout cast basalt tile were chosen to 64× magnification (Figure 12) where it was even better to see the perfect connection of the grout to the tiles, which perfectly mirror the mass of fine unevenness on the surface of the tile. Among other things, at such magnification, the structure of both the grout and the cast basalt paving surface is already well observable. It can be seen in the structure of the grout that the filler is perfectly distributed throughout the epoxy matrix and does not form clumps.

Figure 11. The connection detail of developed grout (30% TOS2) to cast basalt tile (magn. 6×).

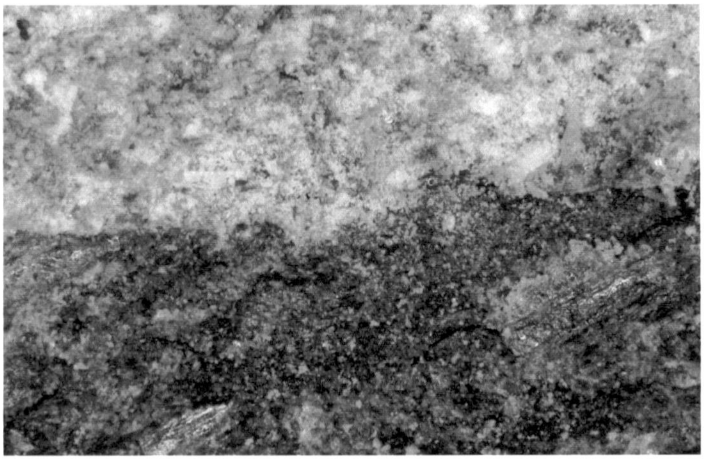

Figure 12. The connection detail of developed grout (30% TOS2) to cast basalt tile (magn. 64×).

4.4. Determination of tensile properties

By examining the results of tensile stress and elongation at break (Figure 13), it is apparent that improved stress exhibited the grout containing the filler prepared by wet granulation, but this difference is minimal. Tensile stress of grout fluctuated values around 18 MPa, and elongation at break of 0.50%. These differences may be caused in the same way as with the compressive strength distribution in the mass of the filler, the shape of the filler particles, and by the amount and distribution of pores in the samples – break was observed always at the point of the largest pores that weaken the cross-section. From the results, it is seen that reference material exhibited a higher tensile strength with a corresponding elongation at break, but since the strength of the difference is minimal, so the filler material based on solidification product appears to be a suitable replacement for primary materials.

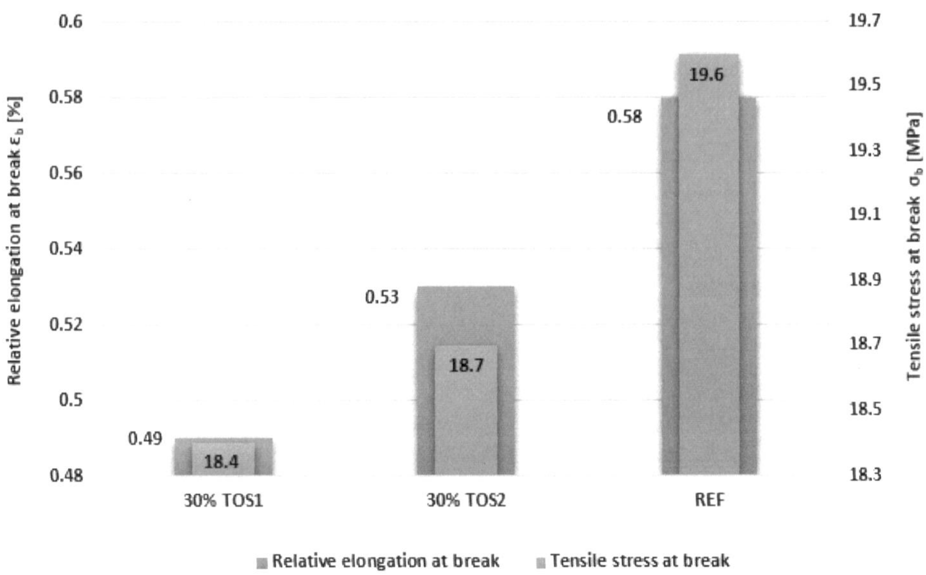

Figure 13. The results of tensile properties of tested epoxy grouts.

5. Conclusion

The research results, which dealt with the use of specially treated hazardous waste (HW), solidification product, as a filler in a polymeric grout, showed that this developed fine material seems very suitable as a filler. A total of two proposed solidification technologies (TOS1, TOS2), namely one wet granulation and one dry homogenisation. Due to the fact that the results of both prepared fillers were very similar, and the kind of solidification technology did not show a significant effect on the resulting physical and mechanical properties of the grout, it appears to be economically more advantageous to use dry homogenisation technology. Utilising solidification product containing HW and appropriate secondary materials in grout contribute to the reduction of landfill materials and reducing environmental burdens.

Acknowledgements

This work was financially supported by the junior specific research project No. FAST-J-19-5784 “Research of special polymer coating using hazardous waste and secondary raw materials“.

References

[1] Prolongo S G, Rosario G D and Ureňa A 2006 Comparative study on the adhesive properties of different epoxy resins *International Journal of Adhesion & Adhesives* (26) pp. 125–132.
[2] Meshram P, Sahu S, Ansari M Z, Mukherjee S 2018 Study on mechanical properties of epoxy and nylon/epoxy composite *Materialstoday: Proceedings* (5) pp. 5925–5932.
[3] Lee H and Neville K 1967 *Handbook of Epoxy Resins* (New York: McGraw-Hill).
[4] Hodul J, Drochytka R and Hodná J 2017 Experimental verification of utilization of fly ash from the glue gas denitrification process as a filler to epoxy patching mortar *Procedia engineering* (195) pp. 134–141.
[5] Bone B D, Bernard L H, Boardman D I, Carey P J, Hills C D, Jones H M, MacLeod C L and Tyrer M 2004 Review of scientific literature on the use of stabilisation/solidification for the treatment of contamined soil, solid waste and sludges. (Bristol: Environment Agency).
[6] Hodul J, Dohnálková B and Drochytka R 2015 Solidification of hazardous waste with the aim of material utilization of solidification products *Procedia Engineering* (108) pp. 639–646.
[7] Lopéz F A, Alguacil F J, Rodríguez O, Sierra M J and Millán R 2015 Mercury leaching from hazardous industrial wastes stabilized by sulfur polymer encapsulation *Waste Management* (35), pp. 301–306.
[8] Leistner CH, Hartmann S, Wittrock J, Bode K 2018 Shrinkage behavior of araldite epoxy resin using Archimedes´principle *Polymer testing* vol 67 pp. 409-416.

Study of temperature influence of cement and water on the fresh cement paste consistency

R Figmig[1] **and A Estokova**[1]

[1] Technical University of Kosice, Faculty of Civil Engineering, Institute of Environmental Engineering, Kosice, Slovakia

E-mail: robert.figmig@tuke.sk

Abstract. The consistency of a fresh cement paste, which is formed after mixing Portland cement with water, depends significantly on the water to cement ratio (w/c), the cement fineness and its composition, but also on the other parameters as temperature of cement and water. The paper is focused on researching the influence of increased temperature of cement and water on the fresh cement paste consistency. Various cement pastes with different w/c ratio (0.28, 0.32 and 0.36) and different temperatures of water and cement (20±2°C, 40°C, 60°C) were investigated. The consistency of cement paste was measured by Vicat roller penetration method. The tendency of losing the consistency of the fresh cement paste with increasing of its temperature was observed and exhibited as linear correlation with the coefficient of determination $R^2(0.28) = 0.9695$ and $R^2(0.32) = 0.9147$ respectively.

1. Introduction

The properties of the cement paste are crucial for the overall properties of concrete in fresh as well as in hardened state [1]. The consistency is one of the rheological characteristics of non-Newton liquids, what the cement paste is [2]. The consistency of a fresh cement paste, which is formed after mixing Portland cement with sufficient amount of water, depends significantly on the water to cement ratio (w/c), the cement fineness and its composition, the temperature and the plasticizer type [3-8]. First, the cement particles are uniformly dispersed in the liquid phase and flocculation of the cement particles occurs within minutes of mixing, which is associated with an increase in paste viscosity. There is a formation of aggregates of cement particles which trap part of the mixing water so that they are unable to participate in the paste flow. At this stage, weak van der Waals forces between molecules and particles are applied. The paste viscosity also increases due to the cement hydration process, leading to a gradual increase in the existing solid/liquid ratio and roughening of the cement particle surface due to the formation of hydrates. At this stage, the amount of hydrated material rapidly increases and the liquid phase volume decreases. Between the individual cement particles on which the hydrates have already been occurred are formed chemical bonds that are stronger than van der Waals and electrostatic forces that are responsible for flocculation. As chemical bonds increase, a continuous three-dimensional solids network develops in the paste, coupled with loss of paste deformability. As the hydration progresses, the amount of hydrated material increases at the expense of the porous space filled with water, leading to a gradual increase in the strength of the cured paste. [9-11]

The consistency of cement composites characterizes its workability and therefore the need of labour during the casting. For cement, there is the technical standard [12] which provides methodology

IOP Conf. Series: Materials Science and Engineering **549** (2019) 012031 doi:10.1088/1757-899X/549/1/012031

for the cement paste testing. The parameter „normal consistency of cement paste“ is defined as water need to reach the specific depth of the Vicat roller penetration (the distance between plate and roller end must be 6 ± 2 mm) and is input information for testing the initial and final setting time of the cement. In noted technical standard, the testing conditions are specified for temperature for cement (20 ± 2 °C) as well as for water (20 ± 2 °C). These are often in a conflict with real conditioning on delivery of the cement, when the temperature of cement often reaches temperature in the range of 20 to 60 °C, rarely up to 80°C. Also, temperature of water in concrete production can reach temperature up to 60°C, mainly in winter for RMC or in precast industry, when there is the need of achieving minimum concrete temperature. Higher temperature of cement paste could also affect initial and final setting time. Considering that it is not usual for concrete producer to measure current cement temperature, there is not water temperature correction providing, but consistency of fresh concrete is mostly adjusted by adding the mixing water, what leads to decreasing the mechanical properties as well as the durability [13-16].

The objective of the paper was to investigate the impact of cement and water temperatures, used in cement pastes preparation, on the fresh cement paste consistency.

2. Materials and methods

Totally 16 cement pastes were prepared for the experiment, using cement CEM II/A-S 42.5 R from locality Turňa and tap water. The pastes preparation was designed as a combination of three different w/c ratios (0.28, 0.32 and 0.36) and three temperatures of water (T_w) – 21, 40, and 60 °C and three temperatures of cement (T_c) – 22.5, 40, and 60°C (Table 1). The cement as well as water were heated prior the experiment in electric oven. The initial water/cement ratios were chosen as 28%, then 32% and finally 38% for control for T_c = 40 and 60 °C and T_w = 60°C.

Table 1. Labelling of the samples (CPw/c - T_w - T_c).

w/c	T_w (°C)	T_c [°C]		
		22	40	60
0.28	21	CP28 - 21 - 22		
	40	CP28 - 40 - 22		
	60	CP28 - 60 - 22		
	21		CP28 - 21 - 40	
	40		CP28 - 40 - 40	
	60		-	-
	21			-
	40			-
0.32	21	CP32 - 21 - 22		
	40	CP32 - 40 - 22		
	60	CP32 - 60 - 22		
	21		CP32 - 21 - 40	
	40		CP32 - 40 - 40	
	60		CP32 - 60 - 40	
	21			CP32 - 21 - 60
	40			CP32 - 40 - 60
	60			CP32 - 60 - 60

Table 1. Completion

0.36	**21**	-	-
	40	-	-
	60	CP36 - 60 - 40	CP36 - 60 - 60

For every batch, 500g of cement and calculated amount of water according to the selected w/c ratio were weighed and mixed according to the technical standard (90s on low speed of blade - 140±5 rpm in rotation and 62±5 rpm in planetary movement, 30s pause and another 90s on low speed). Immediately after mixing, temperature of fresh cement paste (T_{cp}) was recorded and consistency (d) on Vicat apparatus was determined in time of 30s. Considering that already at T_c = 40°C and T_w = 40°C was reached stiffens at w/c = 28% (could be supposed that the Vicat roller did not penetrate the paste), testing at this w/c ratio for T_c = 60°C was not performed.

3. Results

The results of final cement paste temperature (T_{cp}) and of the height between base plate and the bottom of the Vicat roller (d) are given in Table 2.

Table 2. Measured parameters of cement pastes.

Sample	T_{cp} (°C)	d (mm)
CP28 - 21 - 22	25.7	11.0
CP28 - 40 - 22	32.6	24.5
CP28 - 60 - 22	34.4	32.5
CP28 - 21 - 40	31.4	25.0
CP28 - 40 - 40	35.5	32.5
CP32 - 21 - 22	24.8	1.0
CP32 - 40 - 22	33.0	9.0
CP32 - 60 - 22	35.4	11.0
CP32 - 21 - 40	29.8	1.0
CP32 - 40 - 40	35.2	12.5
CP32 - 60 - 40	40.4	15.0
CP32 - 21 - 60	33.4	9.0
CP32 - 40 - 60	37.3	12.5
CP32 - 60 - 60	41.9	18.0
CP36 - 60 - 40	40.2	0.0
CP36 - 60 - 60	44.0	3.0

The lowest d for w/c = 0.28 (28%) equals 11mm and was obtained for the lowest temperature of cement paste (T_{cp} = 25.7°C), when testing was performed at laboratory conditions. The highest d for the same w/c ratio equals 32.5mm and was obtained for 2 cases, when T_w = 60°C and T_c = lab (ca. 22,5°C) while T_{cp} = 34.4°C, respectively T_w = 40°C and T_c = 40°C while T_{cp} = 35.5°C.

Using higher w/c ratio, the lowest d for w/c = 0.32 (32%) equals 1mm and was obtained for the lowest temperature of cement paste (T_{cp} = 24.8°C), when laboratory conditions were retained. The d equals 1mm was also obtained for T_{cp} = 29.8°C, while T_w = lab. and T_c = 40°C. Considering other results within the table, at this w/c, the d started to raise at higher temperature of cement paste or that for the first two cases there was too much water in cement paste and temperature did not have significant impact. The highest d for the same w/c ratio equals 18.0mm and was obtained for T_w = 60°C and T_c = 60°C while T_{cp} = 41.9°C.

At the w/c = 0.36 (36%), for T_w = 60°C and T_c = 40°C was T_{cp} = 40.2 and d = 0 mm, what means, that there was excessive amount of water in cement paste. For T_w = 60°C and T_c = 40°C was T_{cp} = 44.0°C and d = 3.0 mm, what is decrease of 15.0mm for the same material (water, cement) temperature conditions with w/c raising of 0.04 (4%).

The dependences of d on T_{cp} for w/c = 0.28 and 0.32 are rendered in Figures 1 and 2.

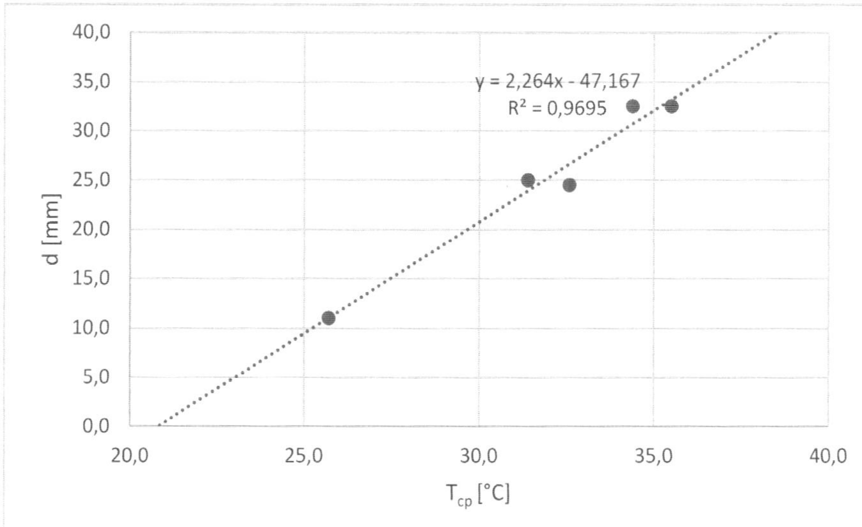

Figure 1. The correlation between d parameters and the temperature of the cement pastes of w/c = 0.28.

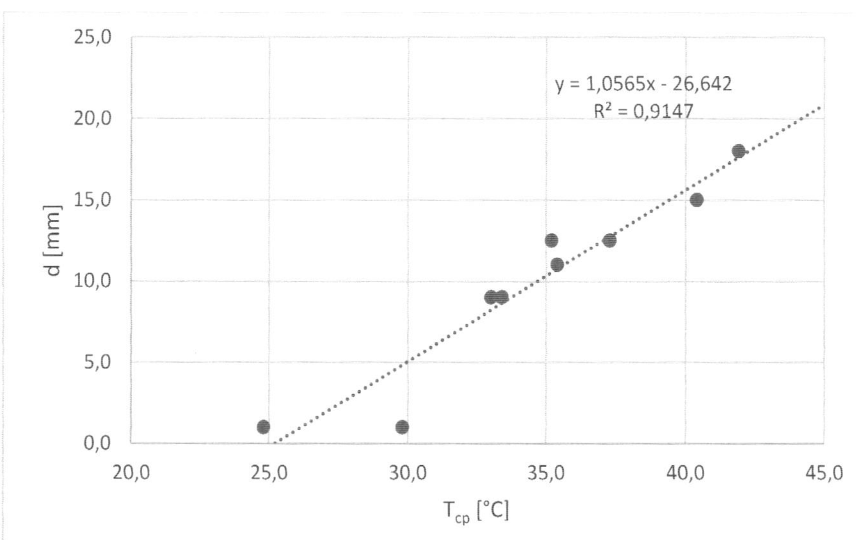

Figure 2. The correlation between d parameters and the temperature of the cement pastes of w/c = 0.32.

Based on results, it could be seen the obvious tendency to lose consistency of cement paste with increasing of its temperature. According to figures 1 and 2, it could be concluded, that the tendency for this experiment with given cement has linear character with the strong positive correlation with the coefficient of determination (R^2) 0.9695 for w/c = 0.28 and 0.9147 for w/c = 0.32. There is also the tendency obtaining the similar consistency at the same temperature of cement paste within the same w/c ratio.

4. Conclusion

The experimental study was focused on temperature influence to cement paste consistency. Based on the results, it can be deduced, that consistency of cement paste (CP) is direct depended on its temperature, when increasing the temperature of CP caused increasing its resistance to Vicat roller penetration. Then, adjusting the consistency is often performed by adding plasticizer or water. In case of water, it could be connected with the decreasing of mechanical properties as well as durability of cement composite. Adding plasticizer could lead to deceleration of setting, decrease of early compressive strength as well as hardness of CP / modulus of elasticity of early age concrete. It can be also assumed, that temperature has also impact on another rheology properties as well as other parameters of fresh CP. Considering the results of water need to achieve normal consistency of CP using standard procedure conditions and the real conditions on the plant, the standard method seems to be inadequate for concrete producers.

Acknowledgment
This research was funded by VEGA - Scientific Grant Agency of the Ministry of Education, Science, Research and Sport of the Slovak Republic and the Slovak Academy of Sciences, grant number and 1/0648/17.

References

[1] Mehta P K and Monteiro P J M 2006 *Concrete: Microstructure, Properties and Materials* (New York: McGraw-Hill)
[2] Roussel N 2012 *Understanding the rheology of concrete* (Sawston: Woodhead Publishing) pp 3-22
[3] Kasinska M, Kempinksi J and Swierzko R 2016 *Acta Sci. Pol. Form. Circ.* **15** 71-81
[4] Rubio-Hernandes F J 2018 *Fluids* **106** 1-16
[5] Vicar H 2009 Influence of temeprature, cement and plasticizer type on the rheology of paste *Proc.of the Sec. Intern. Symp. on Design, Perf. and Use of Self-Cons. Conc.* pp 1-10
[6] John E and Gettu R 2014 *ACI Mat. Journal* **111** pp 67-76
[7] Choi M, Park K and Oh T 2016 *Int. J. Concr. Struct. Mater.* **10** pp 65-74
[8] Nehdi M and Martini S A 2009 *Cem. Concr. Res.* **39** pp 1007-1016
[9] Taylor H F W 1990 *Cement chemistry* (London: Academic Press)
[10] Hewlett P 2004 *Lea's Chemistry of cement and Concrete* (Oxford: ButterWorth Heinemann)
[11] Kurdowski W 2014 *Cement and Concrete Chemistry* (Dordrecht: Springer)
[12] EN 196-3: 2017 *Methods of testing cement – Part 3: Determination of setting times and soundness*
[13] Dyer T 2014 *Concrete Durability* (Boca Raton: CRC Press) pp 246-249
[14] Wasserman R, Katz A and Bentur A 2009 *Mater. Struc.* **42** pp 973-982
[15] Kim Y Y et al **2014** *Adv. in Mater. Scien. and Eng.* pp 1-10
[16] Mullick A K 2018 Ind. Concr. J. **92** pp 12-21

The issue of plastic waste in the environment and its possible uses as a substitute for filler in lightweight concrete

M Ledererová[1], Z Štefunková[1], V Gregorová[1], D.Urbán[1]

[1] Slovak University of Technology, Faculty of Civil Engineering, Radlinského 11, 810 05 Bratislava, Department of Material Engineering, Slovakia

E-mail: miriam.ledererova@stuba.sk

Abstract. The specific nature of plastics causes specific problems in waste management. Plastics are relatively inexpensive and versatile with the potential for versatile use in industry, which has led to their rapid development over the past century, which still persists. The aim is to look for the possibility of reusing plastic waste as one of the lightweight filler fillers. It used plastic recycled material recycled from EVA plastic packaging and polystyrene recycled material in various proportions. All are among the medium-recyclable plastics according to their recycling rate. The experimental part of the work is devoted to the verification of the effect of the filler dose on the monitored physical and sound absorption parameters.

1. Introduction

Plastics are materials whose essence consists of macromolecular organic compounds, which are produced by organic synthetic reaction or made from natural materials as oil, coal and natural gase. Dating of the plastics industry starts as early as 1907 and can now be found in almost every industry, with worldwide annual production of around 245 million tonnes. The main raw material of plastics is oil, which is a non-renewable resource.

The basic characteristics of plastics include durability, low volume density as well as low production costs, which can be characterized as excellent material for use and application in everyday life as well as in construction practice. However, the greatest negative result of this material is its extremely problematic waste. In addition to the above mentioned statistical annual production of plastics, it is also necessary to mention the amount of 25 million tonnes of annually produced plastic waste in the EU. Of these, 38% are landfilled, 36% are recovered and only 26% are recycled [1,2].

In the environment it can last for hundreds of years, annually 10 million tons of plastic waste ends up in the seas and oceans. Despite the technological difficulty of separation and recycling, it should be our duty to make greater use of this waste, to explore the possibilities for reuse, but above all to perceive it as a feedstock. Plastic waste is generally a major environmental problem, using up to 94% of raw materials, oil and energy can contribute to the use of secondary raw materials.

One way to use plastic recycled material is to incorporate it into lightweight concrete or directly produce lightweight concrete based on polyurethane binders. By choosing suitable filler combinations (expanded polystyrene, expanded polypropylene, recycled cable, recycled electrical cable, etc.) and binder, it is possible to use such building material not only for new construction, but also for the reconstruction and rehabilitation of existing buildings [1,2,3].

2. Materials and methods

For experimental testing in research of the recycled waste polystyrene (PS) (C1, L1) and ethylenevinyl acetate (EVA) (C3, L3), was used as shown in figure1. Materials and components of mixtures were mixed in following proportions: fillers 100% EVA (C1), binder Conipur 360, fillers 100% EVA (L1), binder Conipur 360, fillers 50 % PP + 50% EVA (C2) binder Cement 360, fillers 50 % PP + 50% EVA (L2) binder Conipur 360, fillers 100 % PS (C3) binder Cement, 100 % PS (L3) binder Conipur 360.

a) b)

Figure 1. Waste plastic Group Size:

a) ethylene-vinyl acetate 4/8 mm b) polystyrene - 4/8 mm

Ethylene-vinyl acetate (EVA) is the elastomeric polymer with high elasticity [3]. The recycled material of the polystyrene (PS) has good thermal insulation properties, low volume density and good workability [4].

The PS and EVA were crushed and the fraction 4/8 mm was screened wherein the bulk density of PS 14,27 kg/m^3, EVA 112,45 kg/m^3. LC were produced with the use of a filler, or as a combination of two materials according to the percentage share: 100%, 50:50 % (table 1 and table 2).

As a binder was used cement (table 1). The binder dose in all mixtures was also a constant of 0.7 kg per one mixture. The water coefficient was constant at 0.5 for all mixtures.

As a binder was used undiluted one-component polyurethane adhesive Conipur 360 (table 2). This adhesive is used in the implementation of the polyurethane surfaces. It has a firming effect; it is resistant to abrasion and very easily workable. Dose of glue in all mixtures was also a constant of 0.4 l per one mixture. Considering different surface and material – water content was moistened to 2%, which regulates the surface tension.

Table 1. Composition of the individual mixtures with binder cement.

Test samples	Binder (kg) Cement	Graining (l) EVA	PS	(l) Water
C1	0.7	4	0	0.35
C2	0.7	2	2	0.35
C3	0.7	0	4	0.35

Table 2. Composition of the individual mixtures with binder polyurethane adhesive.

Test samples	Binder (l) Conipur 360	EVA	Graining (l) PS
L1	0.3	4	0
L2	0.3	2	2
L3	0.3	0	4

Samples in the shape of 100×100×100 mm were made from homogeneous mixtures. These samples were treated 48 hours in the laboratory environment. Subsequently, they were removed from the mold

IOP Conf. Series: Materials Science and Engineering **549** (2019) 012032 doi:10.1088/1757-899X/549/1/012032

and stored in laboratory environment for adhesive (20 ° C, humidity = 50%) and cement (20 ° C, humidity = 90%) for the binder. Subsequently, the physical properties (bulk density, heat –technical parameter and thermal conductivity) were measured. Samples for sound absorption tests were 100mm and 35mm in diameter, 60mm high. The sound absorption was tested on the EVA sample, the recycled polystyrene and a 50:50% EVA/PS, Conipur 360 binder combination. Measurements of bulk density and heat as technical parameters were carried out in the laboratory conditions of the STU.

Figure 2. Samples with binder cement.

Figure 3. Samples with binder polyurethane glue.

a) EVA diameter 10/3.5 cm b) PS diameter 10/3.5 cm

Figure 4. The reference sample (cylinders with diameter 10 cm /3.5 cm and length 6.5 cm).

3. Results

Physical and thermophysical properties are shown in (Figure 5,6,7). While the bulk density of the polyurethane glue samples (L1-L3) ranged from 220 to 120 kg.m^{-3}, the bulk density of the samples (C1-C3) increased, by changing the binder used, up to 430 kg.m^{-3} (Figure 5). By changing the binder and filler (L1 - L3), the thermal conductivity coefficient of the test samples decreased (Figure 6) and conversely, when using a cement binder and changing the filler, the coefficient of thermal conductivity of the test samples increased (C1-C3) (Figure 7).

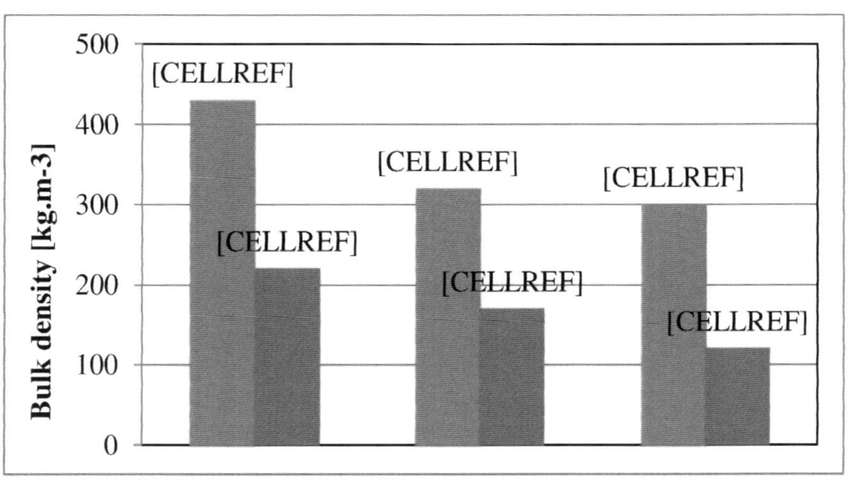

Figure 5. Comparison of the bulk density of mixtures.

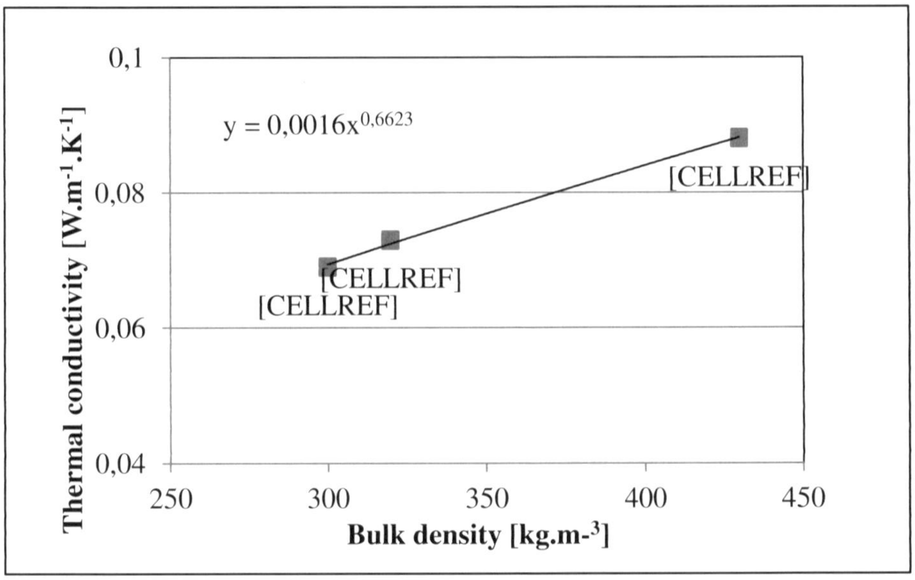

Figure 6. Relationship between thermal conductivity and bulk density of mixtures – cement binder.

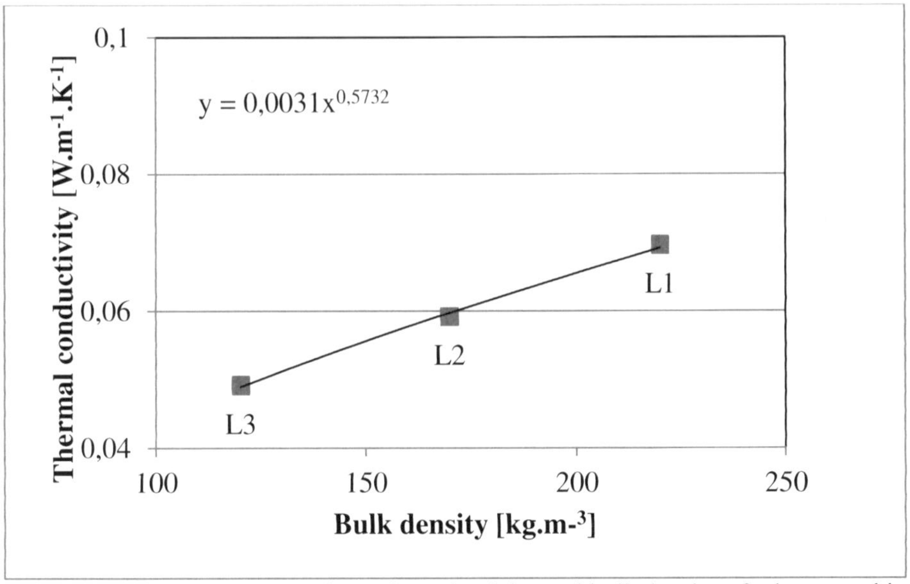

Figure 7. Relationship between thermal conductivity and bulk density of mixtures – binder polyurethane glue.

4. Sound absorption

Knowledge of the sound absorption of building materials, opens up the possibility of their applications also in terms of dampening the sound as in the interior (generally in diffusion audio field), as well as in the outdoors (free sound field) [5,6]. The experiment was aimed at measuring of sound absorption determination. For this purpose, modified measurement set of impedance tube was prepared, process for preparation of this measurement EVA and PS samples were prepared. There have been large differences in the results of two different measurements methods.

For the verification of sound absorption were used samples on the base of ethyl-vinyl acetate (EVA) and polystyrene (PS), in combination with the binder Conipur 360 (polyurethane glue). The aim was to set up a measurement assembly, to perform a reference measurement of a stable measurement protocol. The samples were EVA and PS. They were 6.5 cm thick. Diameter of cylinder of the sample was 3.5 cm and 10.0 cm. The measurement and the used measuring assembly operates on the principle described in the standard ISO 10534-1 , ISO 10534-2 [7,8].

a)

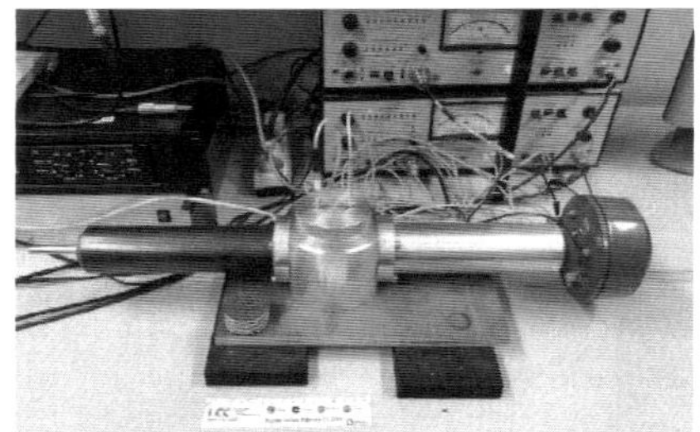

b)

Figure 8. The measurement of the reference sample EVA
a) impedance tube STUBA b) impedance tube KU Lueven.

5. Description of sound absorption measurement

As was mentioned above for the comparative test was used measurement technique explained in standard ISO 10534 [9]. The big advantage of the impedance tube measurement techniques (also so called the fast measurement techniques) in general is the fact, for sound absorption properties determination just relativity small size measurement sample is needed (the size of specimen can be smaller than wavelength of the sound waves in frequency spectrum of interest). Therefore, impedance tube measurement methods are often used for absorption material development. However, the objective parameter, sound absorption coefficient, is determined just for perpendicular incidence of acoustic plane wave.

The measurement principle by means of impedance tube and standing wave method based on sound pressure level measurement inside of the tube is based on seeking for nodes, or superposition, of standing wave in tube (maximal, and minimal value of pressure in impedance tube, also so – called the Kundt´s tube). The standing waves are generated by acoustic plane wave source (loudspeaker) mounted on the one end of the tube, on the other side of the tube is mounted the specimen under investigation [10]. This measurement technique is used in practice since 40ties of 20th century. Measurement performed can be divided to two laboratories, the comparative test between 2 labs STUBA (STU CI Bratislava, Slovakia), KU Lueven (The Katholieke Universiteit Leuven, Belgium). Measurement were performed in frequency range from 100- 1800Hz. The excitation signal was time stationary sine signal tuned in of 1/3rd frequency generated by external sine generator.

The results achieved by interlaboratory tests of sound absorption measurement were compared. Three EVA and PS based composite materials were distributed between two laboratories. In both cases, the excitation measurement signal was sine signal generated by sine tone generator, the tone was gradually tuned 1/3rd octave step. The average values of sound absorption coefficient spectra in frequency range from 100 to 1800Hz are compared in Figure 9. one can see the similar trend of absorption spectra shape. Measurement from both laboratories verified the significant sound absorption in spectrum from 400 to 700Hz. However, there is deviation in results in low and high frequency range. The deviation is usually caused by background noise, tightness of specimen mounting in the device but also by used filter. It is not unusual to have deviation in low frequency spectra. Difference in interlaboratory results comparison was up to $\Delta\alpha = 0.2$.

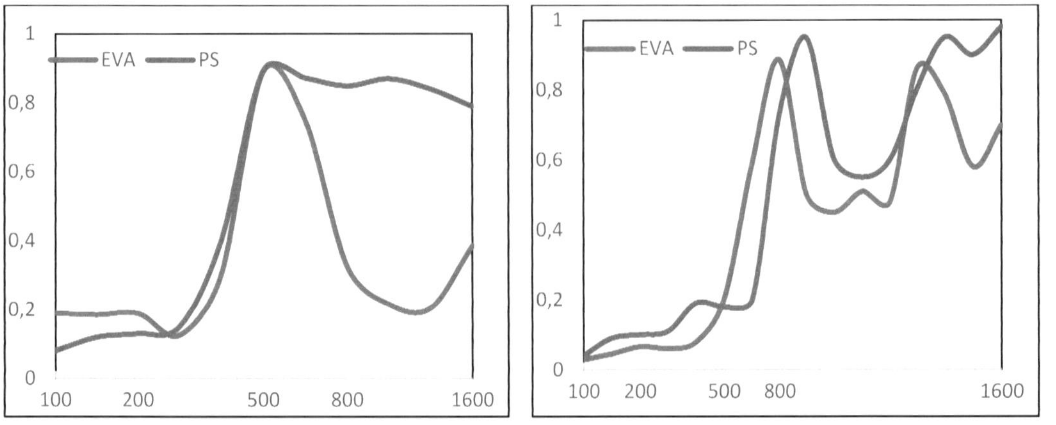

a) diameter 10.0 cm b) diameter 3.5.cm

Figure 9. The interlaboratory results comparison of sound absorption (average results for EVA/PS cylinders with length 6.5.cm in rage 100Hz to 1600Hz).

6. Conclusion

Based on the results obtained, we can conclude that samples of lightweight plastic-based concrete have favourable properties with different fillers. The best heat transfer properties were achieved with samples equipped with 100% polystyrene (P3, L3), which also had the lowest bulk density depending on the type of binder. Its density was increased by the addition of EVA, which also increased the conductivity of lightweight concrete. Most of the differences in sound absorption were in the combination of EVA and polystyrene and 100% EVA.

From the results comparison of sound absorption is clear, that the differences occurred between laboratories measurements. Generally, most sensitive are low frequencies. There was also shown, if the same measurement device and specimen is used, and the same data are processed in the different ways, the results can differ. The measurement presented here were performed on samples with same of size diameter 10 cm and 3.5 cm, but their material properties were not comparable.

For unambiguous conclusions, it would be necessary to perform a larger number of measurements on the samples in order to say more clearly which method is more accurate and comparable with the measurement in the reverberation chamber [10].

The right choice of recycled plastic waste, in combination with particular binder, can have significant influence in decreasing the bulk density as well as improving thermal conductivity. Results of measurements of sound absorption suggest that this material mixtures can be used in floor insulation, noise walls (barriers) and other components where sound absorption is crucially important.

Acknowledgements
This article was created with the support of the Ministry of Education, Science, Research and Sport of the Slovak Republic within the Research and Development Operational Program for the project "University Science Park of STU Bratislava", ITMS 26240220084, co-funded by the European Regional Development Fund and by the Slovak Research and Development Agency under the contractNo. APVV-15-0681.

References

[1] Kuraš M, Dimera V, Slivka V and Březina M 2008 *Odpadové hospodářství Vodní zdroje spol.s.r.o (Chrudim)* p 143
[2] Gregorova V 2015 *Proc. Int. Conf. on for PH.D Students and Yong Researchers (Vilnius)* pp 2484–2489
[3] Chandra S and Berntsson L 2002 *Lightweight aggregate concrete Science Technology and Applications Noyes Publications* p 450

[4] Sicakova A and Sad J 2009 Build. Mater. **5** 36-40
[5] Beranek L L, Tichý J, Břeský M, Kaplan J and Wanke J 1965 *Snižování hluku (Praha)*
[6] Hopkins C 2012 *Sound insulation. Routledge (London)* p 648
[7] ISO 10534-1:1996 *Determination of sound absorption coefficient and impedance in impedance tubes - Part 1: Method using standing wave ratio*
[8] ISO 10534-2:1998 Acoustics - *Determination of sound absorption coefficient and impedance in impedance tubes - Part 2: Transfer-function method*
[9] ISO 354:2003 Acoustics – *Determination of sound absorption in a reverberation room*
[10] Vigran T E 2008 *Building Acoustics CRC Press* p 364

IOP Conf. Series: Materials Science and Engineering **549** (2019) 012033 doi:10.1088/1757-899X/549/1/012033

Metakaolin/carbon black geopolymer with enhanced electrical properties

C Mizerová[1], I Kusák[1], P Rovnaník[1], P Bayer[1]

[1] Brno University of Technology, Faculty of Civil Engineering, Veveří 95, 602 00 Brno, Czech Republic

E-mail: mizerova.c@fce.vutbr.cz

Abstract. Alkali-activated binders represent a promising material for composites with self-sensing properties both for their lower environmental impact and ability to transfer the electric charge through ionic mobility. Metakaolin geopolymers with carbon black conductive admixture up to 4% were produced to test their electrical properties, mechanical properties and microstructure. Electrical capacitance, resistance and relative permittivity were determined via impedance spectroscopy analysis. Notable enhancement of all mentioned electrical parameters was registered from 2% carbon black content, especially at lower frequency range. Another carbon black addition was rather unreasonable due to only slight improvement in electrical properties but significant drop of compressive strength and unfavourable changes in pore volume and binder microstructure determined by mercury intrusion porosimetry and electron microscope.

1. Introduction

Intrinsic self-sensing concrete is a novel type of construction composite capable to serve both as the structural material and a sensor monitoring its own stress, strain or damage occurrence. The self-sensing property is related to changes in electrical conductivity and other parameters that are affected by mechanical loading, temperature and other external stimuli. Concrete incorporated with functional filler, for instance carbon fibres, carbon nanotubes, carbon black, graphite powder or steel fibres, forming the conductive network within the matrix, thus does not require additional monitoring devices that are in most cases expensive, less effective and limited in durability [1]. Apart from self-sensing, conductive composites can be applied in various advanced engineering solutions including modern traffic detection systems, self-heating concrete for deicing and heating or cathodic protection of steel [2]. Especially carbonous materials improve the concrete ability of electromagnetic shielding [3].

Alkali-activated binders drew considerable attention in present efforts to search for an alternative binder to traditional Portland cement production being one of the major sources of CO_2 and other pollutants released into the atmosphere [4]. Emissions related to their production can be similar or a way lower comparable to cement, maximal use of local and waste materials along with the lowest amount of alkaline activator needed represent an applicable approach in their wider adoption in practice [5]. The technology is based on the reaction of highly reactive aluminosilicate precursors, frequently slag, fly ash, metakaolin and other products or natural materials, with alkaline activator solution. A strongly alkaline environment causes dissolution of the source materials and formation of new binding phase [6]. Since the use of precursors containing almost exclusively aluminosilicates enables to form highly coordinated structure of the final product, the term "geopolymer" was introduced [7]. Good mechanical properties of alkali-activated/geopolymer cements and concretes have been proved in lots of structural applications around the world [8].

Geopolymers are usually manufactured from low-calcium source material, i.e. fly ash or metakaolin [9]. Metakaolin is produced by calcination of kaolinite clays at 500–900 °C. High reactivity of metakaolin particles in alkaline solution is achieved by optimal calcination temperature providing desired chemical and mineralogical composition. As long as costs of metakaolin production are higher in comparison to industrial by-products commonly used in alkali-activated binder technology, use of metakaolin geopolymers is predominantly studied in niche applications. Excellent fire resistance and durability in aggressive chemical environment [10] predetermine the use of metakaolin geopolymer in refractory applications or protective structures.

Carbon black, an amorphous form of elementary carbon, is produced by the incomplete combustion of hydrocarbon materials and used as a low-cost rubber reinforcing filler or black pigment. Colloidal carbon black particles exhibit high conductivity with chemical and thermal stability, therefore they can be successfully incorporated in conductive composites [11]. Mingqing et al. [12] used carbon black to manufacture cement mortar with enhanced conductivity that was laid as an indoor electrical heating slab. Lamuta et al. [13] studied metakaolin geopolymers and described the direct piezoelectric effect in their structure related to the migration of hydrated Na^+ cations causing a charge imbalance and generation of local dipoles. Increased availability of mobile ions supplied by the alkaline activator thus causes lower electrical resistance of alkali-activated binders in comparison with cement-based materials [14]. Concerning the application of conductive admixtures in metakaolin geopolymers, MacKenzie et al. [15] states that the electrical conductivity of potassium-activated composites containing the same amounts of graphite was generally lower than those of the SWCNT containing samples.

2. Materials and methods

The main source materials for geopolymer matrix synthesis was metakaolin composed of almost exclusively aluminosilicates (chemical composition is given in Table 1) and water glass solution, commercially sold sodium silicate with molar ratio SiO_2/Na_2O = 1.6 (43 wt.% of dry mass). Conductive admixture - carbon black Vulcan 7H was added in dosage up to 4.0% of solid aluminosilicate precursor along with dispersing agent, 2% aqueous solution of non-ionic surfactant Triton X-100. Mortars were produced using standardized quartz sand as aggregate.

All mixtures were mixed according to proportions listed in Table 2 with water content adjusted to the same flow diameter on the flow table (EN 1015-3: 1999 [16]). Mixing procedure included following steps: first mixing the water glass with carbon black, dispersing agent and initial amount of water in laboratory mixer. After the homogenization, the quartz sand and remaining water were added. Right after mixing, geopolymer mortars were cast into prismatic 40 × 40 × 160 mm moulds, covered with a plastic sealant to protect the binder from moisture loss and left to set in laboratory conditions. After demoulding, the specimens were stored in laboratory till the age of impedance spectroscopy testing (40 days). Before the impedance spectroscopy, samples were dried (5 days at 105 °C) to reduce the influence of moisture on assessed electrical properties.

In order to perform the impedance spectroscopy, specimens were placed between parallel brass electrodes 30 × 100 mm (so that a distance between electrodes was 40 mm) and subjected to analysis in the range of 40 Hz to 1 MHz using an Agilent 33220A sinusoidal signal generator and an Agilent 54645A dual-channel oscilloscope. The output voltage of the signal generator was 5.5 V and the input values of the oscilloscope 13 pF and 1 MΩ for the electrical capacity and the resistance, respectively. These measurements were fully automated. At higher range of frequencies, 100 MHz up to 3 GHz, R&S ZNC vector analyzer with DAK 12 coaxial probe (Speag) has been used to measure the electrical conductivity and relative permittivity as a function of frequency. Seven days after the beginning of the electrical measurements, the specimens were tested for their mechanical properties (compressive and flexural strength) and porosity by mercury intrusion porosimetry (Micromeritics Poresizer 9310). SEM images were taken using electron microscope Tescan MIRA3 XMU.

Table 1. Chemical composition of metakaolin.

	SiO_2	Al_2O_3	Fe_2O_3	CaO	MgO	S_{total}	Na_2O	K_2O
(%)	55.01	40.94	0.55	0.55	0.14	0.34	0.09	0.06

Table 2. Mix composition of metakaolin geopolymer with carbon black.

	REF	CB-0.5	CB-1	CB-2	CB-3	CB-4
Metakaolin (g)	350	350	350	350	350	350
Sodium silicate (g)	350	350	350	350	350	350
Carbon black (g)	-	1.75	3.5	7	10.5	14
Aggregate (g)	1050	1050	1050	1050	1050	1050
2% Triton X-100 (ml)	-	2	3.5	7	10.5	14
Water (g)	120	120	125	130	137	148

3. Results and discussion

3.1. Electrical properties

Electrical resistance depending on the carbon black concentration is depicted in Figure 1. At lower frequency range, the resistance decreased with higher amount of carbon black and with the frequency as well. The logarithmic scale was applied to distinguish a point of frequency around 50 kHz wherefrom the resistance of the reference material drops and remains the lowest among all samples while the specimens with carbon black exhibited similar behavior and minor differences in electrical resistance.

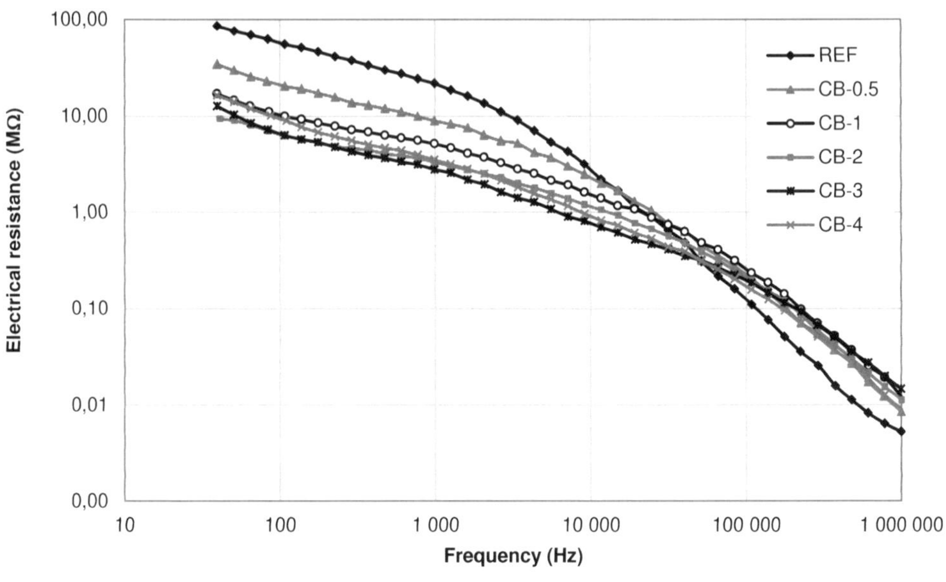

Figure 1. Electrical resistance of metakaolin geopolymer mortars with different carbon black content.

Electrical capacitance curves of samples with carbon black (Figure 2) showed an increment of capacitance, especially at low frequency. There was a steep decline of capacitance in the frequency spectrum up to 1 kHz. Samples CB-2 to CB-4 are close to each other and intersect several times across the spectrum. The reference frequency of 1 kHz was chosen to illustrate the changes of electrical properties, both resistance and capacitance values are shown in Table 3. Enhanced electrical capacitance and reduction of resistance with higher carbon black content is clear, with similar values of CB-2, CB-3 and CB-4.

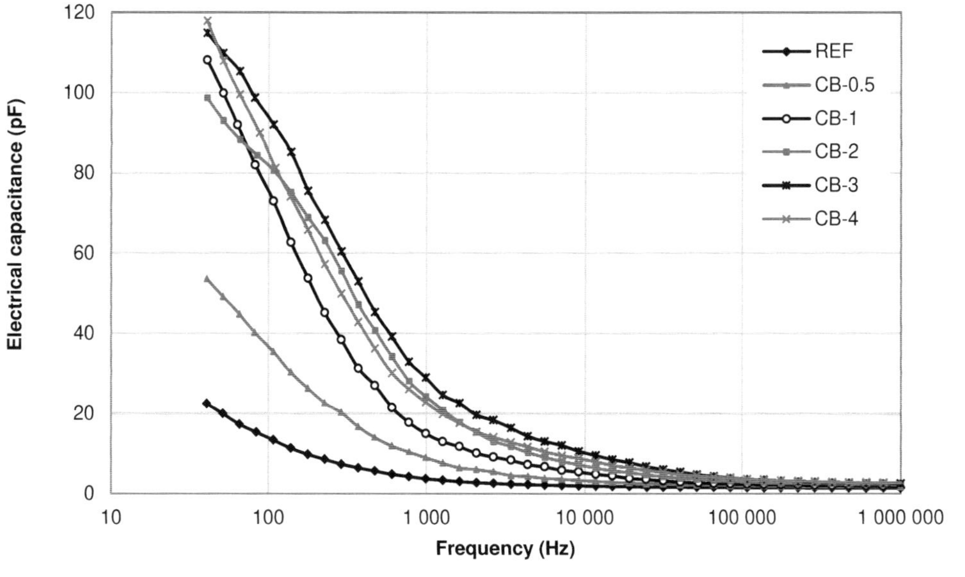

Figure 2. Electrical capacitance of metakaolin geopolymer mortars with different carbon black content.

Table 3. Comparison of electrical resistance and capacitance at reference frequency 1 kHz.

	REF	CB-0.5	CB-1	CB-2	CB-3	CB-4
Electrical capacitance (pF)	3.71	8.97	14.95	24.28	28.99	22.83
Electrical resistance (MW)	21.61	8.87	5.13	3.30	2.75	3.51

Relative permittivity of the reference material without carbon black remains almost constant across the frequency spectrum, as can be seen from Figure 3. Relative permittivity of carbon black-doped material grew, with a slight decrease with frequency. This drop is more apparent with higher carbon black content and at the outset of the frequency applied. At 500 MHz, the relative permittivity increased from 2.8 to 5 in case of maximum carbon black concentration.

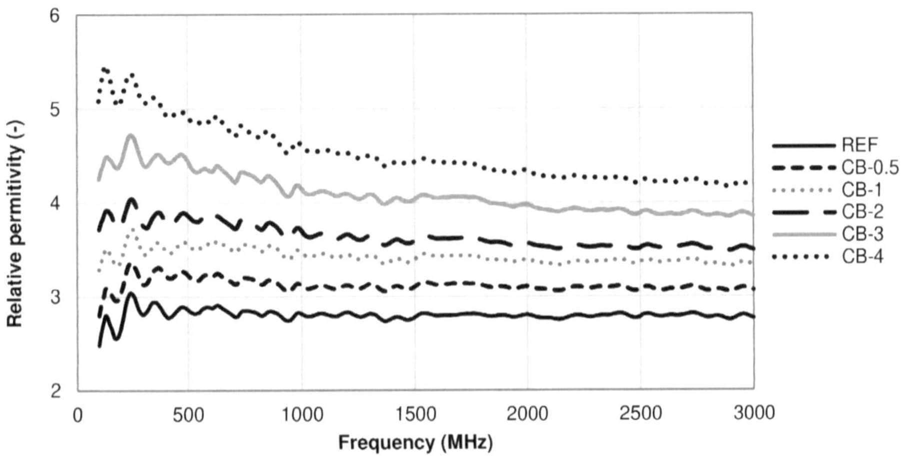

Figure 3. Relative permittivity of metakaolin geopolymer mortars with different carbon black content.

3.2. Mechanical performance

Compressive and flexural strength values of 47 days cured geopolymers are shown in Figure 4. Carbon black content of 0.5% was yet too low to worsen the mechanical properties but higher concentration from 1% caused a drop of strength, especially in compression. Sample CB-4 reached only 11.7 MPa, less than a half of the reference strength (27.5 MPa). Flexural strength was decreasing as well but it was less sensitive to the carbon black content.

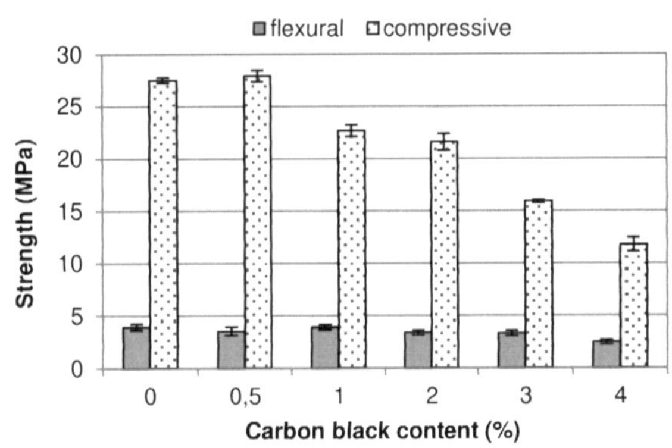

Figure 4. Mechanical performance of metakaolin geopolymer mortars with different carbon black content.

3.3. Microstructure

Comparison of the cumulative intruded volume in Figure 5 shows that the presence of carbon black modified the pore structure of geopolymer matrix. With increasing carbon black content, the total pore volume rose from 0.14 to almost 0.19 $cm^3 \cdot g^{-1}$ and higher number of pores of larger diameter occurred. Similarly, the SEM images in Figure 6 revealed that the microstructure of sample CB-4 (picture B) is porous and clearly less compact compared to the reference sample (picture A). Picture C depicts the agglomerates of carbon black we traced among the binding phase and other products. Bright platelet formations are probably crystals of sodium carbonate crystals found in CB-4 matrix only. The results of porosimetry and electron microscopy correspond well with mechanical strength values achieved.

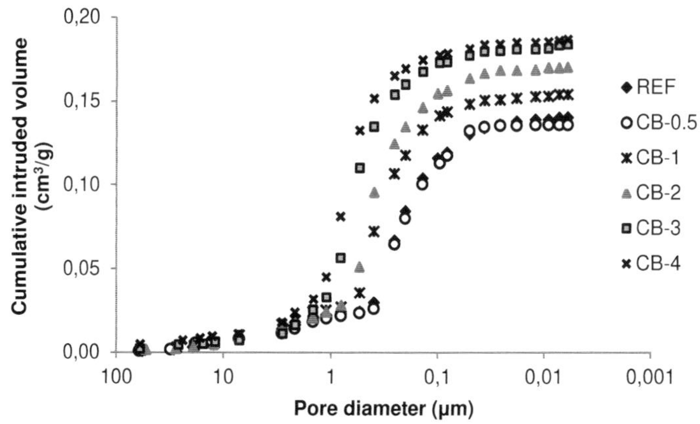

Figure 5. Cumulative intruded volume of metakaolin geopolymer mortars with different carbon black content.

A B C

Figure 6. SEM images. A, B: REF and CB-4 samples (scale: 10 μm), C: carbon soot particles (scale: 2 μm).

4. Conclusions

Alkali-activated binders have been reported to possess potential in fabrication of smart conductive composites. Metakaolin geopolymeric mortars were chosen to investigate the contribution of carbon black (up to 4%) to the enhancement of their electrical properties tested using impedance spectroscopy analysis. Carbon black content from 1% led to drop in electrical resistance across the frequency spectrum up to 50 kHz while the improvement of electrical capacitance was obvious throughout the whole measurement frequency range. The electrical capacitance and resistance of samples with 2, 3 and 4% of carbon black were similar, only the relative permittivity (measured at higher frequency) increased with each carbon black addition. From the results of mechanical performance and microstructural observation it seems that the increase of carbon black concentration from 2 to 4% does not have a considerable effect on the principal electrical parameters but deteriorates the binder quality. Drop of compressive strength is probably associated with multiple factors. Incorporation of nanosized carbon black particles involves additional requirements for wetting and results in increase of pore volume. Moreover, extremely small dimensions of the functional filler make the filler-matrix interface area enormous and poor binding of the matrix to the filler greatly affects the conductivity potential and mechanical performance as well.

Acknowledgements

This outcome has been achieved with financial support by the Czech Science Foundation under the project No. 19-11516S and the BUT student research project FAST-J-19-6016.

References

[1] Han B, Ding S and Yu X 2015 *Intrinsic self-sensing concrete and structures: A review Measurement* **59** pp 110-128
[2] Han B, Zhang L and Ou J 2017 *Smart and multifunctional concrete toward sustainable infrastructures* (Singapore: Springer)
[3] Chung D D L 2012 *Carbon materials for structural self-sensing, electromagnetic shielding and thermal interfacing Carbon 50.9* pp 3342-3353
[4] Juenger M C G, Winnefeld F, Provis J L and Ideker J H 2011 *Advances in alternative cementitious binders Cement and concrete research* **41**(12) pp 1232-1243
[5] McLellan B C, Williams R P, Lay J, Van Riessen A and Corder G D 2011 *Costs and carbon emissions for geopolymer pastes in comparison to ordinary portland cement Journal of cleaner production* **19**(9-10) pp 1080-1090
[6] Shi C, Roy D, Krivenko P 2003 *Alkali-activated cements and concretes* (CRC press)
[7] Davidovits J 2008 *Geopolymer chemistry and applications* (Geopolymer Institute)
[8] Tang Z, Li W, Hu Y, Zhou J L and Tam V W 2019 *Review on designs and properties of multifunctional alkali-activated materials (AAMs) Construction and Building Materials* **200** pp 474-489
[9] Provis L, Van Deventer J S J 2014 *Alkali Activated Materials* (Springer)
[10] Rashad, A M 2013 *Alkali-activated metakaolin: A short guide for civil Engineer–An overview Construction and Building Materials* **41** pp 751-765
[11] Donnet J B ed. 1993 *Carbon black: science and technology* (New York: CRC Press)
[12] Mingqing S, Xinying M, Xiaoying W, Zuofu H and Zhuoqiu L 2008 *Experimental studies on the indoor electrical floor heating system with carbon black mortar slabs Energy and Buildings* **40**(6) pp 1094-1100
[13] Lamuta C, Candamano S, Crea F and Pagnotta L 2016 *Direct piezoelectric effect in geopolymeric mortars Materials & Design* **107** pp 57-64
[14] Lamuta C, Bruno L, Candamano S and Pagnotta L 2017 *Piezoresistive characterization of graphene/metakaolin based geopolymeric mortar composites MRS Advances* **2**(61) pp 3773-3779
[15] MacKenzie K J and Bolton M J 2009 *Electrical and mechanical properties of aluminosilicate inorganic polymer composites with carbon nanotubes Journal of materials science* **44**(11) pp 2851-2857
[16] EN 1015-3:1999 *Methods of test for mortar for masonry. Determination of consistence of fresh mortar (by flow table)*

IOP Conf. Series: Materials Science and Engineering **549** (2019) 012034 doi:10.1088/1757-899X/549/1/012034

Assessment of the resistance of ventilated facade system by vacuum testing

P Schmid[1], P Daněk[1], I Rozsypalová[1], P Holomek[1] and P Juránek[2]

[1] Brno University of Technology, Faculty of Civil Engineering, Veveří 331/95, 602 00, Brno, Czech Republic

[2] Brno Branch, Technical and Test Institute for Construction Prague, SOE (TZÚS Praha, s. p.) Hněvkovského 228/77, Brno, Czech Republic

E-mail: pavel.schmid@vutbr.cz

Abstract. The aim of this research was to determine experimentally the mechanical resistance of preset ventilated facade systems exposed to wind suction. In accordance with the European ETAG 034 Guideline, Part 1, seven facade assemblies were tested. The wind suction load was simulated in a vacuum test in the AdMaS centre using the universal vacuum loading system. The test was performed with a load increasing in increments, where after each level, the specimens were unloaded. The effect of the density of connections of load-bearing plates with the structure of subframes was demonstrated. Further, the resistance to wind suction of specimens with timber subframe was slightly higher than that of specimens with aluminium subframe.

1. Introduction

Ventilated facades have recently been coming into focus because of their use in the residential and industrial construction. The advantages of this system include protection of the building against the climate, assembly without wet processes, potential for use on uneven surfaces, high durability and easy repair of the elements. Mention should also be made of the aesthetic aspect and the variability of application of this type of façade [1].

In accordance with the Guideline for European Technical Approval of Kits for External Wall Claddings [2], these parts of the structure are mainly assessed from the point of view of safety in the case of fire, indoor environment, dampness, outdoor environment, noise protection, energy efficiency and heat retention, durability, operability and safety in use. As far as safety in use is concerned, the evaluated properties include mechanical resistance of connections, resistance to horizontal point load, impact resistance, hygrothermal behaviour, resistance to seismic action and resistance to wind action (suction and pressure) [2].

The aim of this research was to determine experimentally the mechanical resistance of preset ventilated facade systems exposed to wind suction. The tests were performed in accordance with the European ETAG 034 Guideline, Part 1 (Section 5.4.1.1, Wind suction) [3].

2. Specimens

The tests were performed on a total of seven specimens of the ventilated facade system with an area of 3.3 × 2.5 m. The basic difference consisted in the load-bearing subframe, which was made from aluminium in 4 assemblies and from timber in 3 assemblies. Apart from the subfame, the tested specimens differed in the layout of plates on the tested area and in the number of connections. Differences in the individual solutions of the facade systems are given in the following Figure 1. The specific differences in the number of connections are apparent from Table 1, which shows the basic spacings of screws in the load-bearing plates in the longitudinal direction.

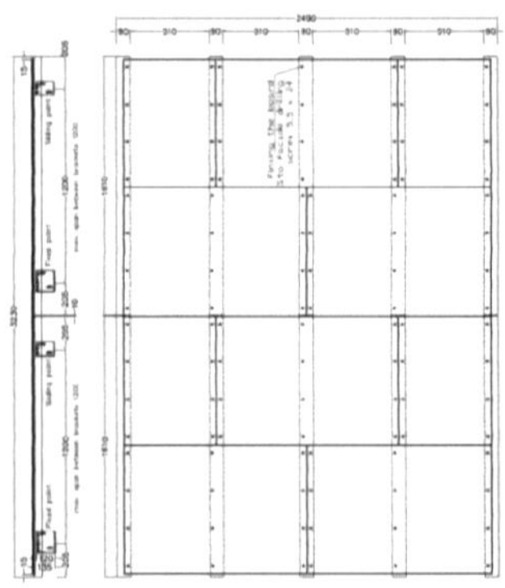

Arrangement 1

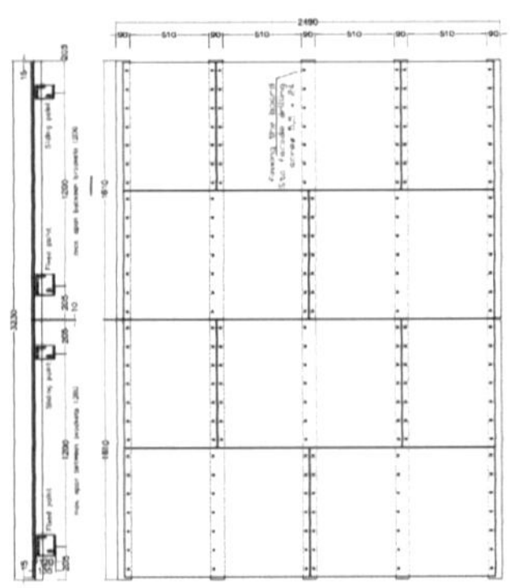

Arrangement 2

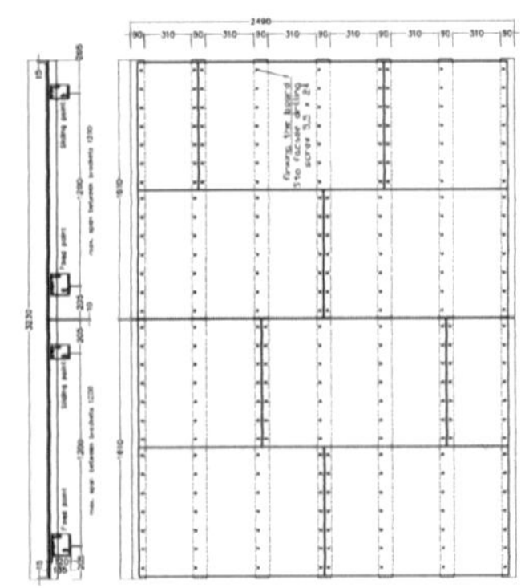

Arrangement 3

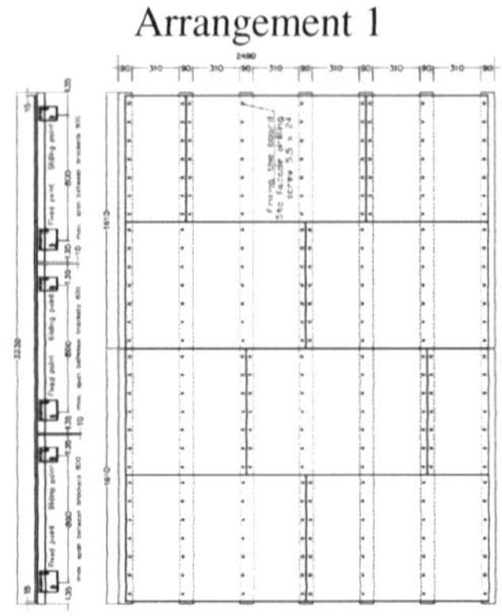

Arrangement 4

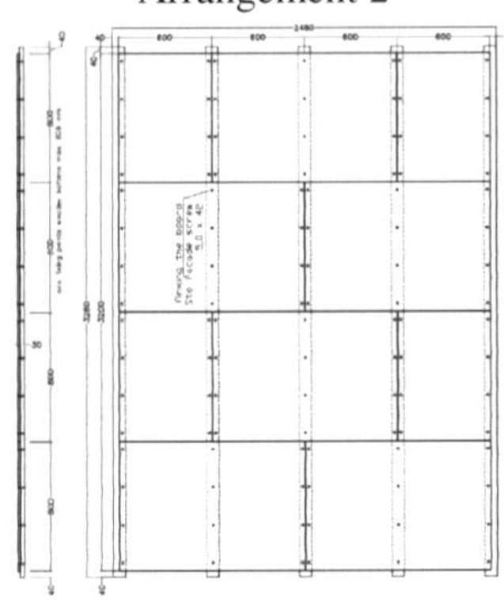

Arrangement 5

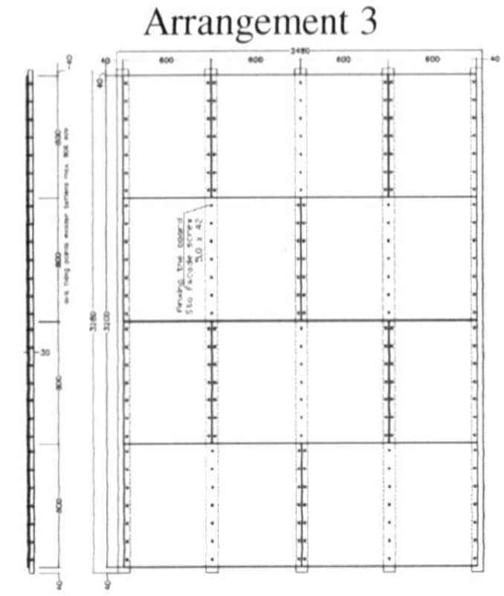

Arrangement 6

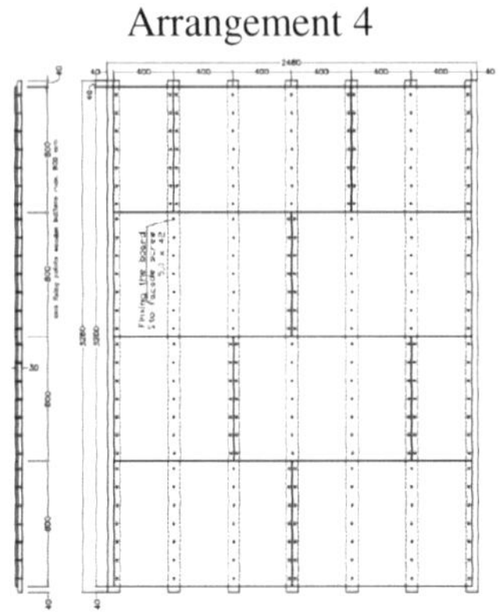

Arrangement 7

Figure 1. Diagrams of facade systems for vacuum testing with layouts of plates and screws.

Table 1. Spacings between screws in the load-bearing plates of the facade system specimens in the longitudinal direction.

Specimen	1	2	3	4	5	6	7
Longitudinal distance of screws [mm]	233.3	117.0	117.0	117.0	233.3	117.0	117.0

The facade systems with aluminium frame (Figure 2) consisted of supporting plates StoVentec Carrier Board Verotec 12 (basic full dimension of 1200 × 800mm) made from glass granulate reinforced with mesh on both sides, the plates are screwed to aluminium subframes made of T-profiles (90 × 52.7 × 2.7 mm) using Sto facade drilling screws (5.5 × 24 mm). The aluminium supporting beams were placed to Sto Wall Bracket GP a FP/GP (both with a length of 120 mm, width of 2.5 mm, stainless steel 1.4301) and fastened with self-tapping screws (5.5 × 19 mm) [4]. These wall brackets were screwed to timber crossbeams (160 × 160mm), which represent the base structure here, using self-tapping screws (5.0 × 42mm). The crossbeams were fastened to the longitudinal timber beams (160 × 160 mm) with self-tapping screws (8 × 280 mm). Unlike setups 1 to 3, the specimen of setup 4 was equipped with two rows of local wall brackets, see Figure 1.

A facade system with a timber subframe (Figure 3) consists of supporting plates StoVentec Carrier Board Verotec 12 (basic full dimension of 1200 × 800 mm) attached to wooden laths (80 × 30 mm, class S 10 according DIN [5]) using Sto facade screws (5.0 × 42 mm).

Figure 2. Preparation of the facade system specimen (including the assembly wooden frame) intended for vacuum testing – aluminium subframe arrangement 1.

IOP Conf. Series: Materials Science and Engineering **549** (2019) 012034 doi:10.1088/1757-899X/549/1/012034

Figure 3. Preparation of the facade system specimen (including the assembly wooden frame) intended for vacuum testing – timber subframe arrangement 6.

3. Methods

The tests were performed in the AdMaS centre on a universal vacuum loading system (see Fig. 4) developed by prof. Melcher and prof. Karmazínová. The dimensions of the air-tight chamber used for this test were 3.7 x 4.1 x 0.5 m. The specimens were placed into the upper part of the chamber and covered by a transparent plastic film. The unused rest of the chamber space was filled with extruded polystyrene and covered with black plastic film. During the test, air was sucked from the chamber using the Mivalt RT 83130 air blower (output 11 kW, suction power 576 m^3h^{-1}). The level of loading was given by a difference between the air pressure in the chamber and the ambient atmospheric pressure. [2]

During the test, deflections were measured using the 1-WA/100 MM-T HBM inductance transducers (with a measuring base of 100 mm) in 11 appropriately selected points of facade element specimens and the deflections of the supporting frame in the middle of its span (see Figure 4). The displacement of the supporting frame was measured by a 1-WA/50 MM-T HBM transducer (with a measuring base of 50 mm). The transducers were attached to an auxiliary independent steel frame using universal magnetic stands, and the deformations were measured from the upper side of the specimen. The inductance transducers were connected to two synchronized QuantumX MX840 systems (each with eight channels). Data from the systems were collected by catman®easy software (V3.5.1) in a computer and subsequently exported into a file in the *.txt format. The pressure in the chamber was regulated by a separate computer with LabVIEW software.

The testing of specimens was carried out in accordance with the procedure described in ETAG 034 [3]. Loading was applied gradually in increments, at first, two increments to 0.3 kNm^{-2}, one increment to 0.5 kNm^{-2}, one increment to 1.0 kNm^{-2} and then every subsequent increment by 0.2 kNm^{-2} higher than the previous one. After each increment, the specimen was unloaded to 0.0 kNm^{-2}. The load was gradually increased to the appearance of apparent irreversible deformations or up to a failure. Each level of loading was maintained for at least 10 seconds and until the stabilization of deformations. Figure 5 shows an example of a real loading of specimen 1 with aluminium subframe.

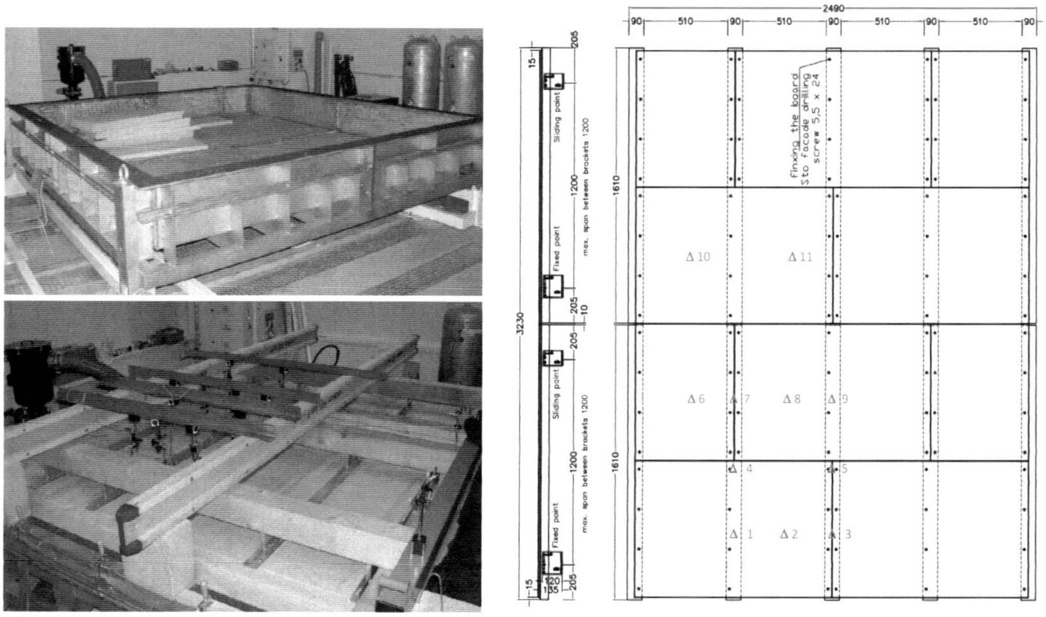

Figure 4. Universal vacuum loading steel chamber (upper left), example of specimen 1 placement on the vacuum chamber incl. location of inductance transducers (lower left), diagram of inductance transducers location of specimen 1 (right).

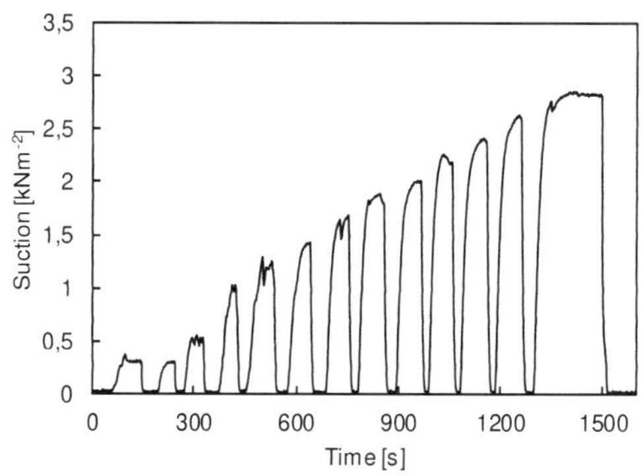

Figure 5. Actual diagram of loading during vacuum testing – Example of specimen 1.

4. Results

The test of ventilated facades by vacuum loading various assemblies using the same load-bearing plates showed differences in the use of load-bearing systems and in their anchoring. The experiment results are summarized in Table 2, which shows the values of suction $Q_{failure}$ [kNm^{-2}] at which the failure occurred, the maximum deflection measured at failure $d_{Qfailure}$ [mm], and it also describes the way of damaging the loaded specimens. In the systems with timber subframes, resistances to wind suction ranged from 3.14 to 7.87 kNm^{-2}. All these assemblies were damaged due to failure of the cladding plates at the point of fixing with a screw. Example of a specimen failure caused by tearing off of the plate at the screw is shown in Figure 6.

The facade systems with aluminium subframes displayed various types of failures, see Table 2. Aluminium T-profiles were tilted, load-bearing facade plates were damaged at the point of plate anchoring to the T-profile, or a combination of both occurred. The load capacities ranged from 2.87 to 7.32 kNm^{-2}. An example of a failure caused by tilting of the aluminium profile and tearing off of the plate at the screw is shown in Figure 6.

Table 2. Suction $Q_{failure}$ at the moment of facade system failure, including the type of failure.

Specimen	Subframe	$Q_{failure}$ [kNm^{-2}]	Failure	$d_{Qfailure}$ [mm]
1	aluminium	2.87	T-profile failure (buckling), cladding elements failure in fixing (tearing the slab off)	35.74
2	aluminium	2.61	T-profile failure (buckling)	15.53
3	aluminium	4.43	T-profile failure (buckling)	30.93
4	aluminium	7.32	cladding elements failure at fixing (tearing the slab off)	9.95
5	wooden	3.14	cladding elements failure at fixing (tearing the slab off)	25.81
6	wooden	4.88	cladding elements failure at fixing (tearing the slab off)	20.73
7	wooden	7.87	cladding elements failure at fixing (tearing the slab off)	7.44

Figure 6. Examples of specimen failures recorded during vacuum testpmg – T-profile buckling of specimen 1 (left), cladding elements failure at a fixing of specimen 1 (middle) and cladding elements failure at a fixing of specimen 6 (right).

5. Conclusion

The research focused on ventilated facade systems and their resistance to the action of wind suction. The loading was simulated by the so-called vacuum test where air is sucked from the chamber and the loading is given by the difference between the pressure in the chamber and the atmospheric pressure. The results of facade system tests showed the dependence of the load-bearing capacity of the elements on the structure of the load-bearing frame of the plates and on the density of connections of the load-bearing plates with the structure of subframes. Specimens with a higher number of screws had a higher load capacity. The load-capacity of specimens with a comparable number of plate anchoring elements to the subframes loaded by vacuum were slightly higher in the case of timber subfame than that of specimens anchored to the aluminium subframe.

Acknowledgement

This paper was created under the support of project No. LO1408 "AdMaS UP - Advanced Materials, Structures and Technologies", supported by the Ministry of Education, Youth and Sports under the „National Sustainability Programme I".

References

[1] Hanzlík P 2014 Ventilated facades - implementation errors (Provětrávané fasády -- chyby při realizaci) *Konstrukce* (Ostrava: KONSTRUKCE Media, sro)
[2] Melcher J 1997 Full-Scale Testing of Steel and Timber Structures: Examples and Experience *Structural assessment: The role of large and full-scale testing* (London: E&FN SPON) pp 301-308
[3] European Organization for Technical Approvals 2012 ETAG 034 Guideline for European Technical Approval of Kits for External Wall Claddings, part 1: Ventilated Cladding Kits Comprising Cladding Components and Associated Fixings (Brussels: EOTA)
[4] Sto Produkt katalog: Fasáda http://www.sto.cz/
[5] DIN 4074-1:2012 Strength grading of wood – Part 1: Coniferous sawn timber (Berlin: German Institute for Standardization)

Test of the tensile load capacity of a timber structure joint with an inserted nail plate

P Schmid[1], P Daněk[1], I Rozsypalová[1] and V Plaček[2]

[1] Brno University of Technology, Faculty of Civil Engineering, Veveří 331/95, 602 00, Brno, Czech Republic
[2] Ostrava Branch, Technical and Test Institute for Construction Prague, SOE (TZÚS Praha, s. p.), U Studia 14, 700 30 Ostrava, Czech Republic

E-mail: pavel.schmid@vutbr.cz

Abstract. The aim of this paper was to determine the axial tensile load capacity of timber structure elements with inserted metal plates with welded nails. The tests were performed on three specimens in the laboratory of the AdMaS centre of the Faculty of Civil Engineering, Brno University of Technology. The tensile force was applied on the specimen by increasing the force up to the specimen failure. The inductive displacement transducers continuously measured the displacement of the inserted joint plates with regard to timber prisms. The tensile load-bearing capacity of the tested joint was determined at 205.0 kN. Due to the advantages, the tested joints with inserted nail plates seem to be appropriate for joining suitable timber elements.

1. Introduction

Elements of timber structures are usually joined by mechanical means - by woodworking joints or by bonding. The basic types of mechanical connectors are divided, according to the principle of force transmission, into pin connectors and plate connectors. Pin connectors (e.g. nails, pins and screws) bend and get pressed into the wood due to the transmission of forces. Plate-type connectors (dowels and nail plates) are inserted or pressed into the elements, and the forces are therefore transferred on the surface of the structural elements. [1] An important improvement in the area of timber structure joints with plates was the shifting of the plate from the surface of the connecting elements into the grooves inside the connected solid timber parts. [2-3]

Joints with inserted plates are usually able to transfer higher loads than joints with outer plates of comparable parameters. The inserted plate contributes to an increase in fire resistance in comparison to the outer plate because the inserted connector is protected from fire by the solid timber. A disadvantage of a structure with inserted plates is the more demanding production process due to the necessity to cut the inner grooves for plates and to secure the sufficiently accurate linkup of the connectors. [1-3]

2. Specimens and method

The purpose of the measurement was to determine the axial tensile load capacity of timber structure elements with inserted nail plates. [4-5] The tests were performed on three specimens marked V1, V2 and V3.

IOP Conf. Series: Materials Science and Engineering **549** (2019) 012035 doi:10.1088/1757-899X/549/1/012035

Each specimen consisted of two timber prisms of 80 × 160 × 830 mm joined together on both ends by an inserted nail plate marked MKD 108×200-110. These connecting plates were 10 mm thick S235 steel plates with rectangular nails welded perpendicular to the plate on both sides. The nails had a perpendicular cross-section of 3 × 4 mm and a length of 50 mm. The wider areas of the nail shafts were profiled to increase the resistance of nails against extraction. The nail shafts were reinforced by a conical shape at the point of their transition to the steel plate, which increased the load capacity and stiffness of the joint. The distance between nails in the direction parallel to timber fibres was 40 mm and perpendicular to timber fibres 12 mm, and alternating arrangement was used. The MKD connecting plates were made by welding nails to a steel plate by a welding machine. The specimens of steel-timber joint were made using a special pressing device in such a manner that the MKD connecting plates were pressed between the two layers of timber, which corresponds to the supposed procedure of girder production, see Fig. 1 and 2. [6]

Figure 1. View of test specimens V1, V2, V3 – specimens of joints with the inserted MKD nail plates between timber prisms.

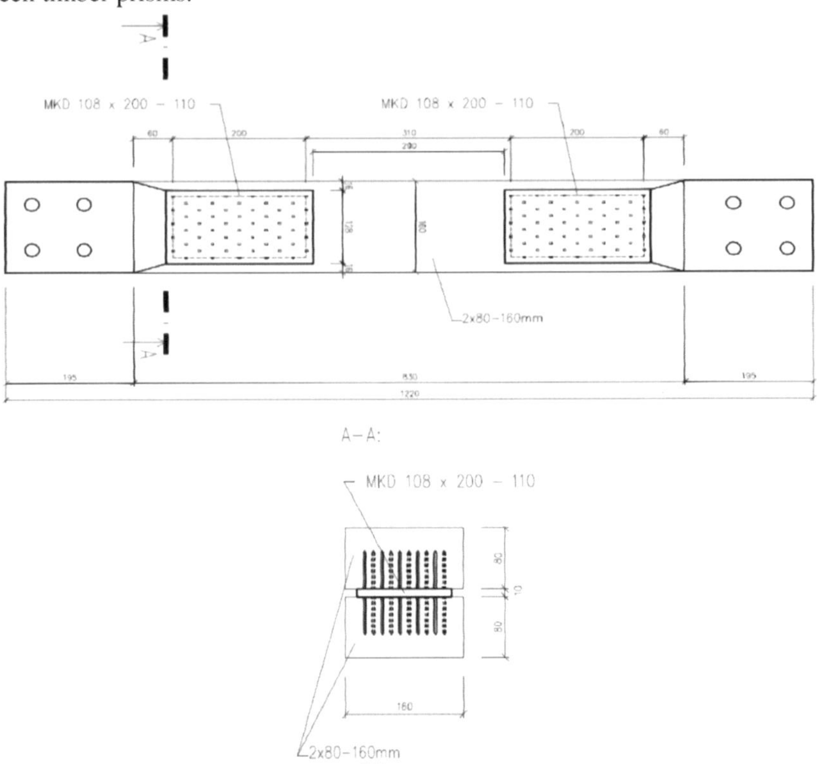

Figure 2. Diagram of the test specimen with an MKD nail plate (above) and the cross-section of the element (below).

The experiments were performed in the laboratory of the AdMaS centre of the Faculty of Civil Engineering, Brno University of Technology. The testing tensile load was applied by a hydraulic cylinder of the automatic electrohydraulic loading system called INOVA with a range of the loading cylinder of up to 500 kN and a strain gauge displacement transducer (displacement of the cylinder). The specimens were anchored in the test frame for static loading tests. The lower part of the test specimen was fastened to the breaking path with an articulated pin joint in the tested joint plates. The upper part of the specimen was fastened to the loading cylinder in a similar way. The tensile force was applied on the specimen by increasing the force of 0.45 kNs^{-1} up to the specimen failure.

In the tensile test, the HBM WA/50mm inductive displacement transducers continuously measured the displacement of the inserted joint plates with regard to timber prisms. Deformations were measured in four measuring bases, by two transducers near the upper nail plate and by two transducers near the lower nail plate. The location of transducers is apparent in Fig. 3.

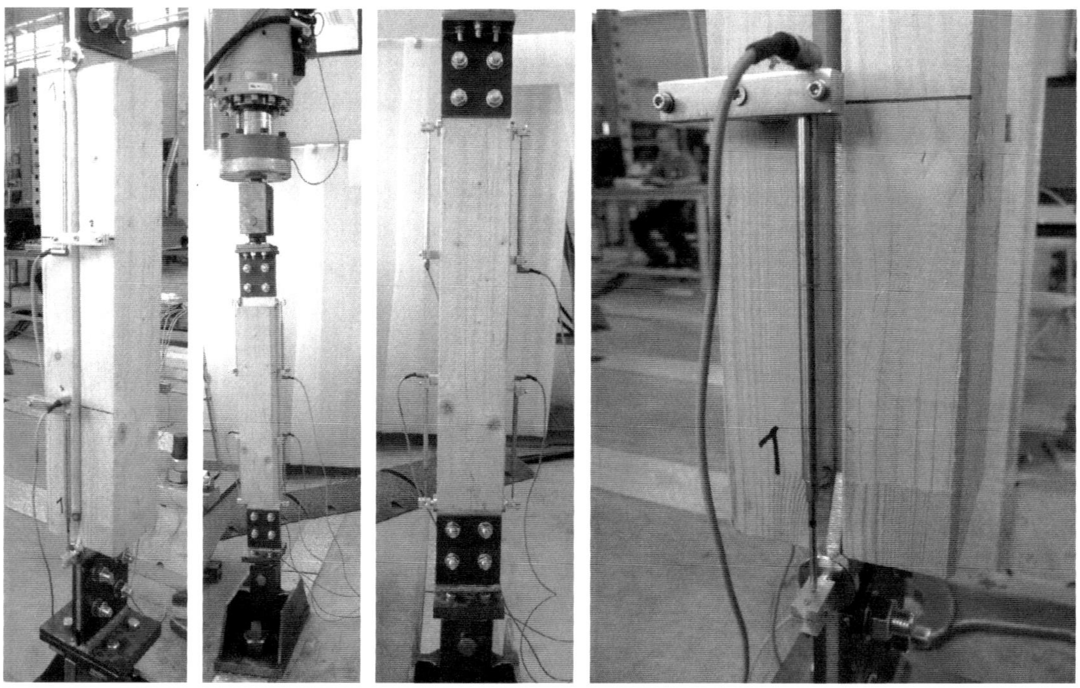

Figure 3. View of the testing setup for determining the tensile load capacity and the detail of the inductive transducer of a selected lower nail plate.

The data were measured by means of the HBM QUANTUM X system and processed in the computer using software for retrieving data called catman Easy version 3.5.1, and subsequently exported to a text file. The record contains data of the time course of the test, of the applied force, and of the displacement of MKD joint plates with regard to timber prisms measured by inductance transducers.

3. Results

The results of experimental measurement of the tensile load capacity are shown in the following working diagrams of joints in Fig. 3, which demonstrate the dependence of the nail plate displacement with regard to the timber prisms on the value of load. The values marked as *displ_up* and displ_bott

IOP Conf. Series: Materials Science and Engineering **549** (2019) 012035 doi:10.1088/1757-899X/549/1/012035

are calculated as average values of the displacement measured on two sides near the upper and lower joint.

The load capacities of nail plate specimens inserted between timber prisms are summarized in Table 1. The tensile load capacity of the tested element $F_{max,m}$ was determined as the arithmetic mean of the forces of the failure limit of the three test specimens. The load capacity of the tested joint was determined at 205.0 kN (with a standard deviation of 13.7 kN and a coefficient of variation of 6.7 %). The character of specimen failure after reaching the tensile load capacity is apparent from the Fig. 4 and 5.

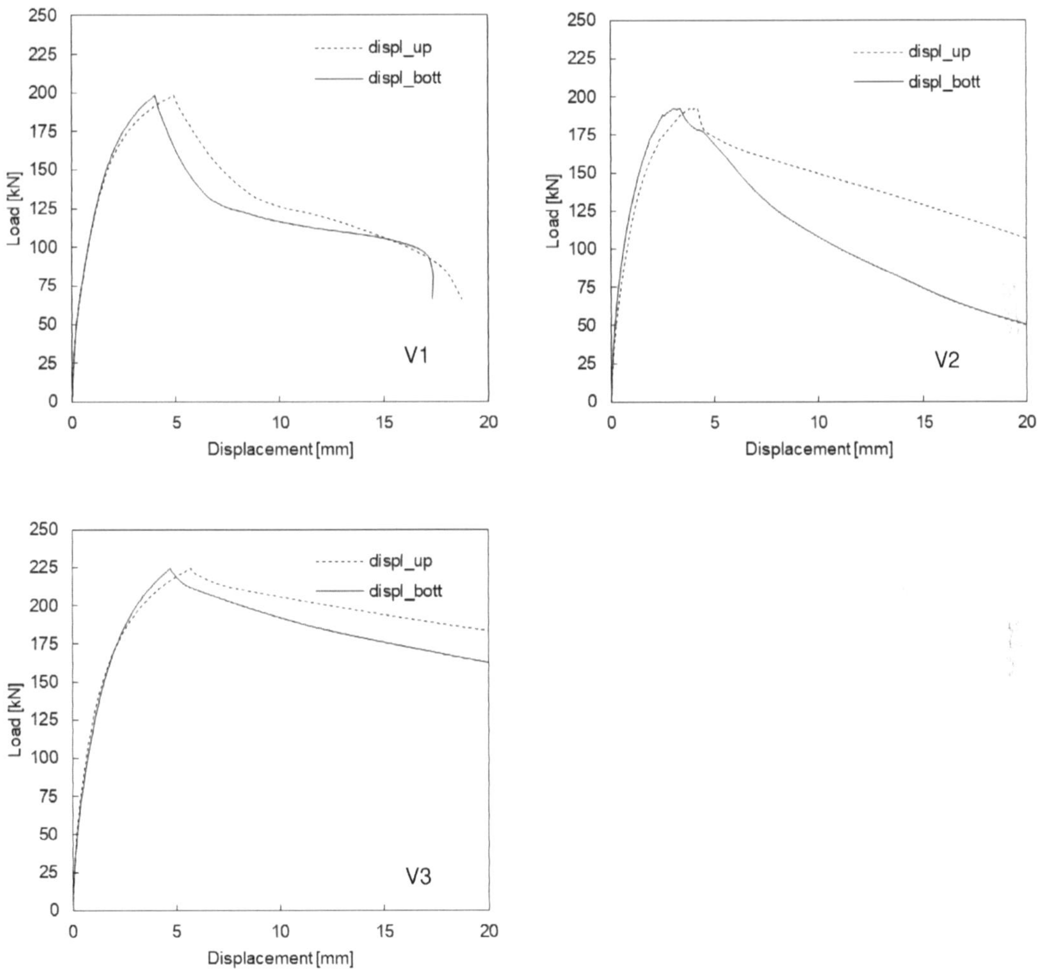

Figure 4. Working diagrams of the specimens of joints V1, V2 and V3 in the tensile test.

Table 1. Tensile load capacities of the examined specimens (incl. diameter, standard deviation and coefficient of variation).

Specimen	$F_{max,i}$ [kN]	$F_{max,m}$ [kN]	σ [kN]	CV [%]
V1	198.2			
V2	192.6	205.0	13.7	6.7
V3	224.1			

IOP Conf. Series: Materials Science and Engineering **549** (2019) 012035 doi:10.1088/1757-899X/549/1/012035

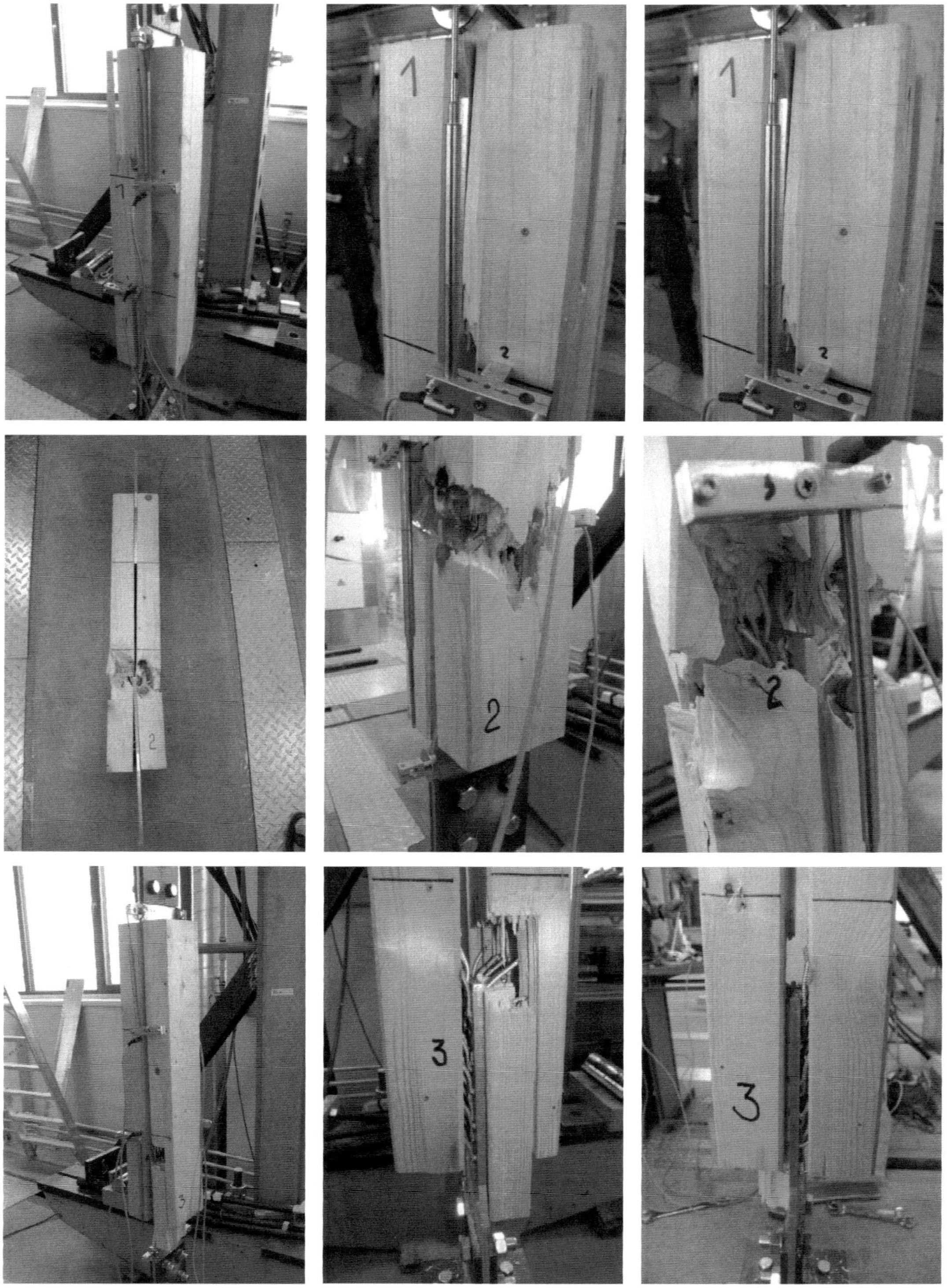

Figure 5. Damaged specimens – overall view and important details of the specimens V1 (1st row), V2 (2nd row) and V3 (3rd row).

4. Conclusion

The paper is a contribution to the knowledge in the field of timber structures and load capacity of joints. The experiment determined the tensile load capacity of a timber structure joint with an inserted metal plate with welded nails. All the three elements tested in the laboratory showed similar behaviour, the average tensile load capacity of specimens was 205.0 kN. All the tested elements behaved in a similar way under tensile loading, which was also supported by the records of deformations of the plates in the area of timber-steel contact. Due to their advantages, the tested joints with inserted nail plates seem to be appropriate for joining suitable timber elements.

Acknowledgement

This paper was created under the support of project No. LO1408 "AdMaS UP - Advanced Materials, Structures and Technologies", supported by the Ministry of Education, Youth and Sports under the „National Sustainability Programme I".

References

[1] Kuklík P 2005 *Dřevěné konstrukce (Timber structures)* (Prague: CTU Publishing House).
[2] Vejpustek Z 2009 *Analýza spojů dřevěných konstrukcí s vkládanými styčníkovými plechy (Analysis of joints in timber structures with inserted joint plates)* dissertation (Brno).
[3] Kuklík P 2005 *Dřevěné konstrukce (Timber structures)*. Prague: Issued by the ČKAIT Information Centre for the Czech chamber of chartered engineers and technicians engaged in construction (ČKAIT), 2005. Technická knižnice autorizovaného inženýra a technika.
[4] ČSN 73 2030 – Loading tests of building structures. Common regulations.
[5] ČSN EN 380 – Tiber structures. Test methods. General principles for static load testing
[6] http://www.strechy92.cz/konstrukce-spoje.html.

Influence of biopolymeric water-retaining admixtures on hydration of Portland cement studied using isothermal calorimetry

T Žižlavský[1], M Vyšvařil[1] and P Rovnaníková[1]

[1] Faculty of Civil Engineering, Brno University of Technology, Veveri 331/95, Brno 602 00, Czech Republic

E-mail: zizlavsky.t@fce.vutbr.cz

Abstract. Viscosity enhancing and water-retaining admixtures based on the cellulose ethers are widely used in the building industry to improve the properties of concrete and ready-mix mortars. The study focuses on the influence of the alternative natural viscosity enhancing admixtures (sodium salt of alginic acid, carrageenan, diutan gum, xanthan gum, and hydroxypropyl derivatives of guar gum and chitosan) in doses of 0.1%, 0.5%, and 1% of the binder weight on the cement hydration. The study is carried out using isothermal calorimetry studying the heat flow and the total heat evolved during the reaction from the very beginning – the water-introduction into system. All of the studied biopolymers showed the set retarding ability as they are all of polysaccharidic origin. The dosage-dependency was significant for the hydroxypropyl guaran and diutan gum, while retarding effect of other admixtures varied only slightly with the amount of biopolymer in the studied range of doses. However the studied admixtures were less effective in the delay of accelerating period, as well as the reduction of total heat generated during the reaction, in comparison with cellulose ethers. Other admixture than hydroxypropyl guar gum or diutan gum should be chosen if the set-retarding ability is not appropriate for the mixture use, otherwise the advantage over cellulose ethers will not be significant.

1. Introduction

Nowadays the use of ready-mix mortars is widely spread due to an easiness of use and reproducibility of the mixture qualities between different mixings. Since the 1960's the biopolymeric admixtures, namely cellulose ethers, are being used to improve the water retention value, adhesion, and workability of cementitious mixtures [1–5]. The cellulose ethers (CEs) used (most common is hydroxyethylmethyl cellulose (HEMC)) undesirably affect early-age compressive strength [6] and noticeably extend setting time [7].

The process of cellulose refining is also problematic with the current environmentally responsible policies, becomes more and more objectionable. The alternative products based on naturally occurring polysaccharides are being studied for their use as replacement of cellulose ethers in the concrete industry [8]. The biopolymers used come from different environments on Earth. The seaweed biopolymers such as carrageenan (CG) and alginates (salts of alginic acid) are obtained by boiling algae in alkaline solution and are furtherly used without additional chemical improvement. Diutan gum (DG) and xanthan gum (XG) are one of the most used alternatives which improve properties of self-consolidating concrete (SCC) notably reducing bleeding and segregation [9, 10]. The DG and XG are microbial polysaccharides, meaning, that they are the product of microbes which feed on the nutrient solution in a process called aerobic fermentation. The XG and DG are also used without further chemical improvements. The biopolymers which, similarly to cellulose, are subjected to further chemical modification are chitosan and guar gum. The biopolymers are modified to achieve better solubility in water and alkaline solutions and to improve their behaviour. Guar gum is a galactomannan form the endosperm of beans of *Cyamopsis tetragonoloba*, a legume widely cultivated in India. Guar gum and its derivatives are widely used mainly in food, textile, and drilling industry [11, 12]. In concrete the addition of guar gum derivatives improves the properties of SCC in similar way to DG and XG [8, 13–15]. Hydroxypropyl guar (HPG) in cementitious mortar increases the water retention value with less efficiency then HEMC [16, 17]. The production of chitosan ethers is the least environmental friendly out of the aforementioned alternatives. Chitosan is mainly obtained from shrimp shells, which are treated with acids to remove mineral constituents (mainly $CaCO_3$) and with alkalis to remove proteins and to deacetylate the chitin. Lately, more ecological alternatives as enzymatic decomposition of the shells have been studied [18]. The chitosan ethers addition also improves the water retention value of mixture [19, 20]. The hydroxypropyl derivatives of both above-mentioned polysaccharides also showed air-entraining function in the lime-based mortars [21, 22].

Overall the influence of these biopolymers on the properties of building materials is scarcely studied with even less studies in the more specific fields such as affection of hydration reaction. The polysaccharides, in general, have retarding effect on the Portland cement hydration. There are several possible explanations of the phenomenon, the complexing ability of sugars with calcium ions [23, 24] or the chelation, which can be important in the adsorption ability [23, 25]. There is sufficient amount of studies concerning the addition of CEs. The authors agree that the CEs affect the hydration of cement particles, in the calorimetric study, the prolongation of induction period, delay of acceleration period, and reduction of the reaction heat in the acceleration period is observed for different derivatives [26–28]. Similar conclusions were made by A. Peshard et al. [29, 30] concerning the conductometric observations. The conductometry alongside with isothermal calorimetry was also used by T. Poinot et al. [31] to study HPG with different degree of substitution, the HPGs showed higher delay of the beginning of acceleration period than HPMC, but still lower than the one of HEC. The addition of chitosan of different molecular weight led to the delay of portlandite precipitation peak with the delay increasing with growing molecular weight as was conductometricaly studied by M. Lasheras-Zubiate et al. [32] Wellan gum, polysaccharide structurally similar to diutan gum, expressed only slight retarding effect in comparison with CEs [33]. Out of other natural polymer-based admixtures studied, the starch ether-based admixture significantly reduced the heat evolution during the acceleration phase with acceptable delay of its beginning [34]. The black gram pulse (one of the India's widely produced legumes) addition had strongly retarding effect in higher doses, postponing the beginning of acceleration phase by several hours, causing acceptable compressive strength difference for cement mortar after 28 days of curing time [35, 36].

The study aims to compare the influence of the biopolymers mentioned in second paragraph (CG, DG, XG, HPG and hydroxypropyl chitosan (HPCH)) on the kinetics of ordinary Portland cement hydration in normal conditions. The same substituent group for both biopolymer derivatives (HPG and HPCH) was chosen to minimize the influence of different substituent groups observed by O.Z. Hua et al [28] on cellulose.

2. Materials and methods

2.1. Materials

The cement used was CEM I 42.5 R class cement according to EN 196-1 (Carmeuse Czech Republic), the chemical and mineralogical compositions of the cement are presented in table 1. The admixtures of dosage 0.1%, 0.5%, and 1% of cement weight were added to the mixture to study their effect on cement hydration. The admixtures used are summarized in table 2.

Table 1. Chemical and mineralogical composition of cement obtained by XRF analysis and calculated by modified Bogue equation according to ASTM C 150-94 and physical properties of the cement.

oxide	content	oxide	content	property	value	mineral	content
CaO	61.48%	MgO	0.86%	LOI[a]	4.17%	C_3S	42.39%
SiO_2	21.26%	Na_2O	0.12%	Blaine fineness	360 m^2 kg^{-1}	C_2S	28.97%
Al_2O_3	5.08%	K_2O	0.91%	specific gravity	3120 kg m^{-3}	C_3A	7.30%
Fe_2O_3	3.64%	MnO	0.07%			C_4AF	11.08%
SO_3	2.42%	TiO_2	0.29%				

[a] LOI = Loss on ignition

Table 2. The list of admixtures used in the study.

admixture	used abbreviation	business name (if given)	Supplier / manufacturer
sodium salt of alginic acid	ALGNA	–	Sigma-Aldrich, Co.
carrageenan	CG	Genuvisco CG-131	Biesterfeld Silcom s.r.o. / CP Kelco
diutan gum	DG	Kelco-crete DG-F	BASF Construction Solutions GmbH / CP Kelco
hydroxypropyl guar	HPG	–	Lamberti SpA
hydroxypropyl chitosan	HPCH	–	Kraeber & co, GmbH
xanthan gum	XG	Kelzan AP-AS	Biesterfeld Silcom s.r.o. / CP Kelco

2.2. Experimental setup, sample preparation

The measurements were executed using TAM Air isothermal calorimeter (TA Instruments). The samples were prepared according to the following procedure: The cement and admixture in specified dose were mixed together and well homogenized. 4 g of dry mixture were precisely measured and put into glass phial. Water to achieve 0.5 water/cement ratio was prepared into specific instrument, which allows mixing and adding water in the closed environment of calorimeter. The ratio of 0.5 was chosen so that even the mixtures with large doses of water-retaining agent can be sufficiently mixed using the implement. The phial with the instrument were put into calorimeter and tempered to 25°C. After the tempering for 24 hours, the water was added to the mixture, it was well mixed and at the same time the measurement was started. The data collection was stopped when the heat flow of all the samples was close to zero.

3. Results and discussion

3.1. Heat flow

The heat flow curves of the cement pastes containing 0.1%, 0.5% and 1% of admixture are presented in figures 1, 2, and 3 respectively. Both peaks, first of initial reaction after introduction of water into cement mixture in the first 30 minutes, and second, the main hydration peak between 2 and 48 hours, are observed in the cut-outs in the figures 1 to 3. The addition of admixture in any dosage with exception of HPG in 0.5% dose decreases the intensity of the heat evolution during this period. Dosage dependency is unclear during this period, only XG with increasing dose decreases the height of the peak. Other admixtures lower the heat evolution the most in certain dose, and in higher or lower dosage the heat evolved is higher. The doses of lowest initial heat evolution are 0.1% for HPG, DG, and CG and 0.5% for HPCH and ALGNA.

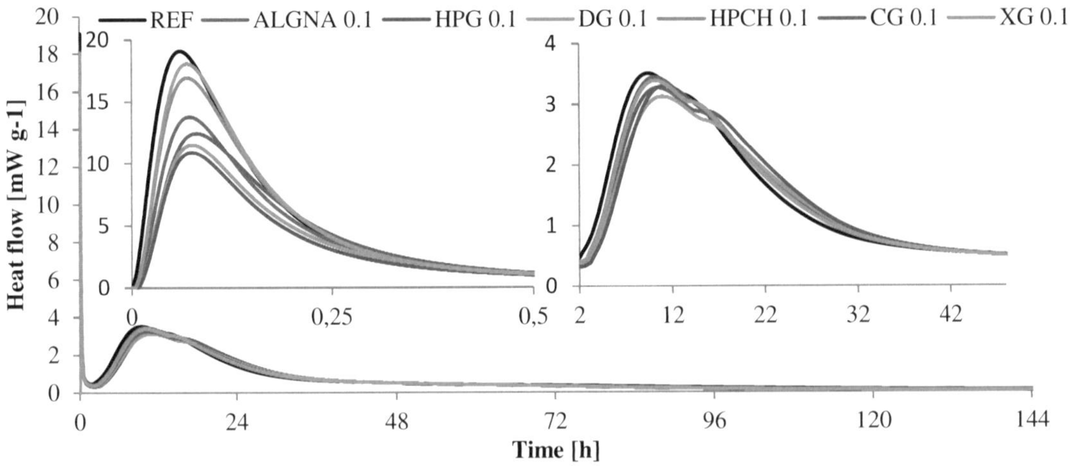

Figure 1. The hydration heat evolution of cement pastes with 0.1% of admixture.

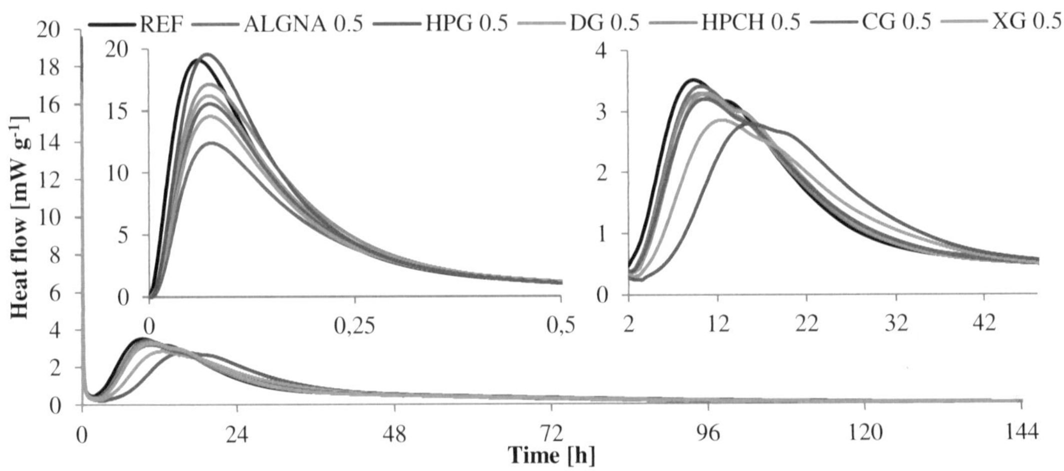

Figure 2. The hydration heat evolution of cement pastes with 0.5% of admixture.

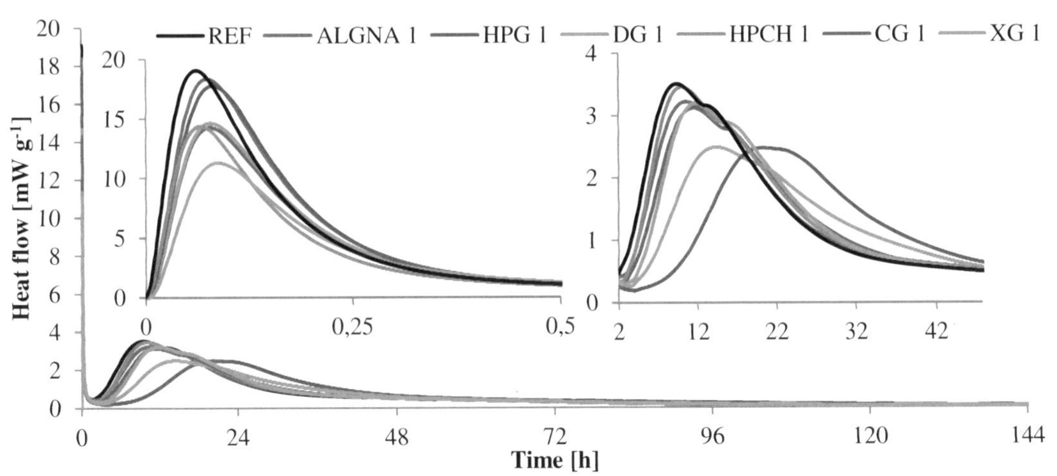

Figure 3. The hydration heat evolution of cement pastes with 1% of admixture.

The evolution of main hydration peak, zoomed in the second cut-out, is crucial for the heat evolution during the concrete hardening in the structure, and it also describes the set-retarding function of the admixuture. All the admixtures tested showed retarding function expectedly, as they are all of polysaccharidic structure [23–25]. The dosage dependency of the admixtures is more clear in the case of main hydration peak. On the figure 4, the most dosage dependent (HPG), and least dosage dependent (HPCH) admixtures are shown. The dosage dependency of HPG has been observed by T. Poinot et al. [31] as a delay of portlandite precipitation time in conductometric curves, thus prolongation of the period of dissolution of anhydrous phases and precipitation of CSH. The dosage-dependency in delay of acceleration phase for HPG and DG is simmilar to the cellulose ethers [26, 27], but the peak height decrease is lower with the tested admixtures. The HPG and DG are a little bit more efficient than the black gram pulse, causing higher heat flow reduction for the main hydration peak [35, 36]. The retarding effect of ALGNA, HPCH, XG, and CG is comparable to the one of starch ether [32] studied by A. Peshard et al.

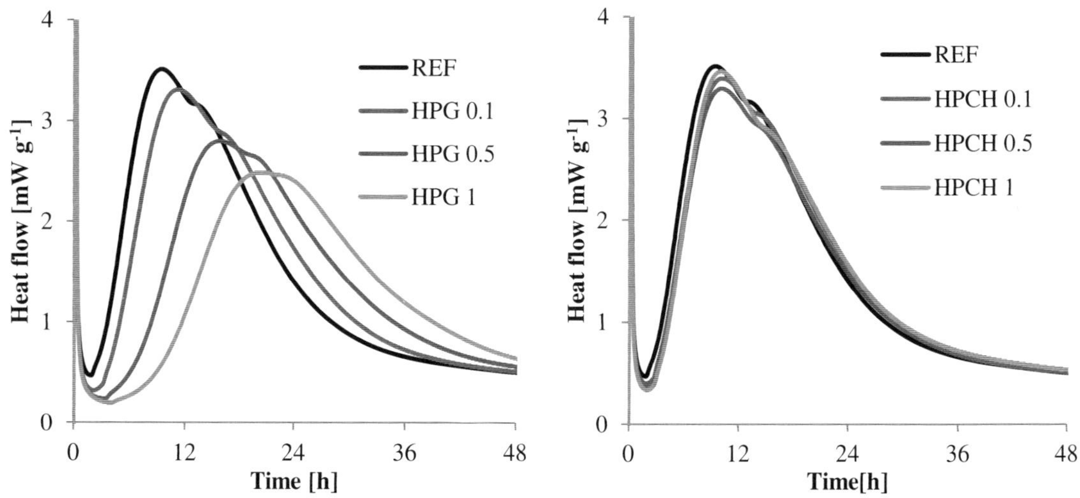

Figure 4. Heat flow curves ilustrating dosage-dependancy of HPG and HPCH addition, numbers representing the dosage in %.

3.2. Total cumulative heat

The cumulative heat evolution for the samples with the highest dose of admixture (1%) is presented in figure 5. The delay of start of acceleration phase of HPG and DG is observed as a delay of onset of the curve. The values at the time-points of 48 hours (the value chosen, because at that time even delayed samples reached similar heat flow, so the main heat-evolution phase is over) and 144 hours are summarized in table 3. The more varying values are seen for the above mentioned biopolymers with retarding function, the other ones like ALGNA or HPCH vary only slightly and the highest difference is observed for the highest dose. If the results are compared with the ones obtained on cellulose ethers, namely HEMC, in the same dosage, the studied biopolymers showed higher heat evolved (not less than 91% of reference value for DG in 0.5% dose in comparison with 84% obtained on HEMC) [26]. The heat evolved in the modified pastes was in average 95% of the heat evolved in the reference paste, the most efficient admixtures HPG and DG did not reach lower than 86.4% and 87.5% respectively for the 48 hours time point. DG is slightly more efficient in comparison with wellan gum (chemically similar polysaccharide with a shorter side-branch), where DG in 0.1% dosage reached 94.9% of heat of reference paste and wellan gum in same dosage and timepoint achieved the value of 97.3% [33].

Table 3. Cumulative heat after 48 and 144 hours of hydration [J g^{-1}].

	ALGNA		CG		DG		HPG		HPCH		XG	
dose\time	48h	144h	48h	144h	48h	144h	48h	144h	48h	144h	48h	144h
0.0%	272	343	272	343	272	343	272	343	272	343	272	343
0.1%	270	344	263	339	258	335	266	342	269	348	268	345
0.5%	268	340	260	334	250	322	250	336	262	336	261	338
1.0%	261	329	257	330	238	316	235	325	269	338	254	325

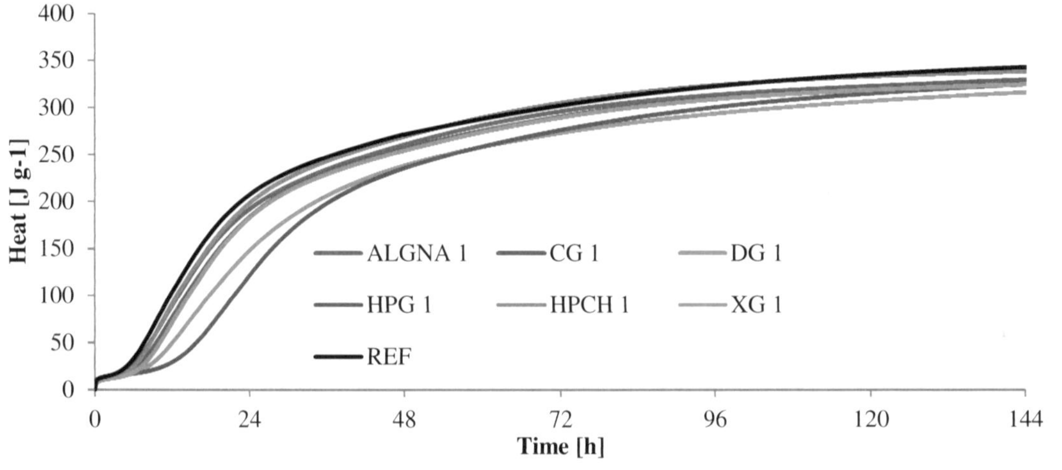

Figure 5. Total cummulative heat of the samples modified by 1% of biopolymeric admixture.

4. Conclusions

The paper studies the influence of addition of biopolymers on the cement hydration process using isothermal calorimetry. All the biopolymers showed retarding effect as expected due to their polysaccharidic nature. The notable dosage dependency of hydroxypropyl guaran and diutan gum has been observed in the range of doses between 0.1% and 1% of cement weight. Other admixtures studied, as hydroxypropyl chitosan, xanthan gum, carrageenan, sodium alginate showed only little dosage-dependency in comparison with the two above-mentioned admixtures.

While compared to the cellulose ethers, the most used water-retaining admixture to cementitious materials, the retarding effect of the hydroxypropyl guaran and diutan gum is comparable to the one of cellulose ethers. The heat flow reduction of all the biopolymers studied is lower than the cellulose ethers. Sodium salt of alginic acid and hydroxypropyl chitosan were the biopolymers the least affecting the hydration of Portland cement in the studied doses.

Any of the studied admixtures is utilisable to Portland cement-based mixtures as a replacement of cellulose-ether based admixture. Depending on the pertinence of the set-retarding ability the appropriate admixture may be chosen taking into account efficiency in fulfilling the primary goal of the admixture use.

Acknowledgements

This work has been financially supported by the BUT specific research project FAST-J-19-5803.

References

[1] Wagner HB 1960 US Patent US2934932A
[2] Wagner HB 1962 US Patent US3030258A
[3] Wang PM, Xu Q, Stark J 2001 *J. Build. Mater.* **4(2)** 122-6
[4] Bülichen D, Kainz J, Plank J 2012 *Cem. Concr. Res.* **42(7)** 953-9

[5] Patural L, Marchal P, Govin A, Grosseau P, Ruot B, Deves O 2010 *Cem. Concr. Res.* **41(1)** 46-55
[6] Ciobanu C, Iluc S, Lazau I, Pacurariu C 2011 *Rom. J. Mater.* **41(1)** 30-41
[7] Singh NK, Mishra PC, Singh VK, Narang KK 2003 *Cem. Concr. Res.* **33(9)** 1319-29
[8] Plank J 2004 *Appl. Microbiol. Biotechnol.***66** 1-9
[9] Isik I E and Ozkul M H 2014 *Constr. Build. Mater.* **72** 239-47
[10] Van Der Vurst F, Grünewald S, Feys D, Lesage K, Vandewalle L, Vantomme J, De Schutter G 2017 *Cem. Concr. Compos.* **82** 190-201
[11] Gross RA, Scholz C 2001 *Biopolymers from polysaccharides and agroproteins* (Washington: American Chemical Society)
[12] Zhang H, Zhou J, Zhou Y, Zhang Y 2013 *Complex hydrocarbon reservoirs* 02
[13] Rinaudo M 2008 *Polym. Int.* **57** 397-430
[14] Khayat KH 1998 *Cem. Concr. Compos.* **20(2-3)** 171-88
[15] Lachemi M , Hossain KMA, Lambros V, Nkinamubanzi PC, Bouzoubaa N 2004 *Cem. Conr. Res.* **34(2)** 185-93
[16] Govin A, Bartolin M-C, Biasotti B, Giudici M, Langella V, Grosseau P 2016 *Constr. Build. Mater.* **122** 772-80
[17] Medić V 2013 *Glas. hem. tehnol. Bosne. Herceg.* **40** 57-60
[18] Chen X, Yang H, Yan N 2016 *Chem. Eur. J.* **22** 13402
[19] Lasheras-Zubiate M, Navarro-Blasco I, Fernández JM, Álvarez JI 2012 *Cem. Concr. Compos.* **34(8)** 964-73
[20] Lasheras-Zubiate M, Navarro Blasco I, Álvarez JI, Fernández JM 2011 *J. Hazard. Mater.* **194** 223-31
[21] Vyšvařil M, Hegrová M, Žižlavský T 2018 *Key Eng. Mater.* **760** 257-65
[22] Vyšvařil M, Žižlavský T 2017 *IOP Conf. Ser.:Mater. Sci. Eng.* **251** 012039
[23] Taplin JH 1962 *Proc. 4th Int. Symp. on the Chemistry of Cement.* Washington **VII-1** 909-26
[24] Tadros ME, Skalny J, Kalyoncu RS 1976 *J. Am. Ceram. Soc.* **59** 344-7
[25] Thomas NL. Birchall JD 1983 *Cem. Concr. Res.* **13** 830-42
[26] Zhang G, Zhao J, Wang P, Xu L 2015 *J. Therm. Anal. Calorim.* **119** 1833-43
[27] Betioli AM, Gleize PJP, Silva DA, John VM, Pileggi RG 2009 *Cem. Concr. Res.* **39** 440-5
[28] Hua OZ, Guo MB, Wei JS 2012 *Constr. Build. Mater.* **33** 78-83
[29] Pourchez J, Grosseau P, Guyonnet R, Ruot B 2006 *Cem. Concr. Res.* **36** 1777-80
[30] Pourchez J, Peschard A, Grosseau P, Guyonnet R, Guilhot B, Vallée F 2006 *Cem. Concr. Res.* **36** 288-294
[31] Poinot T, Govin A, Grosseau P 2013 *Cem. Concr. Res.* **44** 69-76
[32] Lasheras-Zubiate M, Navarro-Blasco I, Fernández JM, Álvarez JI 2011 *J. Appl. Polym. Sci.* **120** 242-52
[33] Zhang Y, Zhang Z, Li X, Li W, Shen X, Wang H 2018 *J. Therm. Anal. Calorim.* **131** 1277-86
[34] Peshard A, Govin A, Grosseau P, Guilhot B, Guyonett R 2004 *Cem. Concr. Res.* **34** 2153-8
[35] Rai S, Kurian S, Dwivedi VN, Das SS, Singh NB, Gajbhiye NS 2009 *Hyperfine Interact.* **188** 19-24
[36] Singh NB, Das SS, Dwivedi VN 2010 *Adv. Cem. Res.* **22(2)** 61-72

IOP Conf. Series: Materials Science and Engineering **549** (2019) 012037 doi:10.1088/1757-899X/549/1/012037

Electronic waste as a filler in concrete

J Junak[1], N Junakova[1] and V Csiszar[1]

[1] Institute of Environmental Engineering, Faculty of Civil Engineering Technical University of Kosice, Vysokoskolska 4, Kosice 04200, Slovakia

E-mail: Jozef.Junak@tuke.sk

Abstract. Electrical appliances are becoming an everyday part of our lives. The more we use them, the more we contribute to the emergence of new waste - electronic waste (WEEE). E-waste is a significant source of plastics, metals and others. Electronic waste (cables) as a volume substitute for natural aggregates (5, 10 and 20%) in our experimental work was used. Flexural and compression was observed on prepared samples, beams 100x100x400 mm. The use of electrical waste as a filler has resulted in a slight reduction in the resulting strengths (about 10%), but the properties of the concrete with dispersed reinforcement have been founded. Flexural strengths ranged from 3.5 to 5.1 MPa and compressive strengths were in the range of 15.2 to 28.2 MPa after 28/60 days of curing.

1. Introduction

In today's ever-evolving world, Electrical and electronical equipment (EEE) is an integral part of everyone's lives, so we do not often realize these advances in the advancement of modern technology that we are replacing older and still functioning EEE with a new and more modern one, although it is often not necessary. However, few of us will ask what is happening with electrical appliances that we throw away, despite the fact that such electrical appliances automatically become waste electrical and electronic equipment (WEEE) or e-waste. Some of them will find their place in the waste dumps and the second part will be handed over to the e-waste processors for further processing. This way leads to the protection of the environment, because there are various materials (plastics, metal and others) in every electrical waste, which can be further evaluated and used, for example as a secondary raw material [1-5].

The basic legislation that represents the European Union's environmental approach to waste electrical and electronic equipment is the EU Directive 2002/96/EC. According to the Directive, manufacturers and importers of electrical and electronic equipment have been jointly responsible for all historical electrical waste since 13 August 2005. The Directive allows producers to pool and collect collectively electronic waste from household and business waste. For the purposes of this Directive, the following definitions shall apply [6]:

- *electrical and electronic equipment* or *EEE* means equipment which is dependent on electric currents or electromagnetic fields in order to work properly and equipment for the generation, transfer and measurement of such currents and fields falling under the categories set out in Annex IA and designed for use with a voltage rating not exceeding 1 000 Volt for alternating current and 1 500 Volt for direct current,

- *waste electrical and electronic equipment* or *WEEE* means electrical or electronic equipment which is waste within the meaning of Article 1(a) of Directive 75/442/EEC, including all components, subassemblies and consumables which are part of the product at the time of discarding.

One part of e-waste is also electrical cables that can be reused as a full-fledged material or are further recovered. The aim of our research was to use electronic waste in the construction industry, specifically in the production of concrete. For this reason, we have decided to use crushed electric cables, similarly to other types of waste [7-15], to replace the natural aggregate in concrete. The crushed electric cables have some properties in common with steel fibres, so we could conclude that the electronic waste will also serve as a scattered reinforcement in concrete.

2. Material and methods

A total of 4 different mixtures (E1-E4) have been prepared and tested in our experiment. Basic mixture composition is shown in the Table 1, while reference mixture (E1) did not contained e-waste cables as a filler substituent.

Table 1. Basic concrete mixture proportion per cubic meter.

Composition	1 m^3
Cement (kg)	350
Aggregates 0/4 mm (kg)	1035
Aggregates 4/8 mm (kg)	405
Aggregates 8/16 mm (kg)	405
Water (l)	180
Plasticizer (%)	0.5

The other mixtures was based on the basic mixture proportion, with the replacement of natural aggregates in the range 5% (E2), 10% (E3) and 20% (E4). The volume substitution of the aggregate was done at a ratio of 3:1:1, where three parts are for aggregates of fraction 0/4 mm and one part for aggregate of fraction 4/8 mm and 8/16 mm.

Beams with dimensions 100x100x400 mm were manufactured as test specimens. After mixing the fresh concrete mixture, the metal forms were filled and placed in a horizontal position under laboratory conditions at 20°C and 50% air humidity. After two days, the test specimens were removed from the forms and subsequently stored in an environment with a humidity of greater than 95%, in the water. Subsequently, the test concrete samples were subjected to compression and flexural strength testing after 28 and 60 days curing.

One type of CEM I 42.5 N cement from the manufacturer Považská cementáreň a.s. Ladce plant was used in the production of individual mixture. This type of cement is characterized by a rapid increase in initial strength, high final strengths and a higher hydration heat generation. The chemical composition of the cement is as follows: SiO_2 (19.87%), Al_2O_3 (4.00%), Fe_2O_3 (3.18%), CaO (64.36%), MgO (4.61%), K_2O (0.32%) and SO_3 (2.60%).

Three fractions of natural aggregate (0/4, 4/8, 8/16 mm) from Calmit spol. s.r.o., Margecany plant was used to prepare the experimental samples. These natural aggregates meet the requirements of STN EN 13242 [16].

The electrical waste (Figure 1) used in the experiment was obtained from V.O.D.S. a.s., Kosice and specifically from the line for processing of electronic waste. The material used is the result of the initial processing of the electrical waste, where the obtained cabling is crushed on the cut shredder. The bulk density of the used e-waste was 1650 kg/m^3 [17].

Figure 1. Used e-waste after the cut shredder.

The MasterGlenium ACE 446 superplasticizer based on polycarboxylate ether was used as a concrete additive at a ratio of 0.5% of the amount of cement used. The additive is optimized for the production of precast concrete and conforms to STN EN 934-2 [18].

3. Results and discussion

The flexural strength test was carried out according to STN EN 12390-5 [19] on beams 100x100x400 mm after 28 and 60 days. Table 2 shows the average maximum loads and the calculated average flexural strengths. All values in Table 2 are still average values calculated from the arithmetic mean of 3 values for each mixture.

Table 2. Flexural strength of tested samples after 28 and 60 days curing.

Mixture	28 days		60 days	
	Maximum force (kN)	Flexural strength (MPa)	Maximum force (kN)	Flexural strength (MPa)
E1	16.5	5.0	21.2	6.4
E2	16.2	4.9	16.9	5.1
E3	14.8	4.5	15.5	4.7
E4	11.6	3.5	12.2	3.7

From the resulting values (Table 2) it can be stated that with increased volume substitution of the aggregate under the influence of electro-waste, the flexural strength of the concrete samples decreases compared to the reference sample. The highest flexural strengths of samples containing electro-waste were achieved mixture E2 with a 5% volumetric substitute, where after 28 days its strength was reduced by only 0.1 MPa, and after 60 days the flexural strength decreased by 1.3 MPa, which represents 20% reduction in strength compared to the reference sample. The lowest flexural strength was found in mixture E4 with a 20% aggregate substitute, where after 28 days its flexural strength was reduced to 3.5 MPa, resulting in a 30% strength reduction and after 60 days the sample strength E4 was 3.7 MPa, where in percentual terms this represents a strength reduction of up to 40% over the reference sample. Furthermore, we can say that samples with electrical waste/cable behaved consistently after the test and the result of this test was a crack where the concrete beam was still connected with parts of the electrical waste and it was not broken into two parts. Thus, from this finding, we conclude that the crushed electrical waste in the concrete also serves as a staggered reinforcement.

The compressive strength test was carried out according to STN EN 12390-3 [20], where fractions of beams were used as test specimens after the end of the flexure strength test. The specimen was placed perpendicularly to the direction of laying and compacting the concrete between the press plates of the test machine, and spacers of 100x100mm were used to maintain the pressure area. The maximum force and average compressive strengths values are given in the Table 3.

Table 3. Compressive strength of tested samples after 28 and 60 days curing.

Mixture	28 days		60 days	
	Maximum force (kN)	Compressive strength (MPa)	Maximum force (kN)	Compressive strength (MPa)
E1	315.3	31.5	363.1	36.3
E2	278.0	27.8	282.3	28.2
E3	198.7	19.9	213.2	21.3
E4	152.1	15.2	160.0	16.0

The highest compressive strengths from samples containing electro-waste were achieved in the mixture E2 with a 5% volumetric substitute, where after 28 days its compressive strength decreased only by 10% and after 60 days by less than 20% compared to the reference sample. The lowest compressive strength was found in the mixture E4 with a 20% volume substitute, where after 28 days its compressive strength was reduced by almost 50% and after 60 days by up to 60% compared to the reference sample. Based on the measured values, we can conclude that with increasing volume substitution of natural aggregate by electronic waste, the compressive strength of tested concrete is reduced.

4. Conclusion

The main aim of the work was to point out the possibility of using electronic waste in construction. For the preparation of tested concrete composites, e-waste after cut shredder (cables) was used as a volume substitute for natural aggregates in concrete. The flexural and compressive strength of the hardened concrete was monitored. In general, the development of strengths was identical for both of the studied types and was directly proportional to the volume of the aggregate replacement - the strength decreased with increasing replacement.

Acknowledgments

This research has been carried out within the project of Slovak Scientific Grant Agency VEGA (Grant No. 1/0524/18).

References

[1] Golev A, Schmeda-Lopez DR, Smart SK, Corder GD and McFarland EW 2016 *Waste Manage* **58** 348-358
[2] Garlapati VK 2016 R*enew Sust Energ Rev* **54** 874-881
[3] Ikhlayel M 2017 *Waste Manage* **68** 458-474
[4] Kumar A, Holuszko M and Espinosa DCR 2017 R*esour Conserv Recy* **122** 32-42
[5] Suchithra S, Manoj K and Indu VS 2015 *Int J Tech Res Appl* **3** 266-270
[6] The european parliament and of the council: 2003 **Waste electrical and electronic equipment**, Directive 2002/96/EC
[7] Cigasova J, Stevulova N, Schwartzova I, and Junak J 2014 **37** *Chemical Engineering*

Transactions 685-690
[8] Junak J, Stevulova N 2011 11th *Int. Mul. Sci. Geo. SGEM 2011 (Albena)* vol 3 (Sofia: STEF92 Technology Ltd.) pp 567–572
[9] Junakova N, Balintova M 2014 *Chem Eng Trans* **39** 637-642
[10] Vaclavik V, Dvorsky T, Simicek V, Ondova M, Valicek J, Kusnerova M and Gola L 2016 **244** *Solid State Phenomena* 77-87
[11] Junakova N, Junak J 2017 *Sustainability* **9** Article number 852
[12] Brenek A, Vaclavik V, Dvorsky T, Daxner J, Dirner V, Bendova M, Harnicarova M and Valicek J 2015 70 *Advanced Structured Materials* 177-188
[13] Junakova N, Junak J 2017 *Procedia Eng* **180** 1292-1297
[14] Stevulova N, Schwartzova I, Hospodarova V and Junak J 2016 **50** *Chem Eng Trans* 367-372
[15] Junak J Sicakova A 2017 *Procedia Eng* **180** 1284-1291
[16] STN EN 13242+A1 Aggregates for unbound and hydraulically bound materials for use in civil engineering work and road construction
[17] STN EN 1097-6 *Tests for mechanical and physical properties of aggregates. Part 6: Determination of particle density and water absorption*
[18] STN EN 934-2+A1 *Admixtures for concrete, mortar and grout. Part 2: Concrete admixtures. Definitions, requirements, conformity, marking and labelling*
[19] STN EN 12390-5 *Testing hardened concrete.Part 5:Flexural strength of test specimens*
[20] STN EN 12390-3 *Testing hardened concrete.Part 3:Compressive strength of test specimens*

Alternative reuse of bottom sediments in construction materials: overview

N Junakova[1] and J Junak[1]

[1] Technical University of Košice, Faculty of Civil Engineering, Institute of Environmental Engineering, Vysokoškolska 4, Košice 04200, Slovakia

E-mail: natalia.junakova@tuke.sk

Abstract. In this paper, the paths of management of the dredged bottom sediments are discussed. Focus is made on the possibility of a recycling method, which consists of the reuse of sediments as a raw material in constructions. The results of selected studies concentrated on the sustainable sediment reuse as a raw material to manufacture lightweight aggregates, bricks and concrete are summarized.

1. Introduction

Long-term and very serious world problem is solid wastes production. Waste accumulation includes a variety of negative consequences and therefore possible solution is to find new paths of management of various types of waste [1-3].

In the center of the sustainable economic approach is the improvement of environmental effectiveness and the effectiveness of the use of materials, which is of specific importance to waste management [4]. Profitability and efficient use of raw materials, recycling and the substitution of conventional raw materials with wastes [e.g. 5-8] become increasingly important. A useful approach for saving natural resources and for promoting alternative solutions to disposal in landfills of by-products or dumping at sea of wastes may be a replacement of raw materials in concrete production by selected wastes [9].

Problem with the solid waste accumulation arises in household, municipal and industrial sphere but also in water management through silting of rivers and reservoirs with sediments [10]. In the world, there are many water structures such as dams and reservoirs that are silted with sediments [11-16]. Reservoirs' siltation causes except for the environmental problems [17, 18] also the problems to water management organizations with disposal of the dredged sediments and their use after dredging [19, 20]. Managers try to find environmentally friendly solutions for these materials. Considering the physical and chemical characteristics of the sediments and their perpetual formation, their use in the construction industry is common practice in the world.

In this paper, the results of selected studies focused on the reuse of sediments in construction materials are summarized.

2. Sustainable reuse of sediment in civil engineering

Several studies have pointed out the possibility of applying dredged sediments as the main or minor material in civil engineering.

2.1 Sediment as raw material to manufacture lightweight aggregate

The sediments can be used as a raw material to manufacture lightweight aggregates (LWA), because the sediments contain high portion of clay particles, which are basic raw material in their production. Several studies [e.g. 21-24] have focused on the reuse of contaminated sediments, whether from harbours or reservoirs, in the production of LWA, as an alternative to landfilling.The results showed that LWA based on the fine sediments meet the requirements of low bulk density for LWA and is also suitable for structural concrete. It is even possible to consider full replacement of raw materials with sediments [22, 23]. The results of selected studies focused on possibilities of sediments' utilization as raw material in the production of lightweight aggregates are summarized in table 1 [25, 26].

Table 1. Sediment as raw material to manufacture lightweight aggregate.

Sediment	Sediment proportion in mixture (%)	Other components (%)	Treatment	Properties
Harbour sediments: Bremen, Germany [21]	60	$Al(OH)_3$, red sludge, bleaching earth, recycled glass/alternatively a by-product of the wool top production	1150°C, 17 min.	- bulk density: 440 - 710 kg/m^3
Harbour sediments: Taichung, Taiwan [23]	100	-	1050–1150 °C, 18 min.	- bulk density: 490-1260 kg/m^3 - water sorption capacity (24h): 2-20%
Reservoir sediments: Shihmen, Taiwan [24]	50, 70, 90	municipal solid waste incinerator fly ash	1070, 1100, 1120, 1150 °C 20 min.	- bulk density: 880–1690 kg/m^3 - water sorption capacity (24h): 7,6-29%
Reservoir sediments: Shihmen, Taiwan [22]	100	-	1175–1200 °C, 30 min.	- bulk density: 622 - 859 kg/m^3 - water sorption capacity (24h): 10,4-12,3%

2.2 Sediment as raw material to manufacture bricks

Due to the high proportion of clay particles, sediments can also be used as a clay substitute for brick production [27] and ceramics [28]. They can be used in the production of bricks as a substitute for quartz sand. Samara et al. [29] used treated sediments as a 15% substitute for quartz sand in bricks. In the experiments, about 15,000 perforated sediment-based bricks were made and the results showed that these produced bricks meet the requirements for clay bricks as well as the relevant environmental requirements. Although the above-mentioned studies and other ones [e.g. 29-31] have shown that the application of sediments in brick production is feasible there are some economic, technical and social limitations. Based on surveys [32], consumers are distrustful of the use of sediments in the production of bricks (lower brick quality, use of contaminated sediments, etc.). Table 2 presents the results of selected studies focused on the reuse of sedimentsas a raw material in the production of bricks.

Table 2. Sediment as raw material to manufacture bricks.

Sediment	Sediment proportion in mixture (%)	Other components (%)	Treatment	Properties
River sediments: Dampremy-Charleroi, Belgium [29]	100	-	1000 °C	- porosity: 15,4% - water sorption capacity (24h): 7,5% - compressive strength: 36 MPa
River sediments: Ria de Aveiro, Portugal [30]	5, 10	plastic red clay, a low plasticity yellow clay	950, 1000, 1050, 1100 °C	- water sorption capacity (24h): 3,5-15,5% - compressive strength: 7,5-35 MPa
Harbour sediments: Santander, Spain [28]	100	-	900 – 1200 °C	- porosity: 12-38% - bulk density: 1600-2450 kg/m^3 - water sorption capacity (24h): 4-22% - compressive strength: 34 MPa
Harbour sediments: Savannah, USA [27]	100	-	900-1000 °C	- compressive strength: 8,3-11,7 MPa
River sediments: Lumsai, Bangkok [31]	80, 85, 90, 95, 100	rice husks	700 °C	- bulk density: 1130-1570 kg/m^3 - water sorption capacity (24h): 17,8-29% - compressive strength: 1,9-9 MPa

2.3 Sediment as a binder in concrete

The unconventional and innovative idea in recent years is the use of dredged sediments as a partial Portland cement substitute in concrete production [16, 33, 34]. In 2008, the technology to produce cement from dredged material was developed by Gas Technology Institute and Unitel Technologies, USA [33]. Since then, the studies in the field of sediments reuse as a substitute of cement in concrete have subsided until 2015 (table 3).

Kazi Aoual-Benslafa et al. [34] studied several cement substitutes (5, 10, 15 and 20% by mass of cement) with sediments dredged from the port of Oran (Algeria). They found that the 5% cement replacement with dredged sediment is the most appropriate, due to the mechanical properties and durability of the composites.

Junakova et al. [25, 26] assessed the possibilities of abiotic reuse of sediments extracted from water reservoirs in Slovakia (Kľušov and Ružin) as a 40% substitute for binder in the concrete production. Composites were prepared with fine-grained sediments without any admixtures and treatment. The sediments were also treated by milling and grinding with the addition of biomass fly ash with the aim of CaO adding in the mixture. The authors investigated the long-term mechanical properties of sediment-based composites as binder. The results showed that the higher (40%) cement replacement with uncontaminated fine-grained reservoir sediments can be considered in the future, especially from the environmental, water management and economic point of view. Treatment of the sediment with short-term grinding with fly ash has not been effective in the long-term strengths' development, but may be optimal in the case of requiring the use of emerging large volumes of sediment and fly ash simultaneously.

Table 3. Sediment as a binder in concrete.

Sediment	Sediment proportion in mixture (%)	Other components (%)	Treatment	Properties
River sediments: Passaic, New Jersey, USA [33]	40	admixtures and mineral modifiers	melting	- compressive strength: 52 MPa - flexural strength: 6,3 MPa
Harbour sediments: Oran, Algeria [34]	5, 10, 15, 20	-	dewatering	- compressive strength: 28-37 MPa
Reservoir sediments: Ružin and Kľušov, Slovakia [25, 26]	40	biomass fly ash	milling	- compressive strength: 19-23 MPa - flexural strength: 4,3-5,3 MPa

3. Conclusions

Reservoir bottom sediments represent a significant environmental, socio-economic and geomorphological resource that is one of the key components of aquatic ecosystems. However, an excessive supply of sediments into the aquatic environment leads to environmental and water management problems. An essential part of maintaining the reservoirs' functionality is sediment extraction, however, there is a problem with their utilization. Depending on properties of extracted materials, they can be used for various abiotic purposes.

With the emphasis on sustainable sediment management, paper presents the results of selected studies focused on the reuse of sediments as a raw material to manufacture lightweight aggregates, bricks and concrete. The results of studies reveal that the application of sediments in construction materials are feasible, however there are some economic, technical and social limitations.This type of incorporation of followed waste - sediment - in the concrete production will allow to reduce the consumption of non-renewable raw material and to extend the extraction duration of quarries.

Acknowledgments
This research has been supported by the Slovak Grant Agency for Science (Grant No. 1/0419/19).

4. References

[1] Junak J, Števulová N 2011 Potential of selected industrial wastes in civil engineering applications *11th International Multidisciplinary Scientific Geoconference and EXPO - Modern Management of Mine Prodhcing, Geology and Environmental Protection, SGEM 2011* **3** pp 567-572
[2] Števulová N, Schwarzová I, Hospodárová V, Junák J 2016 Implementation of waste cellulosic fibres into building materials *Chem Eng Trans* **50** pp 367-372
[3] Junakova N, Balintova M 2014 The study of bottom sediment characteristics as a material for

beneficial reuse *Chem Eng Trans* **39** pp 637-642
[4] Uher TE, Lawson W 1998 Sustainable development in construction *Proceedings of the 14th CIB World Building Congress on Construction and the Environment* **7** pp 7–12
[5] Junakova N, Junak J 2017 Recycling of reservoir sediment material as a binder in concrete *Procedia Eng* **180** pp 1292-1297
[6] Čigášová J, Števulová N, Schwarzová I, Junák J 2014 Innovative use of biomass based on technical hemp in building industry *Chem Eng Trans* **37**(1) pp 685-690
[7] Gola L, Vaclavik V, Valicek J, Harnicarova M, Kusnerova M, Dvorsky T 2015 Drainage concrete based on cement composite and industrial waste *Mech Mater Eng Mod Struct Component Des* **70** pp 155-165
[8] Dvorský T, Václavík V, Šimíček V, Břenek A 2015 Research of the Use of Waste Rigid Polyurethane Foam in the Segment of Lightweight Concretes *Inzynieria Mineralna* **36** pp 51–56
[9] European Commission 2010 *Being wise with waste: the EU's approach to waste management.* Available online: http://ec.europa.eu/environment/waste/pdf/WASTE%20BROCHURE.pdf (accessed on 15 January 2018)
[10] Junakova N, Balintova M 2012 Predicting of soil loss in the Tisovec Catchment, Slovakia*Chem Eng Trans* **28** pp 265-270
[11] Apitz SE, Brils J, Marcomini A, et al. 2007 Approaches and frameworks for managing contaminated sediments—A European perspective. *Processes, Assessment and Remediation of Contaminated Sediments* eds DD Reible, T Lanczos (Dordrecht, The Netherlands: Springer) pp 5–82
[12] Wohl E 2011 A World of Rivers: Environmental Change on Ten of the World's Greatest Rivers (Chicago, IL, USA: University of Chicago Press)
[13] Sedimentation Committee. Sedimentation and Sustainable Use of Reservoirs and River Systems. Draft ICOLD Bulletin, 2009. Available online: http://www.icold-cigb.org/userfiles/files/CIRCULAR/CL1793Annex.pdf (accessed on 22 March 2017)
[14] White R 2001 Evacuation of Sediments from Reservoirs (London, UK: Thomas Telford)
[15] Morris GL, Fan J 1998 Reservoir Sedimentation Handbook (New York, NY, USA: McGraw-Hill)
[16] Junakova N, Junak J 2018 The effect of sediment treatment on strength and water absorption capacity of sediment based concrete*IOP Conference Series: Materials Science and Engineering* **385** pp 1-5
[17] Junakova N, Balintova M 2012 Predicting nutrient loads in chosen catchment *Chem Eng Trans* **26** pp 591-596
[18] Junakova N, Balintova M 2011 Model for predicting the nutrient content in reservoir bottom sediments *11th International Multidisciplinary Scientific Geoconference and EXPO - Modern Management of Mine Producing, Geology and Environmental Protection, SGEM 2011* **3** pp 475-482
[19] Deibel I, Lampe C, Ulbricht J P, Cnudde T, van Dessel G. 2007 Beneficial Use *Sustainable Management of Sediment Resources: Sediment and Dredged Material Treatment* eds P Bortone, L Palumbo (Amsterdam, The Netherlands: Elsevier) pp 119–132
[20] Sumi T, Hirose T 2009 Accumulation of sediment in reservoirs *Water Storage, Transport, and Distribution, Encyclopedia of Life Support Systems* ed Y Takahasi (London, UK: UNESCO) pp 224–252
[21] Hamer K, Hadeler A, Muschalla T, Schroter J, Timmer G 2003 Light Weight Aggregates made from dredged harbour sediments. Leaching behaviour of inorganic pollutants and constructional characteristics *J Soils Sediments* **4** pp 284-291
[22] Chen HJ, Yang MD, Tang CW, Wang SY 2012 Producing synthetic lightweight aggregates from reservoir sediments *Constr Build Mater* **28** pp 387-394

[23] Wei YL, Yang JC, Lin YY, Chuang SY, Wang HP 2008 Recycling of harbor sediment as lightweight aggregate *Mar Pollut Bull* **57** pp 867–872
[24] Hwang CL, Bui LAT, Lin KL, Lo CT 2012 Manufacture and performance of lightweight aggregate from municipal solid waste incinerator fly ash and reservoir sediment for self-consolidating lightweight concrete *Cem Concr Compos* **34** pp 1159–1166
[25] Junakova N, Junak J, Balintova M 2015 Reservoir sediment as a secondary raw material in concrete production *Clean Techn Environ Policy* **17** pp 1161-1169
[26] Junakova N, Junak J 2017 Sustainable use of reservoir sediment through partial application in building material *Sustainability* **9** (5) 1-13
[27] Mezencevova A, Yeboah NN, Burns SE, Kahn LF, Kurtis KE 2012 Utilization of Savannah harbor river sediment as the primary raw material in production of fired brick *J Environ Manag* **113** pp 128–136
[28] Romero M, Andrés A, Alonso R, Viguri J, Rincón J M 2009 Valorisation of contaminated marine sediments to produce ceramic construction materials. *Spanish National Conference on Advances in Materials Recycling and Eco – Energy* pp 45-48
[29] Samara M, Lafhaj Z, Chapiseau Ch 2009 Valorization of stabilized river sediments in fired clay bricks: Factory scale experiment *J Hazard Mater* **163** pp 701-710
[30] Torres P, Manjate R S, Fernandes H R, Olhero S M, Ferreira J M F 2009 Incorporation of river silt in ceramic tiles and bricks *Ind Ceram* **1** pp 5-12
[31] Tangprasert W, Jaikaew S, Supakata N 2015 Utilization of Dredged Sediments from Lumsai Canal with Rice Husks to Produce Bricks *Int J Environ Sci Dev* **6** pp 217–220
[32] Cappuyns V, Deweirt V, Rousseau S 2015 Dredged Sediments as a Resource for Brick Production: Possibilities and Barriers from a Consumers' Perspective *Waste Manag* **38** pp 372-380
[33] Mensinger M C 2008 Cement-Lock® Technology for Decontaminating Dredged Estuarine Sediments *Topical Report on Beneficial Use of Ecomelt from Passaic River Sediment at Montclair State University* (Essex County, NJ, USA: Montclair State University)
[34] Kazi Aoual-Benslafa F, Kerdal D, Ameur M, Mekerta B, Semcha A 2015 Durability of mortars made with dredged sediments *Procedia Eng* **118** pp 240–250

IOP Conf. Series: Materials Science and Engineering **549** (2019) 012039 doi:10.1088/1757-899X/549/1/012039

Comparison of material properties of steel fiber reinforced concrete with two types of steel fiber

Z Marcalíková[1], L Procházka[2], M Pešata[2], J Boháčová[2] and R Čajka[1]

[1] Department of Structures, Faculty of Civil Engineering, VSB-Technical University of Ostrava, Ludvíka Podéště 1875/17, 708 33 Ostrava-Poruba, Czech Republic
[2] Department of Building Materials and Diagnostics of Structures, Faculty of Civil Engineering, VSB Technical University of Ostrava, Ludvíka Podéště 1875/17, 708 33, Ostrava-Poruba, Czech Republic

E-mail: zuzana.marcalikova@vsb.cz

Abstract. The work is focused on the area of mechanical properties of two types of Steel Fibers for reinforced concrete. In both cases, the same concrete mixture is used. The Steel Fibers used differ in shape. The first one is short and straight fiber and the second one is 3D steel fiber. Steel Fiber Reinforced Concrete was prepared at a dosage of 40 and 75 kg steel fibers/m^3. The experiment includes determination of strengths, concretely compressive strength, a three-point flexural test and Splitting Tensile Strength test. The results are summarized; fracture mechanics parameters necessary for structural modeling are also included.

1. Introduction

Steel fiber reinforced concrete is classified as a fiber reinforced composite [1]. Fiber concrete can vary by class of concrete, type of fibers, and its shape, length, diameter and surface finish [2, 3, 4, 5]. For building structures such as floors, foundations of buildings [6, 7, 8] or tunnel linings, it is often more favorable to use fiber reinforced concrete than conventional reinforced concrete or conventional concrete. Reinforcing the concrete by means of fibers increases the tensile strength [9, 10] of the concrete and the ductility. Other advantages of fiber reinforced concrete are reduced shrinkage cracking [11] and reduced concrete deformations, increased toughness and fatigue strength, improved cohesion. For the use of fiber reinforced concrete in construction, it is necessary to know in detail its properties [12, 13, 14], which are most commonly tested in laboratories on test specimens. The research results can be found in a number of recommendations [15] and technical conditions [16]. There are also a number of Steel Fiber Reinforced concrete associations such as RILEM [17], DafStB [18], BS [19] and Model Code 2010 [20]. Contemporary studied topics in the area of reinforced fiber concrete in the Czech Republic include, for example, durability determination and verification of mechanical properties. Also the possibilities of using secondary raw materials, such as cement kiln dust, are monitored.

2. Theory of experimental work

IOP Conf. Series: Materials Science and Engineering **549** (2019) 012039 doi:10.1088/1757-899X/549/1/012039

Tests of steel fiber reinforced concrete include determination of compressive strength on cubes a typical size of 150 x 150 x 150 mm. Testing compressive strength on cubes is always perpendicular to the filling direction.

Other tests included in the experimental program include splitting tensile strength tests. Two variants were chosen: perpendicular to the filling direction and parallel to the filling direction. The reason for this double test is checking homogeneity of fibers in test samples. The diagram of the test is shown in figure 1. The splitting tensile strength test is determined by the following formula:

$$f_{ct,sp} = \frac{2 \cdot P_{\max}}{\pi \cdot l \cdot d} \tag{1}$$

where P_{max} is the maximum load, l is the length of the contact line of the body and d is the chosen cross-sectional dimension of the body.

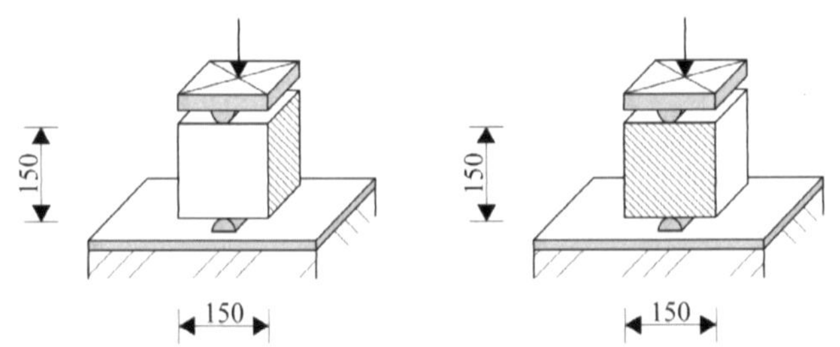

Figure 1. Tests of compressive strength and splitting tensile strength (parallel and perpendicular to the filling direction).

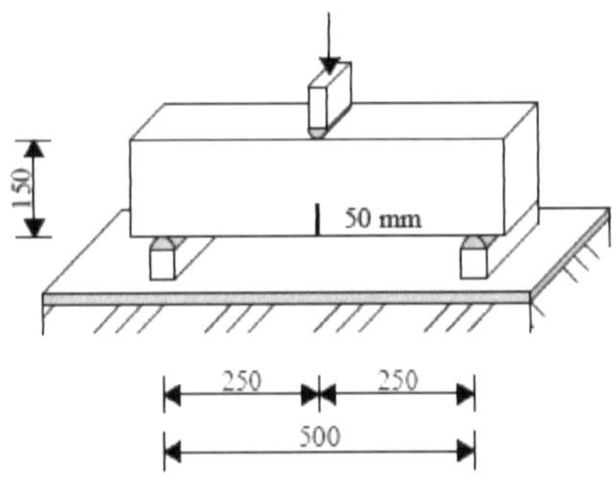

Figure 2. Three-point bend test.

In order to verify the suitability of the test and to determine the fracture mechanics parameters, a three-point bend test in design is performed. The test scheme is shown in figure 2. The flexural strength of concrete in the case of a three-point bend configuration is determined:

$$f_{ct,fl,3B} = \frac{3 \cdot P_{\max} \cdot L}{2 \cdot b \cdot (h - a_0)^2} \tag{2}$$

where P_{max} is the maximum load; L, b and h are dimensions: span, width and height of the cross-section, and the height of notch is a_0.

For wider classification of brittleness of concrete, it is also important to determine the fracture parameters of the concrete, which are also needed to investigate the creation and crack growth. The most critical fracture mechanics parameter is fracture work A. It is the energy that the body takes until it is completely broken. Thee test bodies are provided with a notch in the middle of their span, so that cracks in the middle of the concrete beam could be expected. The cuts are most often performed at 1/3 of sample height. In order for us to correctly perform the three-point bend test, fibers with a maximum length of 60 mm are required for reinforced concrete samples and suitable concrete with a grain fraction of maximum 32 mm is also required. The size of the fraction of aggregate also has a large impact on the value of fracture energy W_f, the larger the fraction of the aggregate used, the higher the value of fracture energy we can expect. From the load diagram, the fracture energy of W_f can be determined. Fracture energy is an integral that is equal to the area under the load diagram or so-called work A needed to create a crack, divided by the unit of surface of the test body without considering the notch depth. The fracture energy is defined as follows:

$$W_f = \frac{A}{h \cdot (b - a_0)} \tag{3}$$

3. Steel fiber reinforced concrete

For the experiment, Dramix OL13/20 [21] and Dramix 3D 55 30BG [21] steel fibers in dosage of 40 and 75 kg/m^3 were selected. The basic characteristics of these steel fibers are listed in table 1. The concrete matrix has the recipe shown in table 2. Samples were prepared in the laboratory.

Table 1. Basic properties of Dramix® steel fibers [21].

Property	Dramix® OL 13/20 (Figure 3)	Dramix® 3D 55/30BG (Figure 4)
Length [mm]	13	30
Diameter [mm]	0.21	0.55
Flexural strength [N/mm^2]	2750	1345
Dosage [kg/m^3]	60	25
Modulus of elasticity [GPa]	200	200

Figure 3. Dramix® OL13/20.

Figure 4. Dramix® 3D 55/30BG.

The concrete mixture used for the tested samples can be classified as conventional concrete C25/30.

Table 2. Basic characteristics of the concrete mix C 25/30.

Description	Mixture [kg/m^3]
Cement	CEM WA-S 42.5
Minimum amount of cement	320 kg
Water-cement ratio w/c	0.625
Aggregates 0/2 DTK Mankovice	525 kg
Aggregates 0/4 DTK Mankovice	420 kg
Aggregates 4/8 Tovačov	150 kg
Aggregates 8/16 HDK	820 kg
Water	200 1
Plasticizer STACHEPLAST	3.2 1

4. Compressive strength of concrete

Compressive strength testing included three samples for each variant. In total, 12 tests were performed. In the case of a 75 kg/m^3 wire dosing variant, the compressive strength was higher. The difference, however, was relatively small. Especially for Dramix OL13/20 steel fibers. The results are summarized in the graph in figure 5. Compressive strength for 40 kg/m^3 fibers was 36.4 MPa for Dramix OL13/20 and 38.5 MPa for Dramix 3D 55/30BG. In the 75 kg/m^3 version, the strengths were very similar to 40.1 for Dramix OL13/20 and 40.9 MPa for Dramix 3D 55/30BG. The aggregate compressive strengths and bulk weights are shown in tables 3 and 4.

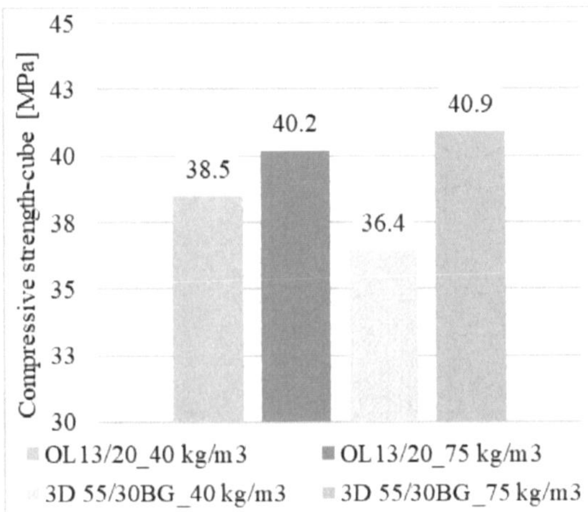

Figure 5. Compressive strength (cube).

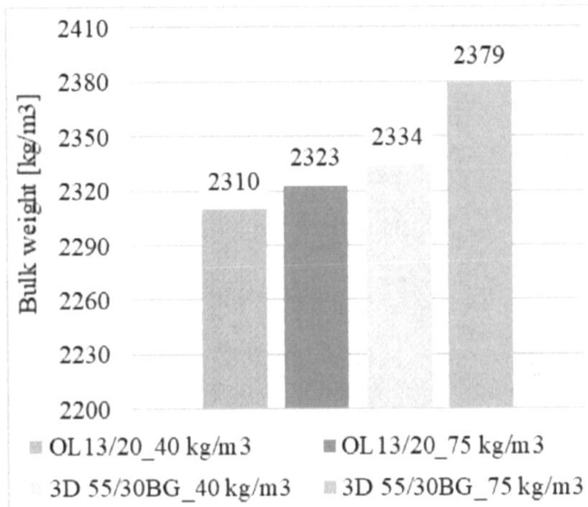

Figure 6. Bulk density.

Table 3. Compressive Strength - Cubic [MPa].

Value	Amount of fibers [kg/m^3]	Type of fiber Dramix OL13	Dramix 3D
MIN	40	37.1	35.5
	75	37.3	39.0
MAX	40	40.8	37.0
	75	42.0	42.1
Average	40	38.5	36.4
	75	40.2	40.9

Table 4. Bulk weight [kg/m^3].

Value	Amount of fibers [kg/m^3]	Type of fiber Dramix OL13	Dramix 3D
MIN	40	2302	2319
	75	2311	2372
MAX	40	2315	2358
	75	2339	2384
Average	40	2310	2330
	75	2320	2380

5. Splitting tensile strength

For the splitting tensile strength test was applied, i.e. 12 samples perpendicular to the filling direction and 8 samples parallel to the filling direction. The weight of these samples was about 8 kg. On the basis of the tests, the splitting tensile strengths computed for the direction perpendicular to the filling direction are shown in table 5 and parallel to the filling direction are shown in table 6. In the case of Dramix OL13/20 fibers, the difference in the dosage of 40 and 75 kg/m^3 was minimal. Flexural strengths were 3.1 MPa. A noticeable difference can be observed for Dramix 3D 55/30BG fibers. In this case, the strengths were 2.8 and 4 MPa. The difference, therefore, is more than 1 MPa. Also interesting are the results for strengths parallel to the filling direction. In all cases, the strength was lower. Smaller difference was detected in Dramix OL13/20 fibers. For Dramix 3D 55/30BG fibers, the difference was 75 kg/m^3 with a difference of more than 1 MPa. For samples with a fiber dosage of 40 kg/m^3, the difference in homogeneity depending on the direction of testing was small. The greater splitting tensile strength was in the direction perpendicular to the filling direction. Larger differences in the strength are for samples with a fiber dosage of 75 kg/m^3. The results are also marked in figure 7.

Table 5. Splitting tensile strength - perpendicular to the filling direction [MPa].

Value	Amount of fibers [kg/m^3]	Type of fiber Dramix OL13	Dramix 3D
MIN	40	2.8	2.6
	75	2.6	3.6
MAX	40	3.3	3.0
	75	3.4	4.2
Average	40	3.1	2.8
	75	3.1	4.0

Table 6. Splitting tensile strength - parallel to the filling direction [MPa].

Value	Amount of fibers [kg/m^3]	Type of fiber Dramix OL13	Dramix 3D
MIN	40	2.7	2.3
	75	2.5	2.7
MAX	40	2.7	2.6
	75	2.6	2.8
Average	40	2.7	2.4
	75	2.5	2.7

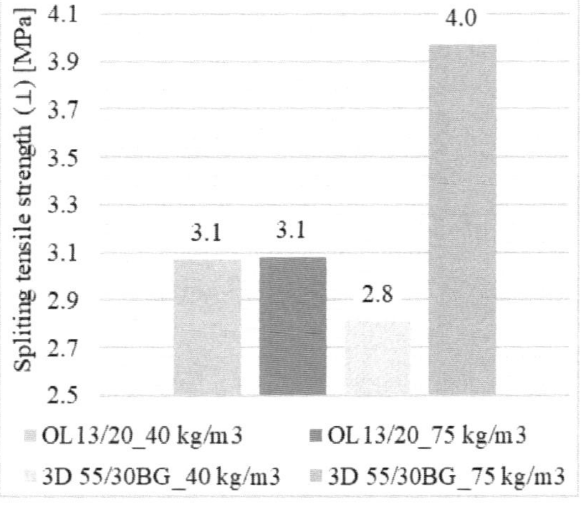

Figure 7. Splitting tensile strength - perpendicular to the filling direction.

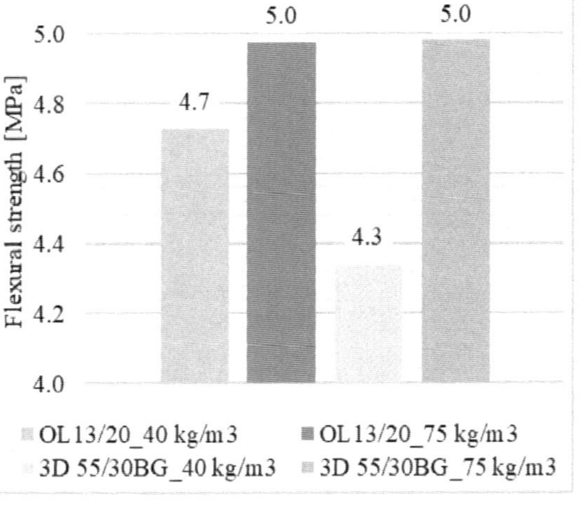

Figure 8. Flexural strength.

6. Three-point bend test

Another test of the concrete was a three-point bend test. A total of 12 beams of 150 x 150 x 600 mm were prepared in the laboratory with a notch in half-span. The depth of notch was 50 mm and the span at the loading test was 500 mm. The weight of the samples was approximately 31 kg. Six beams were reinforced by Dramix OL13/20 fibers, of which 3 beams had a fiber content of 40 kg/m^3 and 3 beams of 75 kg/m^3. Other six beams were reinforced with Dramix 3D 55/30BG fibers, of which 3 beams had a fiber content of 40 kg/m^3 and 3 beams of 75 kg/m^3. After 28 days, the three-point bend test was performed. Based on the results, the strength of all samples was subsequently determined (table 7). The test load was applied until the deflection was 8 mm. The flexural strength was very similar for both types of fiber dosing. The flexural strengths with Dramix OL13/20 fiber were 4.7 and 5 MPa. For Dramix 3D 55/30BG fibers, the flexural strengths were 4.3 and 5 MPa. The results are also marked in figure 8.

Table 7. Flexural strength in a three-point bend test.

Value	Amount of fibers [kg/m³]	Type of fiber Dramix OL13	Dramix 3D
MIN	40	4.4	4.2
	75	4.5	4.8
MAX	40	4.9	4.6
	75	5.8	5.2
Average	40	4.7	4.3
	75	5.0	5.0

Table 8. Fracture energy values corresponding to 8 mm deformation [N/m].

Value	Amount of fibers [kg/m³]	Type of fiber Dramix OL13	Dramix 3D
MIN	40	1189	1699
	75	1329	2973
MAX	40	1518	2956
	75	2043	3603
Average	40	1352	2272
	75	1665	3354

An evaluation of all load diagrams shown in Figures 9 and 10 was also carried out, as well as the determination of the work *A* respectively of fracture energy W_f, for deformations of 8 mm.

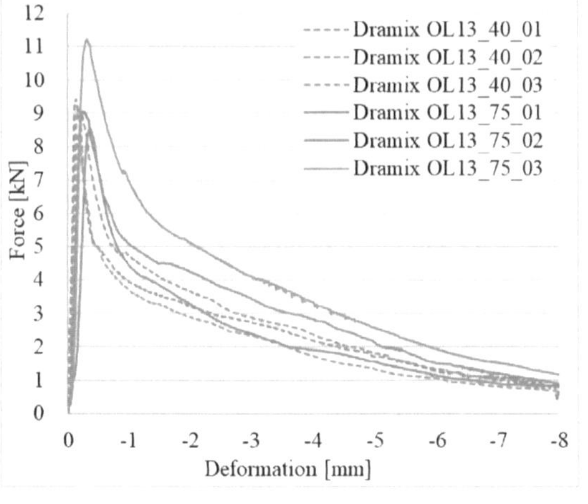

Figure 9. Dramix OL13/20 load test process with fiber dosing of 40 kg/m³ and 75 kg/m³.

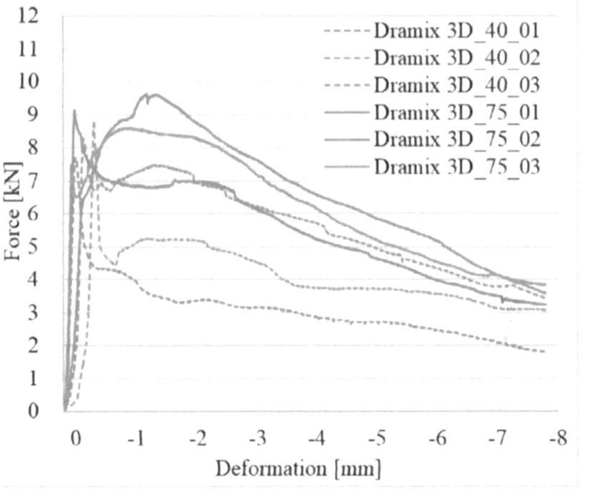

Figure 10. Dramix 3D 55/30BG load test process with fiber dosing of 40 kg/m³ and 75 kg/m³.

The curve of the diagram for both dosages is very similar. Slightly greater residual flexural strength occurs at higher dosages of fibers. For Dramix 3D 55/30BG, the differences are more pronounced. Also, in the case of the 75 kg/m³ dosage, the flexural strengths have two peaks. However, fracture energy needs to be determined for a more detailed description. The average fracture energy values are shown in table 8.

To illustrate the behavior of the fiber-reinforced concrete, a comparison of the working diagrams for the same dosing but different fibers in figures 11 and 12 is also made.

7. Results and discussion

The work dealt with the fracture mechanics parameters of selected two types of steel fiber reinforced concrete that differed in the dosage of fibers, which was 40 and 75 kg/m³. Specifically, Dramix OL13/20 straight fibers and curved fibers Dramix 3D 55/30BG were selected for the experiment. The results have shown effect of fibers on the compressive strength and the flexural strength of the concrete.

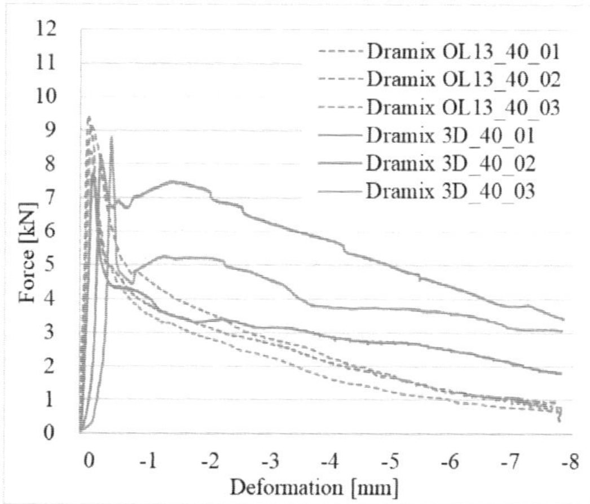

Figure 11. Dramix OL13/30 and Dramix 3D 55/30BG stress tests (bending) with fiber dosing of 40 kg/m^3.

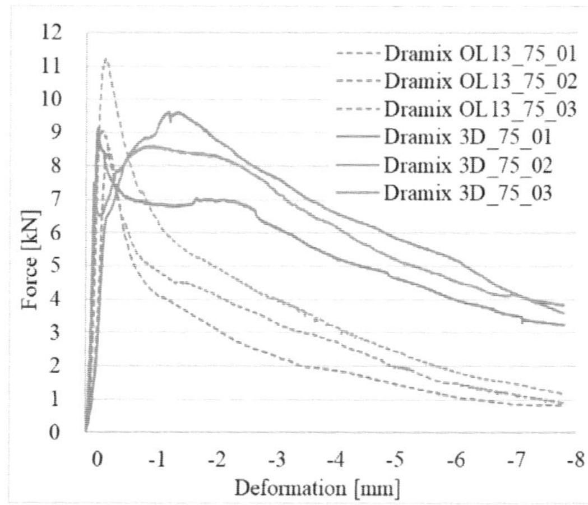

Figure 12. Dramix OL13/30 and Dramix 3D 55/30BG stress tests (bending) with fiber dosing of 75 kg/m^3.

However, the stronger effect of the fibers is in the case of flexural strength and fracture energy. It has also been shown that the flexural strength parallel to the filling direction is lower than in the direction perpendicular to the filling. Smaller differences were for Dramix OL13/20 fibers. The flexural strength difference was small for Dramix OL13/20 fibers for both dosings. However, the fracture energy was higher. Larger differences were for Dramix 3D 55/30BG fibers, both in flexural strength and fracture energy. Also, the curves of the diagrams were different. In the case of Dramix 3D 55/30BG and 75 kg/m^3, two peaks of flexural strength were even achieved in some cases. In the case of Dramix 3D 55/30BG fibers, higher residual flexural strength remains even after the test. Another possibility of expanding the knowledge about the mechanical properties [21] of fiber concrete is to identify detailed fracture parameters [22], [23] suitable for advanced nonlinear analyzes of concrete structures [24], [25] where authors focus on this area in further research.

Acknowledgement

The work was supported by means of the project Utilization of secondary raw material in geopolymers production and by Conceptual development of science, research and innovation assigned to VŠB-TUO by the Ministry of Education, Youth and Sports of the Czech Republic.

References

[1] Brandt A M 2008 *Fiber reinforced cement-based (FRC) composites after over 40 years of development in building and civil engineering Composite Structures* vol **86** (1-3) pp 3–9

[2] Katzer J and Domski J 2012 *Quality and Mechanical Properties of Engineered Steel Fibres Used as Reinforcement for Concrete Construction and Building Materials* vol **34** pp 243–48

[3] Holschemachera K, Muellera T and Ribakov Y 2010 *Effect of steel fibres on mechanical properties of high-strength concrete Materials& Design* vol **31** iss 5 pp 2604-2615

[4] Koniki S and Ravi P D 2018 *A study on mechanical properties and stress-strain response of high strength concrete reinforced with polypropylene–polyester hybrid fibres Cement, Wapno, Beton* iss 1 pp 67-77

[5] Kasagani H and Rao C B K 2016 *The influence of hybrid glass fibres addition on stress – Strain behaviour of concrete Cement, Wapno, Beton* iss **5** pp 361-372

[6] Vaskova J and Cajka R 2018 *Interaction of nonlinear numerical model of SFRC slab and nonlinear numerical subsoil model International Journal of GEOMATE* vol **15** iss 47 pp 103-110 DOI: 10.21660/2018.47.3576

[7] Labudkova J and Cajka R 2017 *3D numerical model in nonlinear analysis of the interaction between subsoil and sfcr slab International Journal of GEOMATE* vol **13** iss 35 pp 120-127.

[8] Sucharda O, Bílek V, Smirakova M, Kubosek J and Cajka R 2017 *Comparative Evaluation Of Mechanical Properties Of Fibre-ReinforcedConcrete And Approach To Modelling Of Bearing Capacity Ground Slab Periodica Polytechnica: Civil Engineering* vol **61** iss 4 pp 972-986 DOI:10.3311/PPci.10688

[9] Kurihara N, Kunieda M, Kamada T, Uchida Y and Rokugo K 2000 *Tension softening diagrams and evaluation of properties of steel fibre reinforced concrete Eng Fract Mech* vol **65** pp 235-245 https://doi.org/10.1016/S0013-7944(99)00116-2

[10] Tipka M and Vasková J 2017 *Tensile properties of structural fibre reinforced concrete of Fiber Concrete 2017 IOP Conf. Ser.: Mater. Sci. Eng.* vol 246 DOI: 10.1088/1757-899X/246/1/012052

[11] Sucharda O and Bilek V *Aspects of Testing and Material Properties of Fiber Concrete 25th Concrete Days 2018* pp 1-6

[12] Zaborski A 2016 *Constitutive model for restricted compression of fiber concrete Cement, Wapno, Beton* iss 1 pp 46-52

[13] Kormanikova E and Kotrasova K 2011 *Elastic mechanical properties of fiber reinforced composite materials Chemicke Listy* vol **105** iss 17

[14] RILEM TC 162-TDF 2002 *Test and Design Methods for Steel Fiber Reinforced Concrete – Design of Steel Fiber Reinforced Concrete Using the σ-w Method: Principles and Application. Materials and Structures/Matériaux et Constructions* vol **35** iss 249 pp 262–267

[15] di Prisco M, Colombo M and Dozio D 2013 *Fibre- reinforced concrete in fib Model Code 2010: principles, models and test validation Structural Concrete* vol **14** iss 4 pp 342-361

[16] RILEM 2011 About Rilem [Online] [Accessed on 4 May 2011]

[17] DAfStb guidelines 2011 *DAfStb-Richtlinie Stahlfaserbeton. Deutscher Ausschuss für Stahlbeton – DAfStb*, Berlin, German. (In German)

[18] BS EN 14721: *Test method for metallic fibre concrete – Measuring the fibre content in fresh and hardened concrete*. BSI, 2005, 2007.

[19] Model Code 2010 – *Final Draft, fib, Bulletin* No 65 and 66. 1-2. 2012

[20] https://www.bekaert.com/en/products/construction/concrete-reinforcement.

[21] Sucharda O, Pajak M, Ponikiewski T and Konecny P 2017 *Identification of mechanical and fracture properties of self-compacting concrete beams with different types of steel fibres using inverse analysis Construction and Building Materials* vol **138**, pp 263-275 DOI: 10.1016/j.conbuildmat.2017.01.077

[22] Sucharda O, Konecny P, Kubosek. Done P. J. *Finite element modelling and identification of the material properties of fibre concrete In Conference: 23rd Conference of Italian Group of Fracture* Location: Sicily, Procedia Engineering, vol. **109**, pp. 234-239, 2015.

[23] Sucharda O, Lehner P, Konecny P and Ponikiewski T *2018 Investigation of Fracture Properties by Inverse Analysis on Selected SCC Concrete Beams with Different Amount of Fibres Ecf22 - Loading And Environmental Effects On Structural Integrity Procedia Structural Integrity* vol **13** pp 1533-1538 DOI: 10.1016/j.prostr.2018.12.313

[24] Sucharda O and Konecny P 2018 *Recommendation for the modelling of 3D non-linear analysis of RC beam tests Computers and Concrete*, vol **21** iss 1 pp 11-20 DOI: 10.12989/cac.2018.21.1.011

[25] Sucharda O, Smirakova M, Vaskova, J, Mateckova P, Kubosek J and Cajka, R Punching *Shear Failure of Concrete Ground Supported Slab International Journal of Concrete Structures and Materials* vol **12** iss 1 Article number 36 DOI: 10.1186/s40069-018-0263-6

Possibilities of evaluation of the masonry wall affected by the explosion

P Schmid[1], J Láník[1], P Žítt[1], L Halašta[2], P Čalkovský[2] and M Bartošková[2]

[1] Brno University of Technology, Faculty of Civil Engineering, Veveří 331/95, 602 00, Brno, Czech Republic
[2] Military Research Institute, s.p. Veslařská 230, Brno, Czech Republic

E-mail: Schmid.p@vutbr.cz

Abstract At present, emphasis is placed on the safety and durability of building structures due to the effects of external loads. Among the various external influences of the load, we can also include loads caused by the blast wave. Structures subjected to extreme loads such as explosion are structurally disrupted in a manner different from that of static or dynamic external loads. In construction practice, we are sometimes faced with the question of how to evaluate the state of a structure after an explosion. The article is focused on the possibility of evaluating the masonry wall affected by the explosion in terms of the expression of residual bearing capacity based on the laboratory tests and inspection in-situ.

1. Introduction

The subject of the assessment is the identification of the extent of the damage with the aim of determining the residual bearing capacity of the masonry wall after exposure to a nearby explosion. The vertical wall masonry structure is made of ten layers of POROTHERM 44 P + D P10 ceramic building materials. There is a masonry mortar MVC 910 of strength class M10. Wall width 2310 mm, height 2470 mm, thickness 440 mm. Layers of masonry mortar in loading joints of approx. 10 mm thickness. The overall condition at the start of the diagnostic work is documented in Figure 1. The test masonry wall was built and tested by a Military Research Institute and subsequently removed to Brno University of Technology.

2. Scope of diagnostic and selected methods

The evaluated structural unit was gradually disassembled manually from the top layer of the masonry (on the layer designation Layer 1) to the last (foundation) layer. The diagnostic itself was performed by a proven combination of testing methods. They are:

- visual inspection,
- manual disassembly of the structure,
- sampling,
- laboratory testing.

IOP Conf. Series: Materials Science and Engineering **549** (2019) 012040 doi:10.1088/1757-899X/549/1/012040

Figure 1. General view of brick wall element damaged by explosion effects (front and back view).

2.1. Visual inspection

The construction is made of ten layers of POROTHERM 44 P + D P10 ceramic building materials. There is a masonry mortar MVC 910 of strength class M10. Wall width 2310 mm, height 2470 mm, thickness 440 mm. Layers of masonry mortar in loading joints of approx. 10 mm thickness [1].

The outer wall portions in contact with the circumferential metal stabilizing frame (layer 1 and 10, vertical edge lines) are heavily damaged by shear and pressure cracks in the inner structure of the billets. The blocks of the blocks are broken by cracks always over the entire height of the building material. These lines are loosely hand-disassembled after mechanical removal of the leveling layers of the mortar to the joints with the metal stabilizing frame. The nature of the damage is related to the bulging of the wall away from the effect of the explosion. The horizontal deformation of the rear wall surface reaches up to 30 mm in the extreme. The face of the wall is damaged by the separation of the ceramic shards of the side peripheral walls of the built-in vertical perforated blocks [2].

2.2. Manual disassembly of the structure

In general it can be stated that the internal defects of the walls (ribs) of the built-in luminaires and the masonry mortar in the joints are so extensive that the construction itself could be dismantled manually.

Sub-findings are listed and commented on the following images.

Figure 2. Loosely removable fragments of built in lumber from layer 1. Pieces of lumps in contact lines with the stabilizing perimeter frame are damaged throughout the wall thickness.

Figure 3. Detail of damage to the internal structure of the ceramic blocks by manually disassembling layer 1.

Figure 4. Detail of the loading joint above layer 5 (overall view of FOTO 8). Total weakening of the wall thickness in the middle part from the original wall width 440 mm to the value of approx. 315 mm (ie 30% weakening).

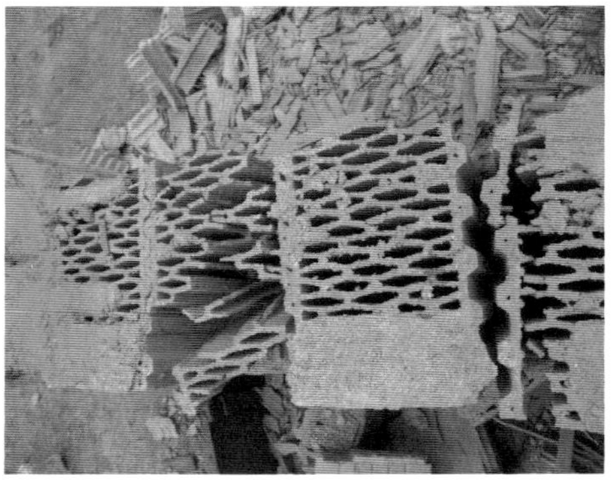

Figure 5. Detail of the internal damage of the built-in pieces of the lower masonry layer during manual disassembly (Layer 10).

Figure 6. Detail of the internal damage of the built-in piece of building materials with stress at the "front" part of the wall section due to arched bulging in the direction of the "rear" face. In the layers at the "back cheek" a slight opening of the joints opens.

Figure 7. Detail of shear damage of built-in blocks in line with stabilizing frame. Furthermore, the damage to the masonry mortar in the bed joint - massive cracks in the contact joints of the building materials.

2.3. Sampling
Fragments, built-in vertical perforated blocks with varying degrees of damage can be removed from the inner wall area (layers 2 to 9) by manually disassembling the wall (light blue contour highlighted in the figure). A total of 12 pieces of built-in construction materials marked V1 to V12 were taken from this area. The collected samples were adapted to be transported to laboratories to diamond circular saws by finishing the lateral damaged wall on regular shaped bodies which are suitable for carrying out laboratory tests.

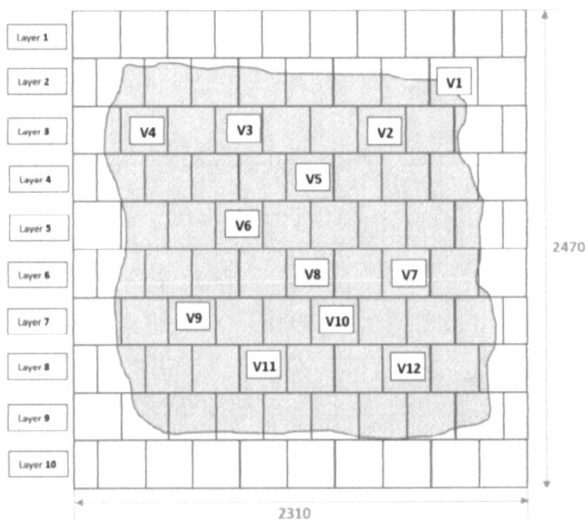

Figure 8. Schematic diagram of taken samples of bricks.

Figure 9. Brick wall after manual disassembly.

The real weakening of the wall thickness from the effects of the explosion is very different on the individual collected building materials. It is a percentage of the total weakening of the original width of 440 mm by 30% to 80%. Some lump construction materials fell apart spontaneously (especially in the case of line joints with the wall stabilization frame), figure 7.

2.4. Laboratory testing
From the inner wall area (layers 2 to 9), fragments of built-in vertical perforated blocks with varying degrees of damage were removable by hand dismantling the wall. A total of 12 pieces of built-in construction materials marked V1 to V12 were taken from this area. The collected samples were transported to the diamond circular saw after transport to the laboratory by finishing the lateral damaged wall on the bodies of regular shapes. Tensile strength tests according to ČSN EN 772-1 + A1 [5] were carried out on the modified test samples.

A total of five "reference" samples were delivered to the laboratory - POROTHERM 44 P + D ceramic blocks, which were not incorporated into the test wall. After taking over to the laboratories, the samples were labeled REF1 to REF 5. Tensile strength tests according to ČSN EN 772-1 + A1 [5] were performed on the supplied samples.

3. Results of the laboratory tests
The aim of the laboratory tests was to assess the impact of the damage of the removed blocks from the masonry of the test wall compared to the reference samples in the aspect of the effect on the final compressive strength of the evaluated piece building materials [4], [5]. The test results are summarized in the following Table 1 and Table 2.

Table 1. Results and evaluation of POROTHERM 44 P + D, reference samples.

Sample marking	Critical force [kN]	Compressive area [mm^2]	Compressive strength [N/mm^2] Individual	Compressive strength [N/mm^2] Average value (coeff. of variation [%])
REF_1	1260	107568	11.7	**10.9** (4.5)
REF_2	1185	106085	11.2	
REF_3	985	92250	10.7	
REF_4	990	95160	10.4	
REF_5	980	93480	10.5	

Table 2. Results and evaluation of POROTHERM 44 P + D blocks, collected fragments.

Sample marking	Critical force [kN]	Compressive area [mm^2]	Compressive strength [N/mm^2] Individual	Compressive strength [N/mm^2] Average value (coeff. of variation [%])
V_2	265	40425	6.6	**6.5** (20.9)
V_3	545	79380	6.9	
V_4	345	64660	5.3	
V_5	460	65636	7.0	
V_6	380	64904	5.9	
V_7	575	71050	8.1	
V_8	505	61250	8.2	
V_9	275	54145	5.1	
V_10	520	59048	8.8	
V_11	245	57096	4.3	
V_12	405	70110	5.8	

4. Evaluation of the survey and recommendations

If the test wall is loaded and modeled the structural design of the masonry vertical structure, then the loaded structure would collapse after the explosion effects.

The built-in model (for example, perimeter walls in the structural support structure) has demonstrated the ability to significantly deform the structural member without overall structural collapse. The effects of the explosion were extensive surface damage by separating the fragments of the side walls of the built-in lumps. The extent of total damage (including the internal structure) and the massive sagging of the compressive strength prove that the real construction would be irreparable in terms of mechanical resistance requirements. It would be necessary to dismantle it (remove) and build a new built-in structure [3].

The "frontal" surface of the wall (face at the blast) is damaged by the separation of fragments of the ceramic body of the side peripheral walls of the built-in vertical perforated blocks of the exterior face of the structure.The vertical deformation of the "rear" wall surface reaches up to 30 mm in the extreme.

5. Conclusion

In view of the findings of the pathological-defectoscopic analysis, the results of laboratory tests on the collected fragments of lumps of construction materials and the supplied reference samples of the supplied blocks can be stated as follows.

The internal defects of the walls (ribs) of the built-in piece of building materials, the overall strongly uneven weakening of the original cross-section (thickness), the damage and depletion of the bonding strength of the bonding bridge between the mortar and the built-in piece of construction in the loading joints show the exhaustion of the mechanical resistance in the test wall after the effects of the explosion on more than 70%.

Acknowledgement

This paper has been worked out under the project No. LO1408 "AdMaS UP - Advanced Materials, Structures and Technologies", supported by Ministry of Education, Youth and Sports under the „National Sustainability Programme I".

References

[1] Technical data sheets of used POROTHRM 44 P + D P10 and MVC 910 M10 mortar.
[2] ČSN ISO 13822 2014 *Principles of Structural Design - Evaluation of Existing Structures*
[3] ČSN 73 0038 2014 *Evaluation and verification of existing structures - Supplementary provisions*
[4] EN 772-16 *Test methods for masonry elements - Part 16: Determination of dimensions*
[5] ČSN EN 772-1 + A1 *Test methods for masonry units - Part 1: Determination of compressive strength.*
[6] Ahmad S et al 2013 *Mater. de Construccion.* **64** 313 e007 p11
[7] Doerr A et al 2012 The effect of near, contact and embedded detonations on masonry structrures, Proccedings 22nd Internationals\ Symposium.
[8] Eamon C D et al 2004 *J of Engin. Mechanics* **130** 9 pp 1098-1106, doi: 10.1061/(ASCE)0733-9399(2004)130:9(1098).

IOP Conf. Series: Materials Science and Engineering **549** (2019) 012041 doi:10.1088/1757-899X/549/1/012041

Biodegradation and its impact on durability of building construction

E Terpakova[1] and A Estokova[1]

[1]Technical University of Kosice, Faculty of Civil Engineering, Institute of Environmental Engineering, Vysokoskolska 4, Kosice 04200, Slovakia

E-mail: eva.terpakova@tuke.sk

Abstract. The durability of building elements is influenced by a number of different factors. Among the most complicated, biocorrosion can be included without exaggeration. From practical experience, we know that biodegradation manifests itself in all kinds of materials and structural elements. Although both primary and secondary protection of a particular building structure is realized, if appropriate living conditions are created for the existence of a particular species of biodeteriogen, this kind of degradation will be initiated. The paper deals with the issue of biocorrosion, which was studied at selected parts of the technical extension in Campus of the Technical University in Košice. Undesirable bio-corrosive symptoms due to the presence of algae and fungi are often visually observed on the surfaces, but their effects also have a negative impact on the internal structure. In practical terms, a solution to identification of these symptom are required which needs to develop methods of monitoring the state of the surface over time and removing them. Diagnostic methods will be presented to check the technical state condition of construction parts as support in identification bio-corrosion symptoms.

1. Introduction

Most of the degradation processes require the presence of water. On the one hand, water is necessary for the performance of technological procedures, but on the other hand, causes disorders construction. Humidity is one of the basic technical parameters which are observed continually from production of building materials up to demolition of building objects. With higher humidity goes plenty of processes which lead to decreasing durability and reliability of construction. Practical determination of humidity of each other materials therefore becomes an important tool of quality control, technological processes while realizing up to diagnostic of corrosion. The determination of presence of water is the basic of diagnostics procedures. There is a particular correlation of humidity and related chemical changes and corrosion processes of building materials [1]. Water within its natural circulation in environment is never chemically pure. When we take this into consideration, the logical conclusion is that because of the presence of water pollutants, in the structure of building materials appears chemical and chemical – physical processes. Polluted water also causes salinity of elements and structures, which is a significantly interconnected degradation processes. The most important physical aspects are phase changes of water (gaseous - liquid - solid state) which are associated with a change in their physical parameters. [2] The main types of such processes in building are soaking, damping and drying. Upon penetration of contaminated water into the structure, the dissolved salts are transported to the whole structure of the elements. If the aqueous phase is evaporated under suitable thermodynamic conditions, the salts may be precipitated. After evaporation of liquid phase are formed new products on the surface or inside of the construction [2]. Their presence is manifested in the form of both crystalline and amorphous new products. Production of new products can be connected with volume changes, which are linked with mechanical consequences (increasing tension, decreasing of strength parameters or even destruction) [2]. In diagnostics the technical state of building elements, salts are very important phenomena and their concentrations are used as assessment criteria of damaged parts of construction. Humidity related degradation processes as well as effect lies in the formation of available conditions to grow of biodeteriogens [3]. Biocorrosion of building materials - stone, concrete, mortar, etc. can be caused by such microorganisms as bacteria, fungi, lichens, algae, and plants [3]. Biofilms are complex communities of microorganisms attached to surfaces of materials generally, or associated with interfaces. Bacterial biofilm formation on inert surfaces is a significant technical and economic problem in a wide range of environmental, industrial, and other areas. Bacterial adhesion is generally a prerequisite for this colonization process and thus, at the same time represents suitable live conditions for another biodeteriogens mostly different types of mosses, moulds, algae's and plants. Multidiscipline methods are available for the study of biodeterioration building materials. Almost of them is bearing to biological part of problem e.g. identification of biodeteriorates, evaluating of birth biofilm or studies of metabolites. An investigation of the effects of humidity and salt crystallization on masonry has been undertaken at chosen part of technical extension of building and its roofing. Masonry has been contaminated with chloride salts above all, which are undergoing cycles of crystallization - dissolution in response to fluctuations in relative humidity. Assessment of degradation masonry consisted of determination of moisture and salinity combination of gravimetric and chemical analysis of samples. Biocorrosion of masonry was also studied by using spectral analysis method.

2. Methods

2.1 In situ determination

In this case study was provided diagnostics of selected building in Campus area of Technical University Košice specially its exterior finish roofing basement. As can be seen from photo documentation, the outer above-ground part of the basement is solved by a combined masonry; from two sides is provided by a concrete fence (Fig. 1), from the north side by a brick wall (Fig. 2). The western part of the fence is directly connected to the facade of building. From a material point of view it can be stated that at this relatively small area has been realized combination of various structural components and materials. In situ was done visual confirmed cracking patterns, efflorescence, and spalling parts. All this knowledge's provide useful information regarding the structure and current situation. To the purpose of biocorrosion study (chemical, physical and structural changes) is important to obtain the representative sample from damaged part. Also, the dimensions and condition of the damaged elements must be taken into account in order to avoid deterioration of their technical condition.

2.2 Methods of sample processing

Samples were taken at atmospheric conditions: relative humidity of air was 65% , air temperature was 15°C. Sampling of external parts of concrete parts was carried out at a height 0.5m, 1.0m and 1.5m measured from the floor of the basement. Sampling of external parts of bricks and mortars (north part) were carried out at a height 0.5 m. All samples were taken carefully from damaged parts - the depth was up to 20-35mm, material was stored in plastic sealable containers, labelled and then processing in laboratory. In the case of a significant crack in the south-oriented part of the structure, the sampling was also carried out from the inside from a depth of approx. 200-300mm.

2.3 Determination of humidity

Humidity of samples was provided by gravimetric method. The samples were weighed in Laboratory using electronic scales Mettler Toledo, to an accuracy of 0.0001g, then dried up to 105-110°C (Laboratory dryer Matest, ITALY) until constant weight. The mass humidity w_{mh} was calculated according to formula (1).

$$w_{mh} = \frac{m_w - m_d}{m_d} \text{ x } 100\% \tag{1}$$

where: m_w = mass of wet sample,
m_d = mass of dry sample,
W_{mh} = mass humidity of sample.

Follow to the standard CSN 73 0610 (*informative Annex A*) [4] masonry mass humidity is classified by percentage mass humidity, follow to criteria in Table1.

Table 1. Criteria for assessment of silicate materials humidity [4].

w_{hm} [%]	humidity
< 3	low
$3 \leq W_{hm} < 5$	very low
$5 \leq W_{hm} < 7.5$	increased
$7.5 < W_{hm} < 10$	high
$W_{hm} > 10$	very high

The obtained values w_{mh} (see Table 2) were different for the individual parts of masonry, partly linked to the fact, that sampling of material was realised after raining weather.

2.4 Chemical processing

For chemical analysis were used a representative parts of samples after their particular homogenization follow to methodical recommendations [5]. Determination of c(Cl) concentration (as selected representative salts), was provided by potentiometric method using selective chloride electrode. Alkality changes of samples were expressed as pH values, the potentiometric method was chosen for their estimation. Table 2 shows the comparison of the average results of silicate samples taken from above mentioned height from the ground. Selected samples of silicates were also subjected to spectral FTIR analysis.

3. Results and discussion

The results of diagnostic assays show (see Table 2), that the samples have followed to criteria [4] higher degrees of humidity. The samples taken from the brick are able to absorb humidity and consequently salts are deposited into pores, their technical state was insufficient.

The presence of salts has also been observed at the surface of the sampling points as efflorescence's. Increased salt concentrations are related to the location of sampling and also in close vicinity of the driveway with parking and their maintenance in the winter months. According to the results in Table 2, it is evident that chloride salts are accumulated at a height of 0.5 m from the basement, with a high degree of salinity according to criteria of WTA directive [6].

Increased concentrations of chlorides which have been determined, indicate degradation processes. At the same time, decreased pH values of silicate samples were observed, indicating the carbonation process [5]. In addition to above mentioned method was made FTIR spectral analysis of sample by ATR FTIR spectrometer Alpha Brucker, in range 4000-360 cm^{-1}, with 24 scans. The results of 2 samples taken from the south part - surface and inner part from crack, are shown in Fig. 5.

Table 2. Results of analytical determination of mass humidity and salinity.

Orientation of sampling	Type / high of sampling	w_{mh} ±0.01%	pH ± 0.01 [-]	c(Cl)± 0.01%	degree of salinity [6]
north	plaster 0.5m	5.64	8.21	0.03	*low*
	mortar 0.5m	8.65	8.45	0.27	*high*
	brick 0.5m	10.64	8.65	0.56	*very high*
east	concrete 0.5m	8.32	8.90	1.30	*very high*
	concrete 1.0m	8.67	9.10	1.70	*very high*
	concrete 1.5m	8.25	9.35	1.40	*very high*
south	concrete 0.5m	16.20	8.85	1.72	*very high*
	concrete 1.0m	14.30	9.15	1.50	*very high*
	concrete 1.5m	10.15	9.14	1.67	*very high*

Figure 1. General view of the southern part of the enclosure (concrete support wall).

Figure 2. General view of the northern part of the enclosure (brick masonry with sheet metal overlap).

Figure 3 also shows a significant failure of the crack, which was located along the entire height of the southern part of the roof (its location is also highlighted in Fig.1). After measuring its depth, it was found to be up to 300mm. The presence of ants as a further source of biodegradation of silicates was found in the interior of the crack during its measurements. A representative sample of the damaged concrete was also selected from this site and subjected to spectral FTIR analysis, spectra recording are at Fig. 5. A significant occurrence of algae from the surface parts is documented in Figure 4. As we can see, there were also cracks and leaching of the coating layers due to biocorrosion.

Figure 3. Determination of wide cracks of damaged concrete parts.

Figure 4. Detail of biocorrosion of masonry due to algae.

Method FTIR is a powerful aid in identifying corrosive changes in building materials. In Fig. 5 are characteristic FTIR transmittance spectra of 2 concrete samples obtained from surface and inner part of crack (approximately from 200mm of depth). In both spectra a weak broad band of approximately 1100 cm^{-1} is an indication of silicate (-Si-O) vibrations, which are overlapped the peak of the anion -SO_4^{2-}. The characteristic peak of calcite (1410 cm^{-1}) is in inner part stronger as in surface parts. The characteristic peaks about 1650cm^{-1}, 1064cm^{-1} respectively at 620cm^{-1} which are an indication of sulphates [2], [3] guiding us to conclude that the samples contain products of biocorrosion processes.

Vibration of -OH groups (1610-1628cm^{-1}) was also partially reflected in both spectra. Surface part has at 3570 cm^{-1} resp. 3398 cm^{-1} stronger OH stretch vibrations, of which we expect to be of the metabolites of present algae. Conversely, at 2920cm^{-1} and 2850 cm^{-1} vibration of CH groups are more pronounced in the interior of the concrete, which could be corresponded to organic acids as a manifestation of biodegradation in the interior of the concrete structure.

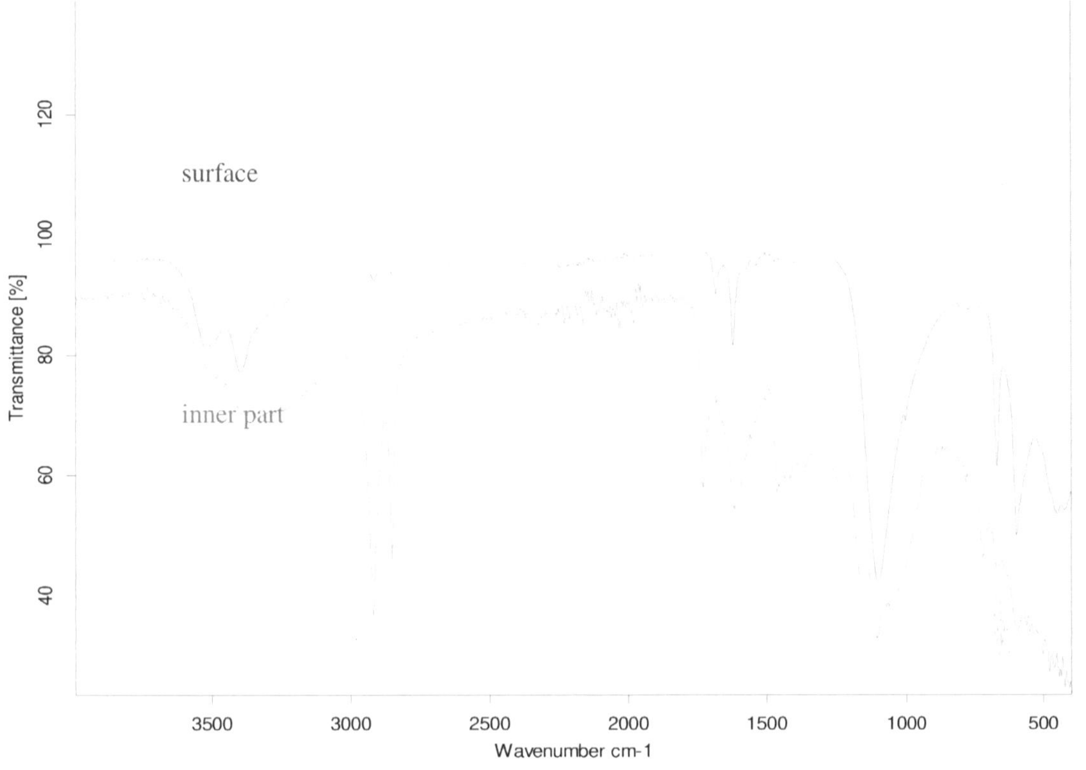

Figure 5. FTIR spectra of concrete samples – south part of construction.

4. Conlusion

Determinations of humidity, corrosion and salinity of silicate materials is from durability and reliability aspects very important, taking into account the need for proper selection of sanitation materials.

The obtained results logically complement and confirm fact, that all parts of the structure under consideration are damaged by bio-corrosion and require reconstruction. Protection against atmospheric conditions would be appropriate, including repair of downpipes of rainwater, revision or addition of waterproofing and surface treatment, etc.. A specific assessment of the samples has unfortunately confirmed the expected state of urgency for repairs.

Acknowledgements
This contribution is supported by Project VEGA 2/0142/19.

References

[1] Dobrý O and Palek L 1988 *Concrete corrosion in building praxis (*Praha: ed. SNTL)
[2] Matoušek M and Drochytka R 1998 *Atmospheric corrosion of concretes (*Praha: ed. Ikas)
[3] Wasserbauer R 2000 *Biological damaged of buildings (*Praha: ed. ARCH a.s.)
[4] ČSN P 73 0610 (2000) *Waterproofing of buildings.* The rehabilitation of damp masonry and additional protection of buildings against ground moisture atmospheric water. Basic provision
[5] Terpáková E 2016 *Diagnostic in civil construction.* (Košice: ed. Technical University)
[6] WTA 4-6-98 (2004) *Sealing of structural elements on contact with soil at a later state*

IOP Conf. Series: Materials Science and Engineering **549** (2019) 012042 doi:10.1088/1757-899X/549/1/012042

Moisture sorption and thickness swelling of wood-based materials intended for structural use in humid conditions and bonded with melamine resin

M Böhm[1], K Kobetičová[1], J Procházka[2], R Černý[1]

[1] Czech Technical University in Prague, Faculty of Civil Engineering, Department of Materials Engineering and Chemistry, Thákurova 7, 166 29 Prague 6, Czech Republic
[2] Mendel University in Brno, Faculty of Forestry and Wood Technology, Department of Wood Science, Zemědělská 3, 613 00 Brno, Czech Republic

E-mail: martin.bohm@fsv.cvut.cz

Abstract. When used in construction, the properties of wood-based materials they are mainly affected by moisture. Moisture sorption and the associated swelling has an impact on the resistance of materials to biocidal attacks, on their mechanical properties or air permeability, and on comfort of use and durability of the structure. Equilibrium moisture content (EMC) and thickness swelling (TS) depending on the relative humidity for plywood (PW), particleboard (PB) and oriented strand board (OSB), for load-bearing purposes in humid environments and bonded with melamine-urea-formaldehyde resin (MUF), are studied in detail in this paper. Equilibrium moisture content and thickness swelling are influenced by the type of material, density, quantity of adhesive composition, and paraffin. The highest values of equilibrium moisture content were found in plywood, whilst the highest thickness swelling was achieved by OSB. A high dependency of nonrecoverable thickness swelling on the equilibrium moisture of the material was also ascertained. The proportion of nonrecoverable thickness swelling increases exponentially, in particular if the material's equilibrium moisture content exceeds 12%.

1. Introduction

In the construction industry, wood-based materials are widely used for ceiling and floor constructions, for reinforcing vertical column structures, for facade systems, roof construction, etc. Wood-based materials may be exposed to increased humidity in all of these application methods.

During moisture sorption, swelling occurs for both wood-based materials and solid wood. While the cross-bonding of the wood elements makes the linear expansion in the board surface low, thickness swelling is significantly higher [1,2]. In addition, during swelling up to thickness, wood-based materials behave differently than solid wood. While solid wood returns to the same dimensions after swelling and re-drying, for wood-based materials, part of the thickness swelling is nonrecoverable [2,3]. As a result of nonrecoverable thickness swelling, the strength of the bonded joints is reduced and consequently the strength of all board materials [4,5] is lost. For OSBs, which are often used in structures as a vapor retarder, their permeability also increases [6]. The factors that most affect the sorption and swelling of wood-based materials primarily include: the density profile of the material, the type and quantity of adhesive used, hydrophobic agents, and the temperature affecting the materials during pressing [7,11].

The uniform density profile and lower compression of the wood elements during pressing reduces both the overall thickness and the proportion of nonrecoverable thickness swelling. However, it also negatively influences mechanical properties, in particular the modulus of elasticity and flexural strength [7]. In terms of the used adhesive, there are big differences not only in their properties but also in price. For the relevant class of material use, MUF adhesives are the most suitable, as they offer relatively good moisture resistance with simultaneous economic advantage. The MUF adhesive is usually added at 6-10%, whereas a higher adhesive deposition reduces thickness swelling while also improving the mechanical properties of the materials [8,9]. Various agents can be used as hydrophobic agents, such as silicones or oils. However, the most effective of these agents is paraffin, which is added to wood chips in the form of an emulsion, usually at 0.5-1.0% [9,11]. The pressing temperature differs in particular with regard to the different types of adhesives, and depends on whether the curing process is based on a temperature increase or pH change. The use of higher temperature and prolonged pressing time generally leads to a reduction in moisture sorption in wood-based materials compared to solid wood [2,7,12]. Achieving lower equilibrium moisture in wood-based materials is generally advantageous, not only because of smaller dimensional changes, but also because increased moisture causes easier vulnerability of wood parts to biotic agents.

The determination of equilibrium moisture, thickness change and proportion of nonrecoverable thickness swelling depending on moisture is therefore one of the most important tasks in evaluating the properties of wood-based materials. It is also important to ensure the use of materials under conditions that do not result in nonrecoverable swelling, and thereby deterioration of their properties.

2. Materials and methods

2.1. Materials

For the purposes of this research, samples of commercially-available materials produced in the Czech Republic were used, which are bonded with the same base adhesive, and the adhesive is cured at a similar pressing temperature. A urea formaldehyde adhesive with the addition of melamine (up to 30%) was used for the production of plywood and particleboard. Since OSBs are only produced in the Czech Republic by one production plant, which mainly uses isocyanates (MDI) for their bonding, it was not possible to evaluate OSBs bonded completely with MUF adhesive. An OSB type was chosen where the centre layer is bonded with MDI adhesive and the surface layers are bonded with MUF adhesive. Previous research has demonstrated [10] that high density surface layers have the largest share in OSB thickness swelling. In addition, the lower-density core layer is bonded with a cross-linked MDI adhesive and only marginally contributes to the nonrecoverable thickness swelling. Therefore, the OSBs were also included for comparison with materials exclusively bonded with MUF adhesive. Basic information on the tested materials is specified in Table 1.

Table 1. Composition of tested load-bearing boards for use in humid conditions

Panel type (Standard)	Thickness (mm)	Wood species	Adhesive	Wax (kg/m^3)	Press. temp. (°C)	Specification
OSB /3 (EN 300)	12.0	Spruce (80%) Pine (20%)	Face: MUF 8.5% Core: MDI 3.5%	1.2	210	Face to Core weight ratio: approx. 50/50
PB P5 (EN 312)	12.1	Spruce (80%) Hardwoods (20%)	MUF 11.5%	2.4	165	Pigments added to the adhesive.
PW 2/S (EN 636-2)	11.9	Spruce (core layers) Beech (face)	MUF 160 g/m2 (approx. 16%)	0	150	Layers: 11 Flour added to the adhesive.

IOP Conf. Series: Materials Science and Engineering **549** (2019) 012042 doi:10.1088/1757-899X/549/1/012042

2.2. Experimental methods

Test specimens were taken from commercially-available materials in accordance with EN 326-1. The deviation from the standard procedure consisted only of the number of selected specimens. The numbers of removed specimens were adjusted to determine the nonrecoverable proportion of thickness swelling. For each material, samples were taken from 10 boards so that 6 samples were randomly selected from each board. The samples were subsequently divided into 6 groups of 10 samples each. In each group, each board was represented by one sample. There were 180 samples in total for three types of materials.

The test samples were oven-dried and then air-conditioned in an air-conditioning chamber at 6 humidity levels at 15, 35, 50, 65, 85 and 100% relative air humidity at 20 °C. Density, moisture content, thickness swelling and nonrecoverable thickness swelling were then determined at all humidity levels.

In order to determine the density, the standard procedure according to standard EN 323 was used (see equation 1). The moisture content was determined according to standard EN 322 (see equation 2).

$$\rho = \frac{m_1}{l \cdot b \cdot t} \cdot 10^6 \quad \left[kg/m^3\right] \qquad w = \frac{m_1 - m_0}{m_0} \cdot 100 \quad [\%] \qquad (1, 2)$$

where m_1 is the mass of the test specimen after air-conditioning in individual humidity stages (g); l, b, t are the length, width and thickness of the specimen (mm) and m_0 is the mass of the test specimen after drying (g).

Thickness swelling (Eq. 3) was ascertained according to standard EN 318 and nonrecoverable thickness swelling (Eq. 4) was ascertained in such a way that after air-conditioning of the samples and ascertaining thickness swelling for the relevant moisture level, a group of samples re-dried at a temperature of (103 ± 2) °C was selected.

$$\Delta T = \frac{t_c - t_0}{t_0} \cdot 100 \quad [\%] \qquad \Delta T_N = \frac{t_{c0} - t_0}{t_0} \cdot 100 \quad [\%] \qquad (3, 4)$$

where t_0 is the thickness of the test specimen after drying (mm); t_c is the thickness of test specimens after air-conditioning at individual humidity levels (mm); t_{c0} is the thickness of re-dried test specimens after air-conditioning at individual humidity levels.

X-ray density profile analyzer DPX300-LTE (Imal, Modena, Italy) was used to ascertain the vertical density profile (VDP) of the tested materials. Measured data evaluation and statistical analyses were performed using the Statistica 13 (TIBCO Software, USA) program.

3. Results and discussion

3.1. Density

The density of the tested materials was first analysed, as the moisture sorption and thickness swelling are affected by density. For solid wood, the higher the density of wood, the more it swells [3] during air moisture sorption. However, this rule may not always apply to wood-based materials. In addition to average density, the moisture sorption of wood-based materials and their thickness swelling is also affected by the shape of their vertical density profile, wood particle size, the used adhesive composition and additives and other manufacturing parameters, as well as their interaction [2,12]. The density of the tested materials, calculated at 20 °C and 65% relative humidity, is shown in Figure 1.

IOP Conf. Series: Materials Science and Engineering **549** (2019) 012042 doi:10.1088/1757-899X/549/1/012042

In the figures with the vertical density profiles of the tested materials, the thickness of the board in millimetres (in this case 12 mm) is shown on the X axis, and the density in kg/m^3 is shown on the Y axis. It can be seen from the figures that the various materials have a different cross-sectional density profile, and that the usually-specified average density data may not be sufficient to assess the causes of thickness swelling.

Both particleboard and OSBs have a typical U-shaped density profile, and they only differ in that the density of the particleboard is slightly higher. However, since the particleboard has been surface treated via sanding, the highest density layers are on the surface of the board. For OSBs that have not been sanded, the layers with the highest density are below the surface of the board and the surface layers have a lower density.

On the vertical density profile of plywood, some variability in the density of individual veneers was observed, as well as increased density in bonded joints. It is obvious from the density profile of bonded joints that the adhesive penetrated the pores in the wood and the bonded joints have a very small thickness.

3.2. Equilibrium moisture content

The measured values of equilibrium moisture of materials at six different degrees of relative moisture are specified in Table 2. The table shows the mean values in % and the coefficients of variation in brackets, also in %.

Table 2. Equilibrium moisture content of studied materials for six different RH.

Material	**Relative humidity**					**100%**
	15%	**35%**	**50%**	**65%**	**85%**	
OSB	2.39	4.68	6.18	8.24	12.56	20.11
	(4.91)	(1.73)	(1.61)	(1.32)	(2.68)	(6.35)
PB	2.60	5.22	6.58	8.34	12.05	19.14
	(3.69)	(2.31)	(2.96)	(1.26)	(1.75)	(4.56)
PW	2.38	5.54	7.11	9.00	12.63	21.85
	(4.16)	(3.12)	(1.88)	(2.44)	(2.32)	(2.63)

All of the tested materials have significantly lower equilibrium moisture than solid wood under the same conditions. Therefore, the equilibrium moisture results were compared to a control sample of spruce wood that was subjected to the same conditions. The equilibrium moisture of solid spruce wood was only 11.68% at 20 °C and 65% relative air humidity at a density of 559 kg/m^3. It is commonly stated in professional literature that the equilibrium moisture of wood-based materials is about 2% lower than that of wood [9]. However, in this experimental measurement, it was found that the equilibrium moisture content of particleboard and OSB is lower by more than 3% at 65% relative humidity compared to wood.

Furthermore, the OSB equilibrium moisture is higher than the particleboard equilibrium moisture at the two highest moisture levels, even if a higher temperature was used during OSB pressing. This is likely due to a smaller coating of paraffin.

The highest equilibrium moisture was achieved by plywood due to the large proportion of solid wood in the form of veneers.

The moisture sorption is shown in Figure 5.

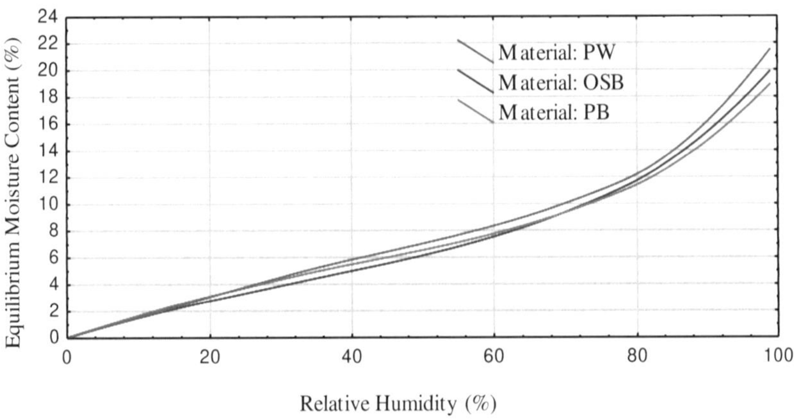

Fig. 5. Dependence between relative air humidity and equilibrium moisture content of materials.

Third degree polynomials were used to graphically depict the profiles of the measured equilibrium moisture values. The moisture sorption is very similar for all of the tested materials. The highest equilibrium moisture was achieved by plywood, but compared to the other OSBs, this difference is less than 10%. Adversely, the changes in thickness swelling are quite significant, see Tab. 3 and Fig. 6.

3.3. Thickness swelling

Table No. 3 shows average thickness swelling values in %. The numbers in brackets represent coefficients of variation, also in %.

Table 3. Thickness swelling of the studied materials.

Material	**Relative humidity**					
	15%	**35%**	**50%**	**65%**	**85%**	**100%**
OSB	0.61	1.58	2.24	4.09	8.52	15.32
	(24.65)	(18.28)	(10.26)	(13.35)	(13.85)	(11.08)
PB	0.71	1.84	2.27	3.35	5.63	11.24
	(14.07)	(8.42)	(5.72)	(6.52)	(2.77)	(4.96)
PW	0.38	1.69	1.75	2.20	3.89	7.22
	(19.61)	(7.26)	(7.96)	(7.87)	(6.76)	(5.40)

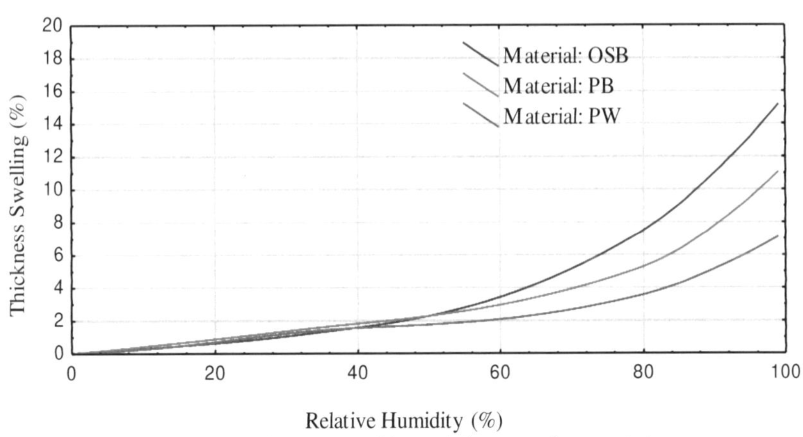

Fig. 6. Thickness swelling of the tested materials.

The moisture sorption is very similar for all the studied materials up to a relative humidity of 40%. However, at a higher relative humidity, the differences between the materials start to increase. OSBs achieved the highest thickness swelling, where 15% was the maximum thickness swelling value. Particleboard had less swelling (11.2%) and plywood (7.2%) had the least. As the proportion of nonrecoverable thickness swelling has a greater impact on material properties than overall thickness changes, nonrecoverable thickness changes were analysed in more detail.

3.4. Nonrecoverable thickness swelling

Non-linear regression was used to estimate the relationship between equilibrium moisture content and nonrecoverable thickness swelling. The outputs from the regression analysis are shown in Fig. No. 7.

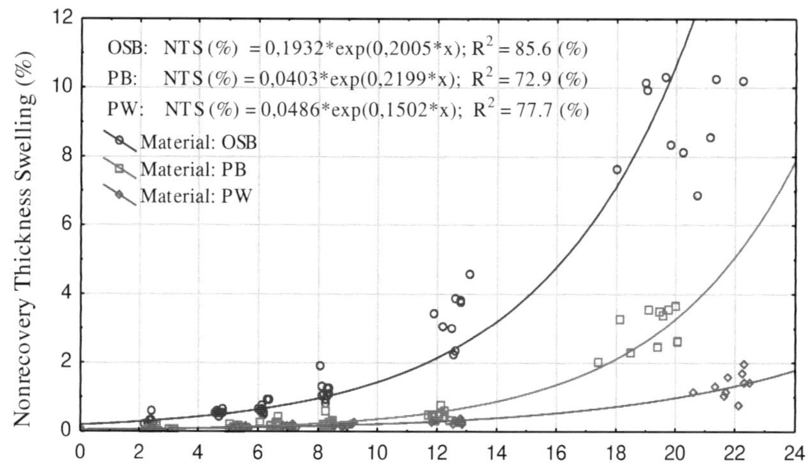

Fig. 7. Relationship between equilibrium moisture content and nonrecoverable thickness changes.

It was found that at low equilibrium moisture, the proportion of nonrecoverable thickness swelling increases very slowly, almost linearly. However, since equilibrium moisture is about 8% for OSB, and about 12% for plywood and fibreboard, the proportion of nonrecoverable swelling is increasing rapidly. The profile of this dependence is best described by the exponential fit. Due to the natural variability of the tested materials, the calculated exponential equation has a relatively close dependence with a determination coefficient exceeding 85% for OSB and 77 and 72% for plywood and particleboard.

The smallest proportion of nonrecoverable thickness swelling is achieved by plywood, which contains only 11 bonded joints per 12 mm thickness. Due to its construction, its thickness swelling consists mainly of recoverable solid wood swelling and the nonrecoverable changes are very small.

Several factors have contributed to a greater proportion of nonrecoverable thickness swelling of OSB versus particleboard. On the one hand, OSB is made of larger flat strands which, when their moisture changes, cause greater shear stress and result in greater stress on the bonded joint and also cause greater unevenness in the density profile. Furthermore, the OSBs are compressed at pressures exceeding 3 MPa, while particleboard at only about a third of the pressure. Due to the larger OSB strands that are more difficult to from and the greater pressure, more residual tension is already accumulated in the OSBs. Wood strands have a greater tendency to straighten, leading to greater overall thickness changes, as well as a higher proportion of nonrecoverable thickness swelling.

4. Conclusions

According to Eurocode 5, all of the tested materials belong to Use Class 2, where it is stated that the "relative humidity of the surrounding air only exceeds 85% for a few weeks per year". Although this concerns use for protected applications, Eurocode 5 permits short-term exposure of water to these construction materials. It arises from the investigation that for the deterioration of their properties, the important thing is the equilibrium moisture content that the materials achieved, as nonrecoverable thickness changes are strongly dependent on the equilibrium moisture content of the materials.

The equilibrium moisture content that the materials achieve is affected by the amount of paraffin and their density profile. For short-term washing with water, higher-density surface layers may be advantageous, as they slow moisture penetration into the material. On the other hand, higher density of the surface layers appears to be unsuitable for exposure to high air humidity. OSB has the highest nonrecoverable thickness swelling values and has the non-uniform vertical density profile.

Neither a significant increase in relative humidity above the limit of 85%, nor washing with water may be problematic for the proper function and use of materials. However, this increase of humidity must only be short-term, wherein the equilibrium moisture content of materials is not over-increased. On the contrary, even a very slight increase in relative humidity above 85% can have a negative impact on the properties of the tested materials if it is long-term. The higher moisture exposure time depends on the way the materials are applied. In a laboratory experiment, unprotected materials that were accessible to moisture from all sides except the lower edge achieved an equilibrium moisture content of about 12% after 5 days of increasing relative humidity from 65% to 85%.

The results also show that it is particularly important to ensure that the long-term use of tested materials in structures does not increase their equilibrium moisture content by more than 12%, when the proportion of nonrecoverable thickness swelling begins to increase rapidly.

Acknowledgment
This work was supported by the Czech Science Foundation, under project No. 19-02067S and by the project SGS19/143/OHK1/3T/11.

References

[1] Wu Q and Suchsland O 1996 Linear expansion and its relationship to moisture content change for commercial oriented strandboards *Forest. Prod. J.* **46**(11/12) 79-83
[2] Halligan AF 1970 A review of thickness swelling in particleboard *Wood. Sci. Technol.* **4**(4) 301-12
[3] Stamm AJ 1935 Shrinking and swelling of wood *Ind. Eng. Chem.* **27**(4) 401-6
[4] Aydin I, Colakoglu G, Colak S and Demirkir C 2006 Effects of moisture content on formaldehyde emission and mechanical properties of plywood *Build. Environ.* **41**(10) 1311-6
[5] Böhm M 2009 The influence of moisture content on thickness swelling and modulus of elasticity in oriented strand board bending *Wood. Res. Slovakia.* **54**(4) 79-90
[6] Hodoušek M, Böhm M, Součková A and Hýsek Š 2018 Effect of Moisture Content on the Air Permeability of Oriented Strand Boards *Bioresources* **13**(3) 4856-69
[7] Kelly MW 1977 Critical literature review of relationships between processing parameters and physical properties of particleboard *Gen. Tech. Rep. FPL-10* pp 1-65
[8] Stoeckel F, Konnerth J and Gindl-Altmutter W 2013 Mechanical properties of adhesives for bonding wood - A review *Int. J. Adhes. Adhes.* **45** 32-41
[9] Youngquist JA 1999 Wood-based composites and panel products Wood handbook: wood as an engineering material *Gen. Tech. Rep. FPL*-GTR-113 pp 10.1-10.31; 113
[10] Wang S and Winistorfer PM 2003 An optical technique for determination of layer thickness swell of MDF and OSB *Forest. Prod. J.* **53**(9) 64-71
[11] Mantanis GI, Athanassiadou ET, Barbu MC and Wijnendaele K 2018 Adhesive systems used in the European particleboard, MDF and OSB industries *Wood. Mater. Sci. Eng.* **13**(2) 104-16
[12] Frihart CR and Hunt CG 2010 Adhesives with Wood Materials Bond Formation and Performance Wood handbook: wood as an engineering material *Gen. Tech. Rep. FPL*-GTR-190 pp 10.1-10.23; 190

IOP Conf. Series: Materials Science and Engineering **549** (2019) 012043 doi:10.1088/1757-899X/549/1/012043

Influence of concentration change of calcium ions over time on their diffusion through sandstone

K Kobetičová[1], I Medveď[1], M Keppert[1] and R Černý[1]

[1] Department of Materials Engineering and Chemistry, Faculty of Civil Engineering, Czech Technical University in Prague, Thákurova 7, 166 29 Prague, Czech Republic

E-mail: klara.kobeticova@fsv.cvut.cz

Abstract. Diffusion is one of the most important salt transport mechanisms. The knowledge of the rate of transport of ions in the pore system of building materials is thus of primary importance for their service life prediction. In this paper, the transport of Ca^{2+} in sandstone Mšené was studied as a model system; two types of experiment were performed: a) without addition of Ca^{2+} ions for the achievement of original initial concentration and b) continuous addition of Ca^{2+} ions. The experiment was performed under laboratory conditions in a two-side plexiglass box at the ambient temperature of (20±2) °C for a period of 426 hours. Calcium concentrations were analyzed chelatometrically.

1. Introduction

Calcium ions are able to destruct concrete and other building materials. The chemical reactions involved can form complex salts. It is therefore interesting to study the transport of calcium ions (Ca^{2+}) in porous building materials. The transport is most often realized by diffusion that is driven by a concentration gradient of migrating particles. They diffuse through a bulk mobile phase stagnating in pores, so that there is practically no interaction between the particles and the inner walls of the pores. However, particles may be adsorbed on these walls and diffuse on the surface while staying adsorbed. This particle migration in the adsorbed state is called surface diffusion. Surface diffusivity depends both on the diffusing particles and on the substrate and it is more complicated to determine it than the standard (bulk) diffusivity [1]. Diffusion has been mostly studied in gas-solid systems, while liquid-solid systems have been less frequent.

However, both diffusion modes are influenced by many factors as temperature, humidity, biological agents, concentration of diffused ions and their mutual interactions [2-3]. For this reason, there is necessary to study these effects in simple experiments for a better understanding of the intensity of individual factors on diffusion rate. Setting the laboratory conditions under which the experiment is conducted is also important.

In this paper we present a pilot study of through-diffusion experiments of calcium ions (Ca^{2+}) in aquatic solutions for sandstone under different laboratory conditions. We compare the effects of original calcium concentration and equilibrium concentration during the experiment on diffusion effectivity of calcium ions.

2. Studied materials

Sandstone from Mšené, Czech Republic, was used as a model sample. It is siliceous rock, containing 99% of quartz and traces of kaolinite. Fundamental parameters of the rock are shown in Table 1. The sample is unimodal and its pore system if formed by voids between individual clasts. The mean pore diameter is 37 μm. The nature of sample is well observable by SEM (Figure 1).

$CaCl_2$ (99% purity) from Lach-Ner Ltd. was used in the experiment. Powdered $CaCl_2$ was dissolved in distilled water. The concentration (C_0) was chosen on the base of preliminary experiment as 4 g l^{-1}.

Table 1. Basic properties of the studied sandstone.

Quantity	Value
Porosity	28.2 %
Pore volume	0.1393 cm^3g^{-1}
Specific surface area	0.44 m^2g^{-1}
Compressive strength	8.7 MPa

Figure 1. SEM image of Mšené sandstone.

3. Experimental setup

Two boxes made of plexiglass were used in the experiment (Figure 2). The following description is the same for both the boxes. Initially dry sandstone samples of disk shape (d = 5 cm, h = 1 cm) were sealed with silicone in a circular aperture cut in the plexiglass panel splitting the box into two chambers. One of the chambers (chamber 1) was filled with 500 ml of distilled water, while the other one (chamber 2) was filled with calcium solution of a desired concentration. The box was closed by a plexiglass lid. The box was placed on a table at ambient temperature (20±2) °C for a period of 426 hours.

Calcium concentrations in distilled water, as well as the residual concentrations were measured in both boxes. The initial concentration (4 g.l^{-1}) was decreased during the time. However, the calcium chloride solution was continuously adjusted to its initial concentration (4 g.l^{-1}) in only one of the model boxes at each measurement (the experiment b). In the second box (the experiment a) the calcium chloride concentration was allowed to decrease during the whole time of experiment. The calcium concentrations in chamber 1 and chamber 2 were then compared for the entire experimental time.

4. Chelatometrical measurement of calcium concentration

Chelatometrical measurement of calcium concentration was done according to [4]. A 0.05 M solution of chelaton III was used to measure the calcium amount. Namely, 500 µl of a 5 M solution of KOH was added to a specific amount of the monitored sample from chamber 1 or 2 and colored by murexid red indicator [5]. Then chelaton III was dropping to the sample using an automatic byrette. When the titration was finished, the solution had violet color. The amount of titration reagent was used for the calculation of the calcium concentration according to equation 1:

$$c = \frac{V_{ch} M_{ch} M_{Ca}}{V_{sam}}, \tag{1}$$

where c is the calcium concentration ($mg.l^{-1}$), V_{ch} is the volume of used chelaton (ml), $M_{ch} = 0.05$ is its molarity, $M_{Ca} = 40$ $g.mol^{-1}$ is the calcium molar mass, and V_{sam} is the volume of analyzed aquatic sample (ml).

The number of moles of calcium was calculated according to equation 2:

$$n = \frac{cV}{M_{Ca}}, \tag{2}$$

where V is the volume of chamber 1 or 2 (0.5 l).

Figure 2. The plexiglass box (2 pieces) used in the experiment.

5. Results and discussion

The results of the measurement of cumulative calcium concentration in chambers 1 and 2 as a function of time are plotted in Figures 3 and 4, respectively. Diffusion started in both boxes after 20 hours. It is evident, that calcium diffused similarly in both boxes at the start of the experiment. The diffusion was observed during the whole time of the experiment. In the experiment b), after first measurement calcium was added into one box for increasing the actual calcium concentration up to the initial concentration C_0 (4 $g.l^{-1}$). This process was repeated at each measurement when it was necessary. The measurements indicated that diffusion of calcium was again very similar in both boxes with exception to the last measurement (Figure 3). In the experiment a) the initial concentration was decreasing during the time to one half, compared to the original value, at the end of the experiment (Figure 4). These changes did not though have any effect on diffusion.

We can assume that the concentration available for the accumulation processes and the number of filled binding sites was not a principal factor affecting the diffusion rate. Unfortunately, we cannot compare our results with data from any other studies because they are not available.

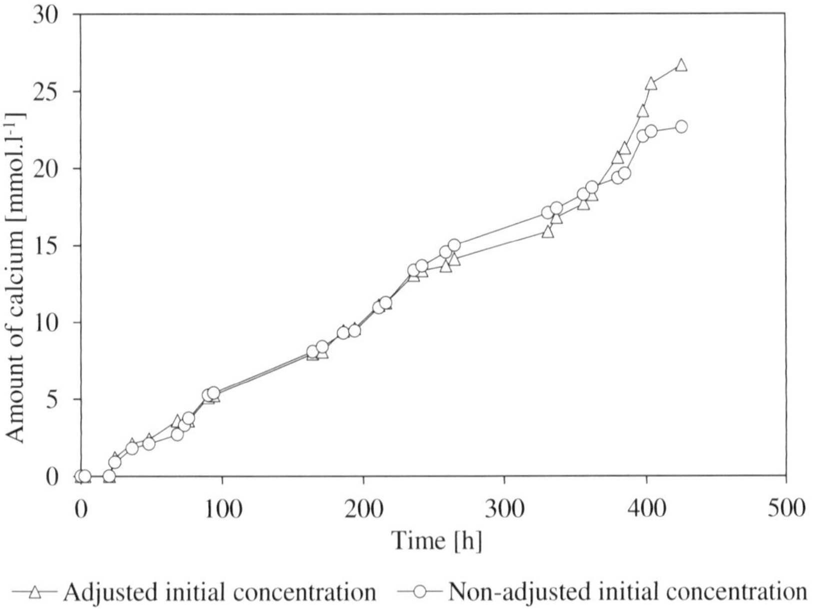

Figure 3. Cumulative calcium concentration in chamber 1 as a function of time.

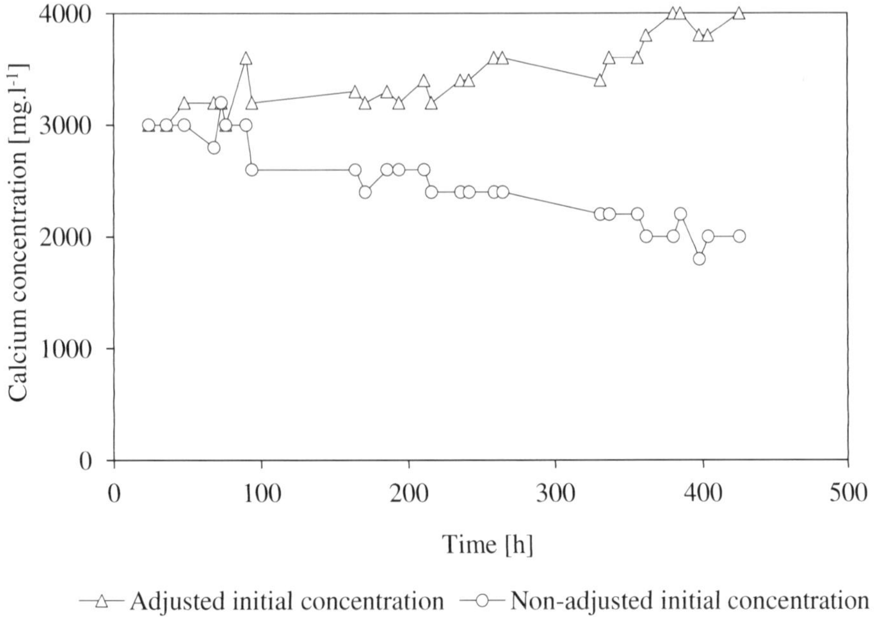

Figure 4. Actual calcium concentration measured in chamber 2 as a function of time.

6. Conclusions

In this paper, through diffusion of calcium in sandstone was studied in two experimental setups (adjusted and non-adjusted initial concentration during the experiment). The amount of calcium ions diffused into distilled water was similar for both model boxes for the whole time of the experiment. Therefore, besides diffusion as the main driving force of ion transport also other processes (e.g. capacity of calcium accumulation in the sample and gradual release of calcium ions from sandstone) had a significant influence on the amount of diffused ions.

Acknowledgement

The research in this paper was supported by the project No. SGS19/143/OHK1/3T/11.

References

[1] Medveď I and Černý R 2011 *Micropor. Mesopor. Materials* **142** 405
[2] Saafi M., Gullane A, Huang B, Sadeghi H, Ye J and Sadeghi F 2018 *Compos. Struct.* **201** 766
[3] McCabe S, Smith BJ, McAlister JJ, Gomez-Heraz M, McAlister D, Warke PA, Curran JM and Basheer PAM 2013 Environ. Earth Sci. 69 1225
[4] Kobetičová K, Medveď I, Keppert M and Černý R 2018 *Thermophys. 2018: Conf. Series* **1988** 020019
[5] Ringbom A, Persan G and Wanninen E 1958 *Anal. Chim. Acta* **19** 525

IOP Conf. Series: Materials Science and Engineering **549** (2019) 012044 doi:10.1088/1757-899X/549/1/012044

AC electrical properties of geopolymers with carbon black admixture

M Petríková[1] and L Fiala[1]

[1] Department of Materials Engineering and Chemistry, Faculty of Civil Engineering, Czech Technical University in Prague, Thákurova 7, 166 29 Prague 6, Czech Republic

E-mail: fialal@fsv.cvut.cz

Abstract. Geopolymers as competitors to the cement-based construction materials are intensively studied in the present. Their competitiveness mainly arises from their high strength, favourable development of hydration heat at early age, good chemical resistance and thermal stability. Further qualitative improvement of functional properties can be achieved by adding electrically conductive admixtures. In an appropriate amount (called percolation threshold), mechanical properties remain reasonable and electrical properties become sufficient to ensure evolution of heat by acting of an external power source (self-heating), to detect material damage (self-sensing) or to harvest thermoelectric energy (energy harvesting). In this paper three geopolymers with different dosages of carbon black (CB) admixture (0 wt. %, 4 wt. %, 10 wt. %) were studied by means of LCR bridge AC measurements. It was observed significant difference in electrical behavior of the studied geopolymers. 0 wt. % geopolymer exhibited highly capacitive character, 4 wt. % geopolymer was slightly shifted to resistive behavior and 10 wt. % geopolymer behaved like resistor even to high frequencies with reasonable resistance which indicates its possible self-heating ability.

1. Introduction

Geopolymers are inorganic materials based on various aluminosilicate-rich precursors that in order to gain sufficient mechanical properties need to be activated by alkalis. Strength can be comparable or even higher than of the cement-based materials when an appropriate precursor and activator is used. Together with good behaviour at high temperatures [1], good durability characteristics [2], and the fact that precursors are very often of waste nature, such materials can be considered as good competitors to the cement-based materials with lower impact to environment [3].

The type of precursor determines material structure that is formed during hydration and influences material properties of the formed geopolymer. In high-calcium systems (blast-furnace slag), typically C-A-S-H structures are formed [4], whereas in low-calcium systems (fly-ash, metakaolin, clay), N-A-S-H type gels appears [5]. However, both, C-A-S-H and N-A-S-H type geopolymers are typical electrical insulators which is a limiting factor for their utilization in smart applications in building industry. Such deficiency can be solved by addition of electrically conductive admixtures (ECA), such as steel fibres (SF), carbon fibres (CF), carbon black (CB), nickel powder (NP) or graphite powder (GP), that in appropriate amount ensures self-heating [6], self-sensing [7] or energy harvesting ability [8] of the enhanced composites.

Research presented in this paper was focused on determination of AC electrical properties of geopolymers based on high-calcium granulated blast-furnace slag electrically enhanced by CB. Such ECA consisting of aggregated spherical particles with high surface area was used in three different dosages in order to prepare composites with different electrical properties. AC electrical properties were determined in the frequency range of 10 Hz – 300 kHz and acquired functional properties were compared in terms of the materials applicability in sophisticated applications.

2. Experimental

The studied geopolymers were based on high-calcium precursor, alkali activator, filler and ECA. Precursor was represented by granulated blast-furnace slag with fineness of 380 $m^2 \cdot kg^{-1}$, SMŠ 380 (Kotouč Štramberk Ltd.) which is an industrial by-product of iron production with high calcium content (39.8 %). Alkali activation was carried out using water glass (sodium silicate solution, molar ratio $Na_2O/SiO_2 = 2.07$). Filler was represented by three normalized CEN fractions of quartz sand (fine aggregates: PG1, PG2, PG3) produced by Filtrační písky, Ltd. complying with ČSN EN 196-1. The reference samples were prepared for comparison of materials properties with the electrically enhanced ones. Enhancement of electrical properties of samples was secured by 5 % and 15 % CB suspension CS CABOT VULCAN 7H in amount of 4 wt. % and 10 wt. %.

VULCAN 7H is manufactured by thermal cracking of a heavy aromatic feedstock in a hot flame. Oil is a typical feedstock that is injected into a hot flame zone in oil furnace where present hydrocarbons are cracked to carbon and hydrogen by means of quenching the flame by water. It is mainly used in tires production and industrial rubber product applications, but for its to high surface area and reasonable high electrical conductivity, it can be effectively used for optimization of electrical properties of polymers [9] and inorganic building materials [10].

Consistence of the fresh mixtures was tested in terms of ČSN EN 1015-3 standard (Determination of Consistence of Fresh Mortar by flow table) and kept at the same, reasonable level to attain good workability of the mixtures. Water-slag ratio of the mixtures with the consistence (average base diameter equal to 160 mm) was equal to 0.44 (CB 0 wt. %), 0.65 (CB 4 wt. %) and 0.88 (10 wt. %). Composition of the studied materials is presented in Table 1.

Table 1. Composition of geopolymers.

Component [g] / Mixture	0 wt. % CB	4 wt. % CB	10 wt. % CB
Slag	1170	1170	1170
Water glass	234	234	234
Sand	3 × 1170	3 × 1170	3 × 1170
Water	520	759.2	1027
Carbon black	0	46.8	117

The bulk density was determined on samples with dimensions (50 × 50 × 50) mm^3 by gravimetric method.

AC electrical properties represented by the absolute value of impedance $|Z|$ [Ω], the phase angle θ [°], the resistance R [Ω] and the reactance X [Ω], were experimentally determined on samples with dimensions (50 × 50 × 50) mm^3 (Figure 1a) by LCR bridge GW Instek LCR-6300. Such device provides precise measurements of AC electrical properties in the frequency range of 10 Hz – 300 kHz. Two opposite lateral sides of the samples were first painted by conductive carbon paint and pasted by copper tapes in order to gain good contact of the samples with measuring device (Figure 1b). Measurements were carried out at voltage level 1V at wide frequency range: 10 Hz, 50 Hz, 60 Hz,

100 Hz, 120 Hz, 1 kHz, 2 kHz, 10 kHz, 20 kHz, 40 kHz, 50 kHz, 100 kHz, 200 kHz, 250 kHz and 300 kHz, respectively.

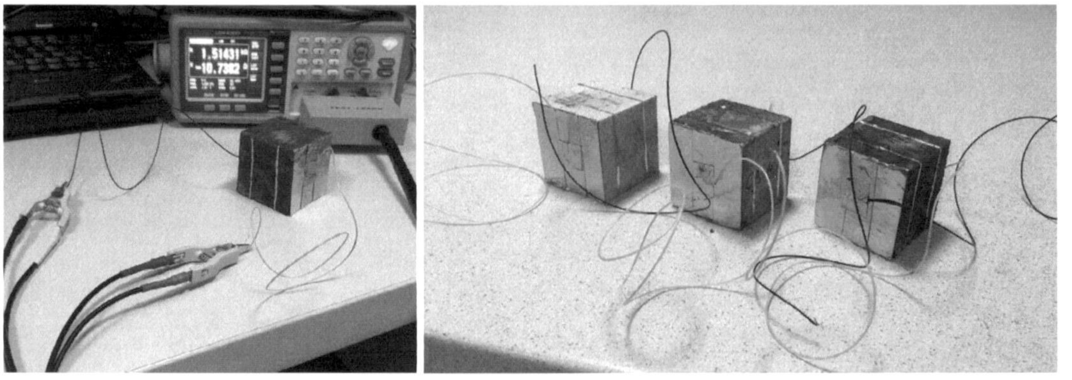

Figure 1. a) Measuring LCR bridge b) Samples for measurements of electrical properties.

3. Results and discussion

The bulk density of the studied geopolymers were equal to 2278 $kg \cdot m^{-3}$ (0 wt. % CB), 2070 $kg \cdot m^{-3}$ (4 wt. % CB) and 2004 $kg \cdot m^{-3}$ (10 wt. % CB), respectively. Decrease in the bulk density with an increasing amount of CB admixture was due to an increase in porosity influenced by increase of the water-slag ratio.

In Figure 2 and Figure 3, the absolute value of impedance and the phase angle dependence on the frequency are presented. It was observed considerable decrease of the absolute value of impedance on increasing amount of CB admixture. Comparing the absolute value of impedance of 0 wt. % CB and 10 wt. % CB, the difference was more than three orders of magnitude for low frequencies and about two orders of magnitude for high frequencies. In case of 0 wt. % CB and 4 wt. % CB, the absolute value of impedance decreased about two and one order of magnitude, respectively, with increasing frequency. 10 wt. % CB did not exhibit any significant change. Concerning the phase angle on the frequency dependence, it was nearly constant for 10 wt. % CB in the whole frequency spectrum with the values close to the zero value which exhibit resistive character of such material. Both, 0 wt. % CB and 4 wt. % CB exhibited increasing capacitive behavior with increasing frequency (- 90° - ideal capacitor, 0° - ideal resistor).

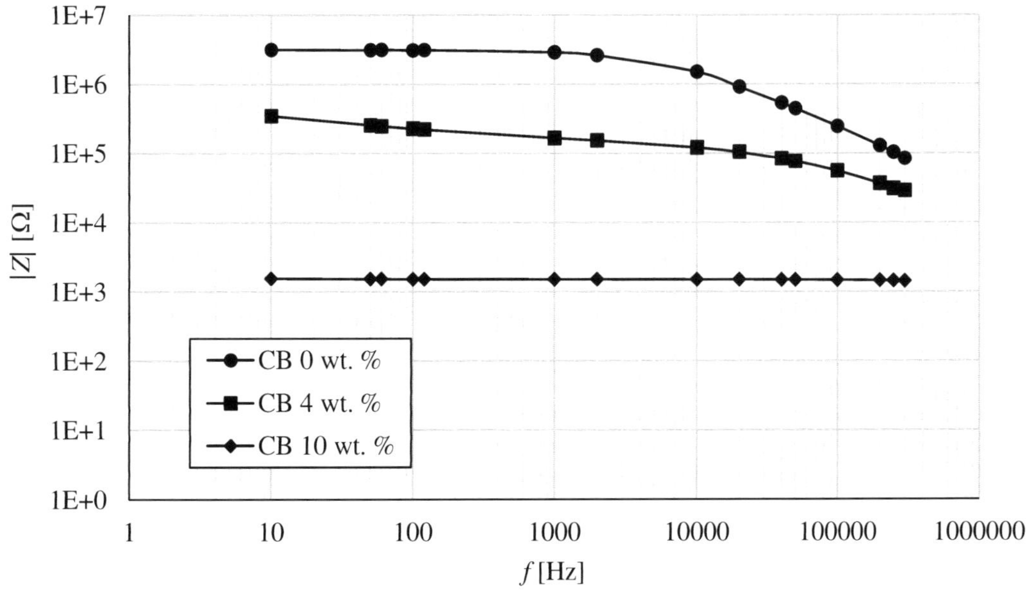

Figure 2. Dependence of the absolute value of impedance on the frequency.

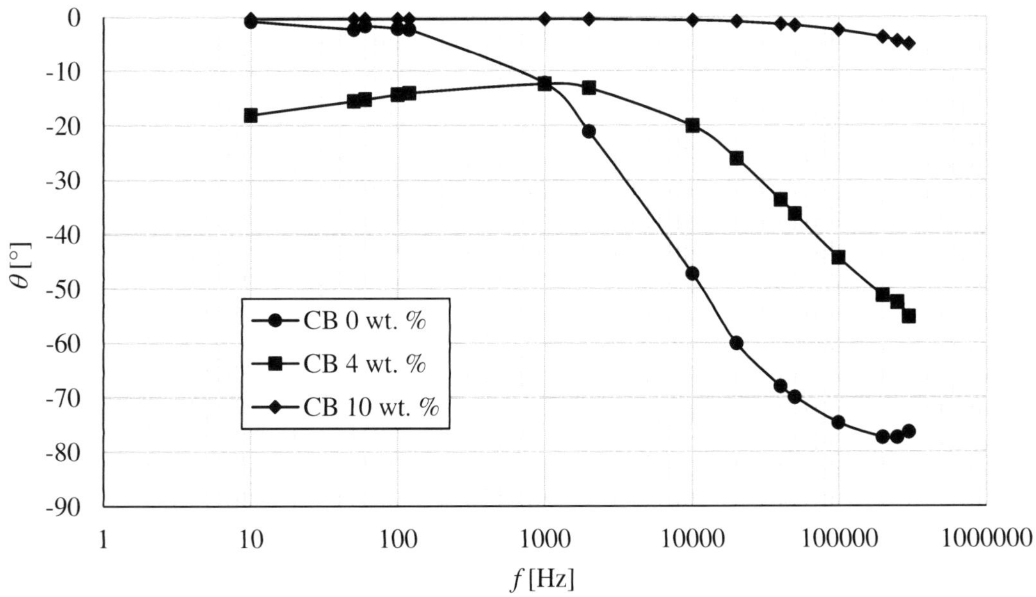

Figure 3. Dependence of the phase angle on the frequency.

In Figure 4 and Figure 5, the resistance and the reactance dependence on the frequency are presented. Results were significantly lower for 10 wt. % CB in comparison with the other two geopolymers. The resistance was almost constant for 10 wt. % CB, whereas for 0 wt. % CB and 4 wt. % CB it was significantly frequency dependent. Reactance which is not participating in heat evolution was in the range from the fourth to the sixth order of magnitude for 0 wt. % CB and 4 wt. % CB which was significantly higher compared to the 10 wt. % CB (first or second order of magnitude). Reactance of all the geopolymers were frequency dependent.

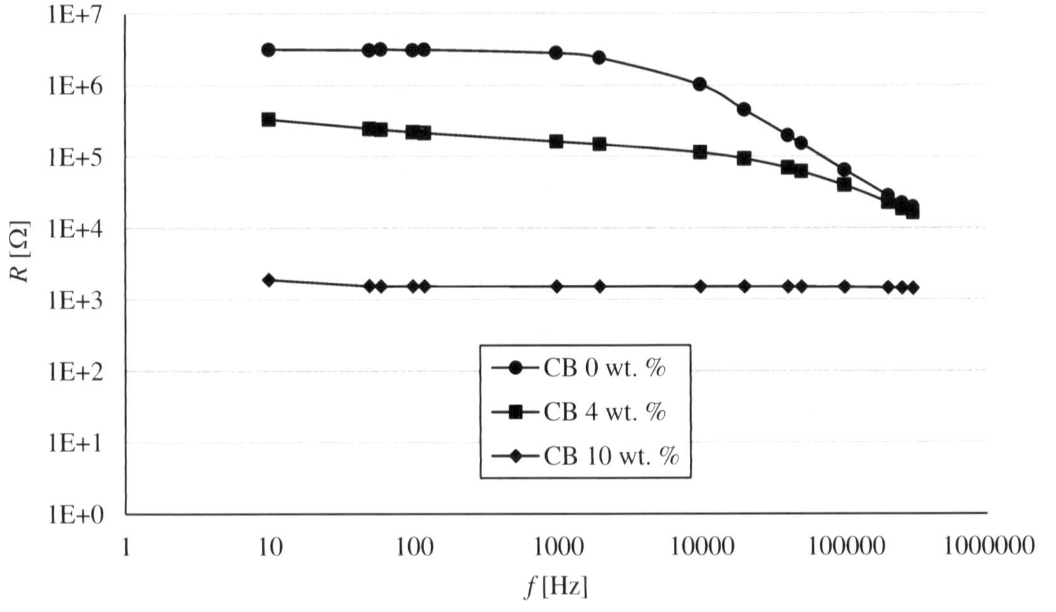

Figure 4. Dependence of the resistance on the frequency.

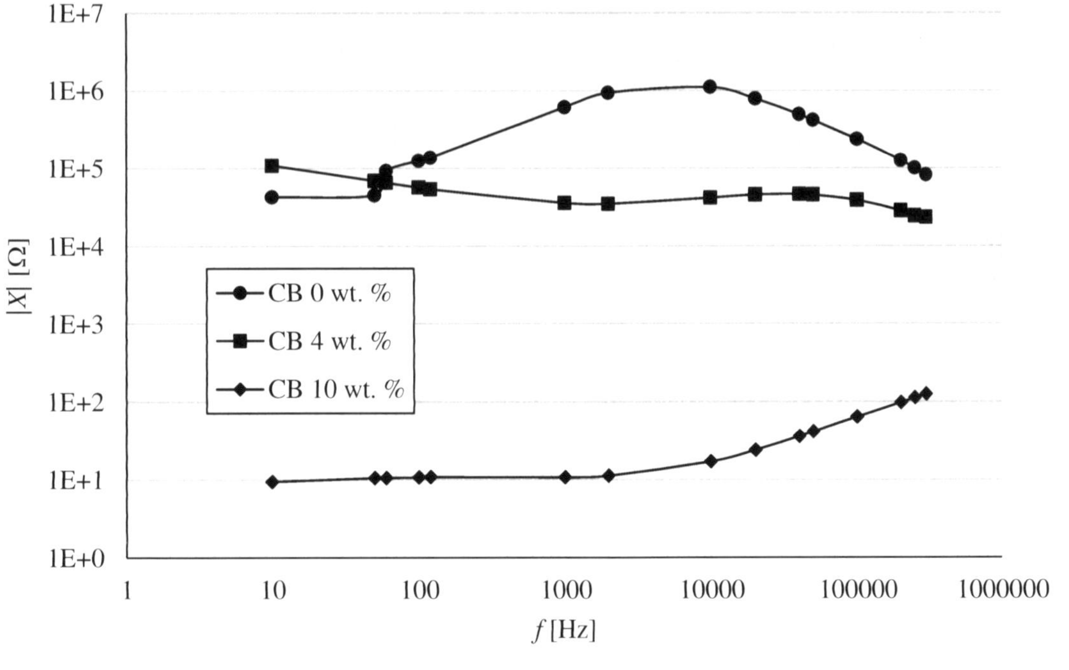

Figure 5. Dependence of the reactance on the frequency.

4. Conclusions

The paper is focused on experimental determination of AC electrical properties of three geopolymers with different dosages of CB admixture (0 wt. %, 4 wt. %, 10 wt. %). Measurements were carried out in frequency range 10 Hz – 300 kHz by LCR bridge GW Instek LCR-6300.

It was observed distinctly capacitive character of the geopolymer without CB admixture, slight transition to the resistive behavior of the geopolymer with 4 wt. % of CB admixture and distinctly resistive character of the geopolymer with 10 wt. % of CB admixture. Capacitive character and high resistance of the reference geopolymer does not allow to evolve heat using the external power source. In case of geopolymer with 4 wt. % of CB admixture, the resistance is slightly lower compared to the reference material but still pretty high which is not a favor assumption for evolution of heat. Geopolymer with 10 wt. % of CB admixture is due to its resistive character and reasonably high resistance promising candidate for self-heating applications.

With respect to the presented results, future observations will be devoted to further electrical analysis of geopolymers with various electrically conductive admixtures together with comparison of their self-heating ability under DC and AC electric field load.

Acknowledgments
This research has been supported by the Czech Science Foundation under project No. 19-11516S and project No. SGS19/143/OHK1/3T/11.

References

[1] Zuda L, Drchalova J, Rovnanik P, Bayer P, Kersner Z and Cerny R 2010 *Cement Concrete Comp* **32** pp 157-63
[2] Firdous R, Stephan D and Djobo J N Y 2018 *Constr Build Mater* **190** pp 1251-63
[3] Rakhimova N R and Rakhimov R Z 2019 *J Non-Cryst Solids* **509** 31-41
[4] Thi Hai Yen N, Tsuchiya K and Atarashi D 2018 *Constr Build Mater* **191** pp 114-24
[5] Davidovits J 2017 Geopolymers: Ceramic-Like Inorganic Polymers *J Ceram Sci Technol* **8** pp 335-50
[6] Fiala L, Rovnanik P and Cerny R 2017 *Cem Wapno Beton* **22** pp 201
[7] Rovnanik P, Kusak I, Bayer P, Schmid P and Fiala L 2019 *Cement Concrete Res* **118** pp 84-91
[8] Wei J, Nie Z, He G, Hao L, Zhao L and Zhang Q 2014 *Rsc Adv* **4** 48128-34
[9] Li Y C, Huang X R, Zeng L J, Li R F, Tian H F, Fu X W, Wang Y and Zhong W H 2019 *J Mater Sci* **54** pp 1036-76
[10] Fiala L, Jerman M, Rovnanik P and Cerny R 2017 *Mater Tehnol* **51** pp 1005-9

Utilization of hydrophilic cellulose fibers for preparation of plaster with enhanced moisture control capability

J Fort[1], J Kocí[1], E Vejmelkova[1], M Dolozelova[1], R Cerny[1]

[1] Czech Technical University in Prague, Faculty of Civil Engineering, Department of Materials Engineering and Chemistry, Thákurova 7, 166 29 Prague 6, Czech Republic

E-mail: eva.vejmelkova@fsv.cvut.cz

Abstract. The relative humidity is one of the most important parameters having a substantial impact on ambient climate comfort. The negative effects induced by undesired levels of relative humidity in building interiors are related not only to a deterioration of materials but also to a potential health risk for building occupants. In the recent paper, a traditional lime plaster is modified by biomass fly ash and hydrophilic cellulose fiber admixture to improve material durability and provide material with enhanced capability for the moderation of ambient relative humidity. Within the experimental analysis, the basic material properties, as well as mechanical strength of designed plaster, is determined. The moisture buffer potential of all designed plasters is evaluated according to the Nordtest test procedure. Obtained results show that the application of cellulose fibers has a synergic effect on both improvements of the potential of studied plaster for passive adjustment of interior relative humidity and mechanical strength.

1. Introduction

In the recent period, several concerns related to the maintenance of the ambient environment can be distinguished. Besides the issues associated with insufficient thermal performance, problems connected with indoor humidity management represent a substantial task for materials engineers and architects. Maintaining the optimal level of relative humidity is related not only to buildings with specific environmental condition requirements such as museums or galleries, but also the indoor air quality of residential buildings need to be secured. In light of the recent findings, the unsatisfactory level of relative humidity in modern passive houses can be responsible for various health problems [1]. A limited response of used building materials to outside weather variations has a negative effect on respiratory discomfort, eye and skin irritation, allergies. Moreover, the condensation of water within the building structure was identified as one of main triggers for degradation of interior materials and furniture [2].

To avoid the problems mentioned above, additional HVAC devices are usually employed, especially in the case of rooms with special requirements on environmental conditions. However, the application of such devices is connected with increased energy consumption and substantial economic costs. In order to overcome the moisture related issues and maintain an optimal range for indoor relative humidity level, the development and application of moisture responsive materials became an important task [3]. The modification of commonly applied building materials can substantially improve the interaction between seasonal or diurnal relative humidity fluctuations and thus improve the quality of indoor air. The characterization of building materials to modulate the interior climate in terms of adjustment of humidity in the room was described by Rode and Grau [4] who introduced moisture buffer value as the integrated indicator.

The modification of the porous space was found as an efficient way to promote passive, moderation of indoor RH. Across the literature is possible to distinguish several attempt aimed at fulfillment of this task, for example by application of light-weighted admixtures such as perlite, charcoal or vermiculite; use of porogene substances (aluminum powder) or utilization of superabsorbent polymers used for control of hydration heat and shrinkage of concrete mixtures [5-7]. However, most of the performed research concluded that incorporation of particular admixtures could provide only limited success due to an adverse effect of applied admixture to material durability. Some drawbacks by the meaning of the loss of workability, shrinkage, and segregation were also observed [8].

An elegant solution for overcoming of described issues can be found in the utilization of natural hydrophilic cellulose fibers. Additionally, the application of such material meets with the requirements associated with the renovation of historical objects. Incorporating natural fibers can significantly improve both the moisture buffering and mechanical strength [5], however application of hydrophobized cellulose fibers resulted in the improvement of mechanical parameters, and more favorable results can be achieved by application of hydrophilic fibers.

In the performed study, the moisture buffering potential of plasters modified by hydrophilic cellulose fibers is studied. The moisture buffering is evaluated according to the Nordtest method in order to access material response to diurnal relative humidity fluctuations. Obtained results from dynamic loading of tested plasters are compared with moisture buffer value (MBV_{ideal}) calculated from steady-state values.

2. Materials and methods

2.1. Materials

The biomass fly ash (BFA) together with cellulose fibers as a hygroscopic material were used to increase the functionality of studied plasters. The BFA was originated during the combustion of biomass in the electricity plant. The chemical composition of BFA given by X-ray fluorescence analysis is shown in Table 1. A substantial part of BFA is formed by SiO_2 (51.3 %), CaO, (15.4 %) and Al_2O_3 (14 %), while the rest is composed by Fe_2O_3, MgO, K_2O, Cl^-, SO^{3-} and P_2O_5.

Hydrophilic cellulose fibers with length ranging between 2 and 4 mm were incorporated for reinforcing of a plaster structure. Hydrated lime CL 90-S (Mokrá Plant of Carmeuse Czech Republic) was used as a base binder for plaster design. In order to improve the durability of plasters, the BFA was used as 20% replacement of lime. The amount of applied cellulose fibers varied from 1 to 4 %. In order to provide a material with the same workability, the flow table test was carried out. Water/binder ratio was modified according to the results of the flow table test. The detailed composition of the designed plasters is shown in Table 1.

Table 1. Composition of designed plasters.

Mixture	Lime (kg)	Aggregate (kg)	Cellulose (kg)	BFA (kg)	Water (kg)
RP	3	9	-	0.9	3.20
CBP1	3	9	0.036	0.9	3.25
CBP2	3	9	0.072	0.9	3.30
CBP3	3	9	0.108	0.9	3.40
CBP4	3	9	0.144	0.9	3.50

2.2. Experimental methods

2.2.1. Basic physical properties

IOP Conf. Series: Materials Science and Engineering **549** (2019) 012045 doi:10.1088/1757-899X/549/1/012045

Basic physical properties of studied plasters were characterized by measurements of the bulk density, matrix density, and total open porosity. Performed measurement of the bulk density was done on five cubic samples of 50 mm side and determined from the measurement of sample sizes (using digital caliper) and its dry mass. The matrix density was accessed by helium pycnometry using apparatus Pycnomatic ATC (Thermo Scientific).

2.2.2. Mechanical properties

Compressive strength (MPa) and flexural strength (MPa) as the main mechanical parameters were determined by employment the device VEB WPM Leipzig having a stiff loading frame with the capacity of 3000 kN. The strength was determined for 28-days cured samples.

2.2.3. Moisture buffering

The experimental determination of moisture buffer values using the step-response method when the sample with the known exposed area is continuously weighting during the exposure to various relative humidity levels. For this purpose, DVS (Dynamic Vapor Sorption Device) was used to maintain cycles of 8 hours at high RH (70%) and 16 hours at low RH (30%). The sample mass was recorded for 4 cycles in order to reach a dynamic equilibrium where the final mass at the end of the cycle and initial mass vary by less than 5%. The practical Moisture Buffering Value (MBVpractical) was consequently calculated by taking into consideration the maximum moisture uptake after 8 hours of adsorption phase according to used the RH levels. Here, the procedure presented by McGregor et al. [9] was adopted. Employment of this methods is beneficial for determination of material response to diurnal relative humidity fluctuation and can be viewed as an indicator of material suitability for application in specific indoor climate conditions.

The moisture buffer capacity of materials is described on the basis of the heat-mass transfer analogy. The moisture effusivity is derived from the thermal effusivity, by a description of the material ability to absorb and release moisture as is given in Eq. 1.

$$b_m = \sqrt{\frac{\delta_p \cdot \rho_0 \cdot \dfrac{\partial u}{\partial \varphi}}{p_s}} \quad . \tag{1}$$

where δ_p (kg/msPa) is the water vapor permeability, ρ_0 (kg/m³) the dry density of the material, u (kg/kg) the moisture content, φ (-) the relative humidity, and p_s (Pa) the saturation vapor pressure at a temperature of the experiment.

Ideal moisture buffer value (MBV_{ideal}) was obtained from the theory of moisture transport; the estimation of surface moisture flux over time related to sample exposure to relative humidity variations. The accumulated moisture uptake $G(t)$ (kg/m^2), respectively the moisture release that both happen within the time period t_p, can be described by integrating the moisture flux over the surface $g(t)$ as in Eq. (2)

$$G(t) = \int_0^t g(t)dt = b_m \cdot \Delta p \cdot h(\alpha)\sqrt{\frac{t_p}{\pi}} \,, \tag{2}$$

where

$$h(\alpha) = \frac{2}{\pi}\sum_{n=1}^{\infty}\frac{\sin^2(n\pi\alpha)}{n^{3/2}} \approx 2.252[\alpha(1-\alpha)]^{0.535} . \tag{3}$$

where α (-) is the fraction of the time period where the humidity level is high. For the 8/16 hours scheme, α=1/3, which makes $h(\alpha)$=1.007, the accumulated moisture uptake can be given in a simplified form:

$$G(t) \approx 0.568 \cdot b_m \cdot \Delta p \cdot \sqrt{t_p}\ . \tag{4}$$

MBV (g/m^2 %RH) can be expressed by normalization of accumulated moisture uptake by the change in surface relative humidity. Therefore, the moisture buffer value is proportional to the moisture effusivity b_m times the square root of the time period, t_p½ (s^{½}). The MBV can be therefore defined as follows:

$$MBV_{ideal} \approx \frac{G(t)}{\Delta RH} \approx 0.00568 \cdot p_s \cdot b_m \cdot \sqrt{t_p}\ . \tag{5}$$

Based on the described experimental arrangement, the moisture effusivity can be determined from the steady-state experiments. On the other hand, the moisture buffer value represents a dynamic characteristic. Additionally, the ideal conditions can be barely maintained, so Eq. (4) represents an approximation, and $MBV_{practical}$ (g/m^2 %RH) was calculated from the DVS experiment.

3. Results and discussion

3.1. Basic physical properties

Determined basic physical properties of studied plasters are given in Table 2. Looking at the results, one can see a gradual increase in the total open porosity induced by a decrease in the bulk density in line with the increasing amount of applied cellulose fibers. Namely, the total open porosity of RP about 37.6 % were shifted up to almost 41 % for CBP4 mixture. This factor represents important information since the materials with a higher level of porosity were found as more perspective in the sense of moisture buffering.

Table 2. Basic physical properties of studied materials.

Mixture	Bulk density (kg/m^3)	Matrix density (kg/m^3)	Total open porosity (%)
RP	1596.7	2558.6	37.6
CBP1	1578.0	2571.9	38.6
CBP2	1562.3	2566.8	39.1
CBP3	1548.1	2549.0	39.3
CBP4	1518.1	2572.1	40.9

3.2. Mechanical properties

Results of mechanical properties are given in Table 3. Here, despite the increase in the total open porosity, application of cellulose fibers was found as beneficial for strength development. Obtained results of compressive strength were improved from initial 1.76 MPa for RP to 2.19 MPa for CBP4. The flexural strength was shifted even more significantly from initial 0.73 to 1.35 MPa. Revealed findings point to successful incorporation of applied fibers which reinforced the plaster matrix thanks to the synergy between the structure of the material matrix, tensile strength of cellulose fibers and created bonds between fibers and material matrix.

IOP Conf. Series: Materials Science and Engineering **549** (2019) 012045 doi:10.1088/1757-899X/549/1/012045

Table 3. Strength of studied materials.

Mixture	Compressive strength (MPa)	Flexural strength (MPa)
RP	1.76	0.73
CBP1	1.86	0.79
CBP2	1.99	0.89
CBP3	2.13	1.11
CBP4	2.19	1.35

3.3. Moisture buffering

The record of moisture buffer measurement given in Fig. 1 proved a beneficial influence of applied cellulose fibers in designed plasters. As one can see, a higher content of cellulose fibers shifted absorption capability of studied plasters during moisture loading (represented by dotted line). Despite the minor changes in material porosity, plotted curves of samples mass record revealed distinct differences which can be assigned to the hygroscopicity of applied cellulose fibers. Looking at the results given in Table 4, MBV_{ideal} was found as slightly lower compared to $MBV_{practical}$. Thus the calculation of moisture buffering from the steady-state experiment can be viewed as satisfactory. Calculated $MBV_{practical}$ from the performed dynamic moisture buffering experiment was only about 5 % higher on average. Namely, $MBV_{practical}$ was gradually shifted in line with increased cellulose fiber dosage from initial RP was about 1.44 obtained for RP to 2.13 for SBC4. The achieved improvement can be partially assigned to changes in porosity, but more significant influence can be assigned to a higher amount of applied hydrophilic cellulose fibers with a great moisture absorbability.

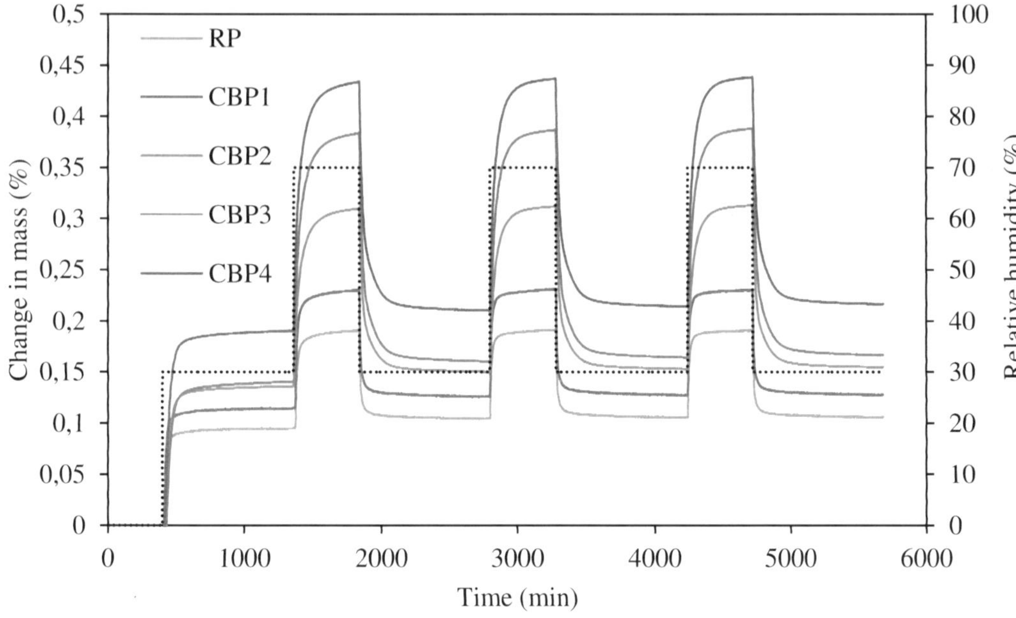

Figure 1. Record of MBV measurement.

Table 4. Moisture buffer values of studied materials.

Mixture	MBV_{ideal}	$MBV_{practical}$
RP	1.28	1.44
SBC1	1.31	1.37
SBC2	1.58	1.68
SBC3	1.76	1.91
SBC4	2.01	2.13

4. Conclusions

In the present study, the moisture control capability of modified plasters by cellulose fibers was studied. At first, the designed plasters were characterized by the meaning of determination of the bulk and matrix density, and porosity. The effect of incorporated hydrophilic cellulose fibers resulted in a slight increase in the total open porosity assigned to a loss of bulk density. This finding was not accompanied by a decrease in strength, and both compressive and flexural strength was improved thanks to the reinforcing the effect of applied cellulose fibers. The main aim of the performed work consisted of the determination of moisture buffering of modified plaster according to the Nordtest protocol by 16/8 h loading schema at 70/30 %RH. Obtained findings point out the synergic effect of applied cellulose fibers which substantially increased the moisture buffer value. Considering the classification of building materials introduced by Rode and Grau [4], SBC4 plaster can be labeled as excellent moisture buffering material while RP, SBC1, SBC2, and SBC3 was found as good moisture buffering material. The mixture SBC4 with the highest content of cellulose fibers enhanced moisture buffering about 70 % which substantially increased moisture control capability in case of varying conditions. Such material can be effectively employed in interiors to mitigate a negative effect of undesirable humidity level and improvement of ambient air quality. Comparison between the calculation of MBV_{ideal} given by steady-state experiment and $MBV_{practical}$ from the dynamic experiment revealed only minor differences and proved the applicability of both approaches. Nevertheless, further research focused on the moisture penetration depth and subjecting of materials to testing in varying weather conditions should be done in order to determine the potential impacts of hygroscopic materials on energy consumption related to the building's maintenance.

Acknowledgment

This work was supported by the Czech Science Foundation, under project No 18-03997S and by the project no. SGS16/199/OHK1/3T/11.

References

[1] Shi L Y, Zhang,H B, Li Z X, Man X X, Wu Y, Zheng C R and Liu J 2018 *Build. Environ.* **143** pp 366

[2] Ge H, Yang X J, Fazio P and Rao J W 2014 *Build. Environ.* **81** pp 162-171

[3] Pavlik Z, Fort J, Pavlikova M, Pokorny J, Trnik A and Cerny R 2016 *Energ. Buildings* **126** pp 113

[4] Rode C and Grau K 2008 *J. Build. Phys.* **31** (4) pp 333

[5] Senff L, Ascensao G, Ferreira V M, Seabra M P and Labrincha J A 2018 *Energ. Buildings* **158** pp 721

[6] Arrigoni A, Beckett C T S, Ciancio D, Pelosato R, Dotelli G and Grillet A C 2018 *Resour. Conserv. Recy* **137** pp 11

[7] De Rossi A, Carvalheiras J, Novais R M, Ribeiro M J, Labrincha J A, Hotza D and Moreira R 2018 *Constr. Build. Mater.* **191** pp 39

[8] Goncalves H, Goncalves B, Silva L, Vieira N, Raupp-Pereira F, Senff L and Labrincha J A

2014 *Energ. Buildings* **74** pp 61
[9] McGregor F, Heath A, Shea A and Lawrence M 2014 *Build. Environ.* **82** pp 599

Enhancement of sorption capacity to Sr and Cs of a cement composite by addition of brick powder

E Vejmelkova[1], V Pommer[1], K Kobeticova[1], D Konakova[1], L Scheinherrova[1], P Konvalinka[2], M Keppert[1], I. Medved[1,3] and R Cerny[1]

[1] Czech Technical University in Prague, Faculty of Civil Engineering, Department of Materials Engineering and Chemistry, Thákurova 7, 166 29 Prague 6, Czech Republic
[2] Czech Technical University in Prague, Faculty of Civil Engineering, Experimental Centre, Thákurova 7, 166 29 Prague 6, Czech Republic
[3] Slovak University of Technology in Bratislava, Faculty of Civil Engineering, Department of Physics, Radlinskeho 11, 810 05 Bratislava, Slovakia

E-mail: eva.vejmelkova@fsv.cvut.cz

Abstract. Concrete is a suitable material for secondary casings in multi-barrier systems used in radioactive waste repositories, because it is durable and radiation- and thermal resistant and it has good mechanical properties. The casings should also limit possible leaching of radionuclides in case of failure of inner barriers. An efficient way to tackle this problem is the addition of a sorption material in the concrete mixture. Here mechanical and thermal properties of concrete containing different amounts of brick powder are studied in relation to the concrete mixture porosity. It is found out that, as expected, the bulk density and compressive and bending strengths decreased with an increasing amount of the brick powder. In addition, the sorption of cesium and strontium chlorides into the mixtures was analyzed under laboratory conditions, and their sorption isotherms and distribution coefficients were calculated. The results indicate that brick has higher sorption capacity for Sr than for Cs. Relatively low distribution coefficients indicate, however, that the studied brick is not a very suitable additive material for sorption of Sr and Cs elements.

1. Introduction

The present solutions of radioactive waste disposal are based on multi-barrier systems of protection to maximize the preservation of the environment surrounding waste repositories, including the safety of living organisms [1–3]. The barrier systems are composed of natural, geological barrier (the surrounding rock) and engineered barriers that include [4]

- durable metallic or concrete canisters (containing sealed hazardous waste);
- clay or cementitious backfills (void filling, limiting water infiltration, sorption and precipitation of radionuclides);
- concrete, reinforced concrete, or clay containment barriers (additional protection of waste and physical stability);
- gravel, sand, ceramic, or concrete drains (control of leachate).

Concrete and cementitious composites are conveniently used in radioactive waste repositories due to its good mechanical properties, durability, low corrosion, and low cost. Concrete casings are usually

IOP Conf. Series: Materials Science and Engineering **549** (2019) 012046 doi:10.1088/1757-899X/549/1/012046

reinforced to have sufficient mechanical parameters in order to withstand the manipulation on site and ground pressure after the final emplacement in the repository [5, 6]. In addition to proper mechanical properties, there are requirements on the radiation stability of concrete, its thermal stability and resistance to aggressive water and, perhaps, organisms. Concrete barriers also limit the leaching of radionuclides into the environment in case of an accidental leach from the inner barriers. Natural and synthetic zeolites, presenting pozzolanic activity and having good sorption properties [7, 8], can be used as additives to concrete, taking part in the cement hydration and/or acting as a filler [9, 10].

In this paper we study whether an addition of a brick powder in varying amounts can enhance the capacity of concrete to sorb strontium and cesium. We also investigate changes in mechanical and thermal properties of the concrete in relation to the amount of the brick powder.

2. Materials and methods

2.1. Materials

The principal binder component was Portland cement CEM I 42.5 R (Mokrá Ltd., Czech Republic) with the specific surface area of 408 m^2kg^{-1} and bulk density 3110 kg m^{-3}. It was the only binder component for the reference mixture MCR. In four other mixtures, brick powder (Heluz) was added in varying amounts. Namely, 20% of the cement mass (80 kg in 1 m^3) was first substituted by brick powder. Then the addition of brick powder was dosed increasingly by 40 kg m^{-3}, while the cement dosage was kept constant. Brick powder served as an inert mineral additive and took part in the hydration of the binder system. This material has the pozzolanic activity of 840 mg($Ca(OH)_2$) g^{-1}, its specific surface area is 402 m^2kg^{-1}, and its bulk density was 2540 kg m^{-3}. The mixture contained also a fine filler – silicate flour ST6 (Sklopísek Střeleč Ltd., Czech Republic) with the specific surface of 0.4 m^2g^{-1}. Superplasticizer Stachesil 2000 (Stachema CZ Ltd., Czech Republic) based on polycarboxylate was added to achieve good rheology of the fresh mixture, and the amount of water was dosed in order to meet the prescribed slump of the fresh mixture. Table 1 gives the chemical composition of the input materials. The particle size distributions of fine components are shown in Fig. 1. The composition of the studied materials is presented in Table 2.

Table 1. Chemical composition of the input materials (wt. %).

	Portland cement 42.5 R	**Brick powder**	**Silicate flour**
SiO_2	18.7	51.3	99.7
Al_2O_3	4.5	20.0	0.2
Fe_2O_3	3.4	6.0	–
CaO	65.9	11.5	–
MgO	1.3	4.5	–
K_2O	0.8	3.2	–
Na_2O	0.2	1.3	–
TiO_2	0.3	0.8	–
SO_3	4.3	1.0	–
P_2O_5	0.1	–	–

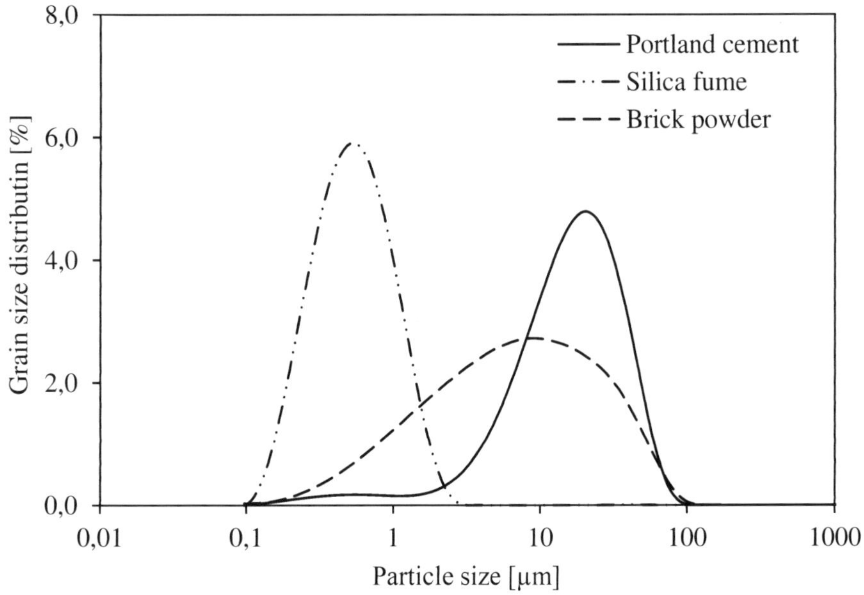

Fig. 1. Grain-size distribution curves of used components.

Table 2. Composition of studied materials (kg m^{-3}).

	RCR	RC1	RC2	RC3	RC4
Portland cement CEM 42.5 R	400	320	320	320	320.00
Brick powder	(0 %)	80 (20%)	120 (30%)	160 (40%)	200 (50%)
Silicate flour ST6	120	120	80	40	-
Superplasticizer – Stachement 2000	6.9	6.9	6.9	6.9	6.9
Water	195	220	220	220	220

2.2. Experimental methods

2.2.1. Basic physical properties
Bulk densities, ρ [kg m^{-3}], were determined from the volume and mass measured for the individual samples.

2.2.2. Mechanical properties
The compressive strength [MPa] and bending strength [MPa] are the main mechanical parameters and were determined according to ČSN EN 12390-3 [11] and ČSN EN 12390-5 [12], respectively. A device EU 40 was used. The strength was determined for 28-days cured samples.

2.2.3. Sorption isoterms
Sorption isotherms were determined by batch experiments. The sorbate solutions of varying concertations were prepared from CsCl and $SrCl_2.6H_2O$ (p.a.). Namely, 100 mg of pulverized concrete and 100 ml of solution of a proper concertation were agitated for 72 hours and 100 rpm. These conditions proved to be sufficient for equilibration of Cs and Sr between sorbent and solution. The sorbed amount, q (in mg g^{-1} of the sorbent), was determined as

$$q = \frac{V(C_0 - C_e)}{m}, \tag{1}$$

where C_0 is the initial concentration in mg l^{-1}, C_e is the equilibrium concentration, V is the volume of the solution in liters, and m is the mass of the sample (sorbent) in grams. The concentration was determined by Atomic Absorption Spectrometer Agilent AA 280 FS with a flame atomization technique (Flame: Acetylene – air). The distribution coefficient, K_d (in ml g^{-1}), was determined as the ratio q/C_e.

3. Results and discussion

3.1. Basic physical properties

The measured values of bulk densities are given in Table 3. The bulk density decreased with the increasing amount of brick powder in the mixture, which is a result of increasing open porosity. This is caused by a higher porosity of the input brick powder. The decrease of bulk density is by about 10% (RC1) to 35% (RC4).

Table 3. Bulk density of studied materials.

	RCR	**RC1**	**RC2**	**RC3**	**RC4**
Bulk density [kgm^{-3}]	2161	1984	1998	1997	1612

3.2. Mechanical properties

Results concerning mechanical properties are given in Fig 2. There is a clear decreasing trend for both compressive and bending strengths. The drop in compressive strength for RC1 compared to the reference concrete is 17% and in the case of RC4 the residual strength is at about 35%. In the case of the bending strength, the decrease for RC1 is lower than in the case of the compressive strength – by 4%. However, for RC4 the bending strength showed less than half the reference value.

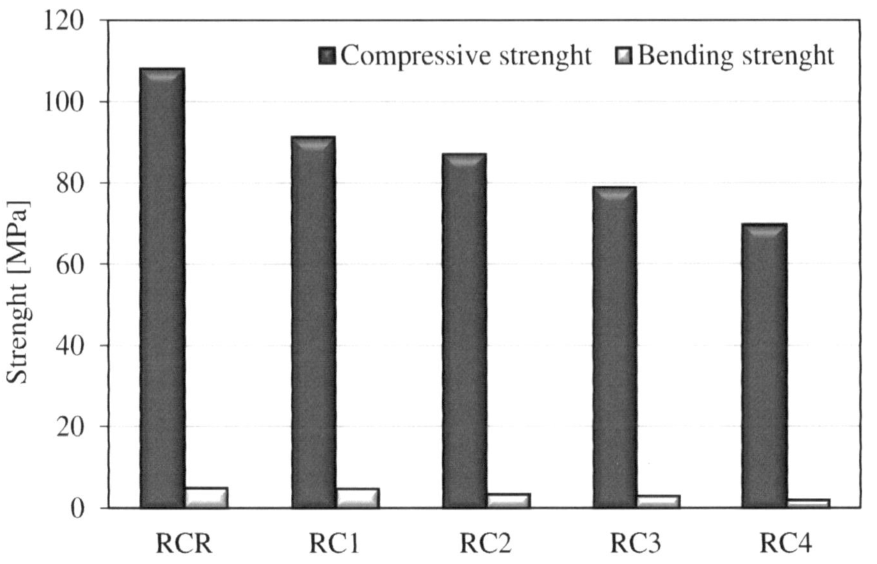

Fig. 2. Mechanical properties of the studied materials.

3.3. Sorption isotherms

The sorption isotherms obtained for the studied samples are shown in Figs. 3 and 4. These results

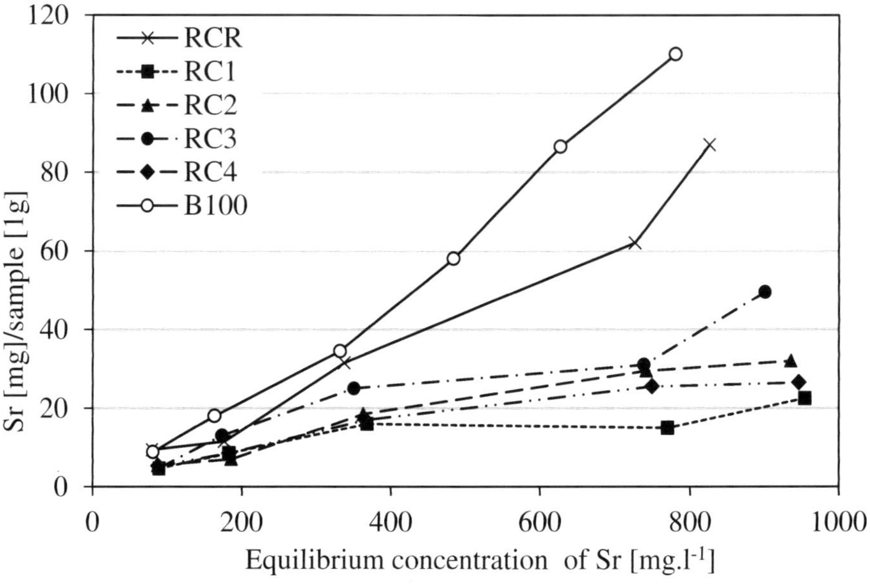

Fig. 3. Relationship between equilibrium concentration of Sr and its amount sorbed into the tested samples.

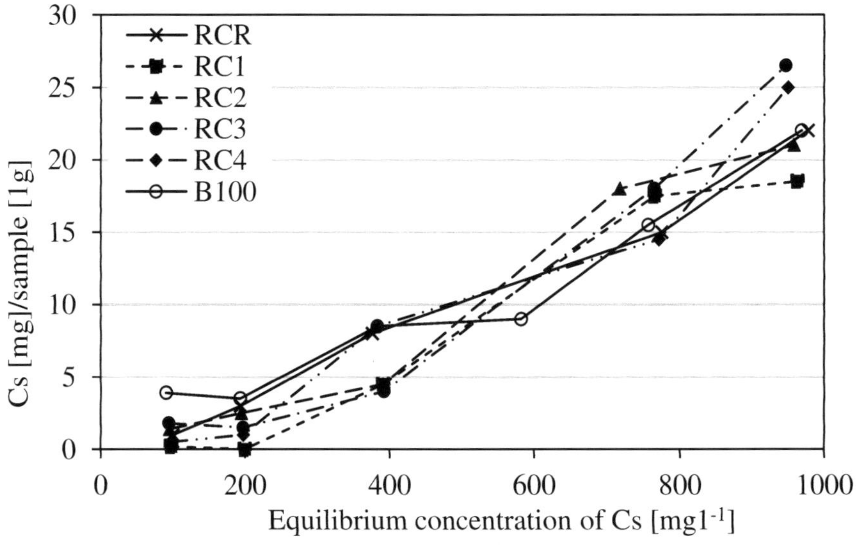

Fig. 4. Relationship between equilibrium concentration of Cs and its amount sorbed into the tested samples.

indicate that the samples containing brick adsorbed strontium to a larger extent than cesium. For samples containing the brick powder, the calculated values of the distribution coefficient K_d (see

Tables 4 and 5) are on an average about twice higher for Sr (from 19.48 mg l^{-1} to 141.03 mg l^{-1}) than for Cs (from 0 mg l^{-1} to 57.88 mg l^{-1}). For the reference samples the range of values of K_d is 50.96 – 117.18 mg l^{-1} for Sr and 0 – 31.91 mg l^{-1} for Cs, which is a more profound difference than for the samples with the brick powder. Thus, in general, Sr is sorbed in the samples in greater amounts than Cs.

Table 4. Distribution coefficient for all tested concentrations of Sr and all samples.

Sample	RC1	RC2	RC3	RC4	B100	RCR
Initial concentration [mg Sr.l^{-1}]			K_d [mg Sr.l^{-1}]			
100	50.66	61.17	51.88	58.66	106.06	117.18
200	46.45	37.63	74.71	46.45	109.76	64.97
400	43.48	50.96	71.43	46.45	104.23	93.47
800	19.48	39.81	42.01	34.05	137.96	50.96
1 000	23.56	34.19	54.94	27.98	141.03	105.33

Table 5. Distribution coefficient for all tested concentrations of Cs and all samples.

Sample	RC1	RC2	RC3	RC4	B100	RCR
Initial concentration [mg Cs.l^{-1}]			K_d [mg Cs.l^{-1}]			
100	2.01	14.40	18.67	5.05	42.30	0.00
200	0.00	12.82	7.61	5.05	18.13	15.46
400	11.51	11.51	10.20	22.19	22.19	31.91
800	22.88	57.88	23.56	18.81	29.10	16.13
1 000	19.21	21.92	27.98	26.32	16.00	11.25

4. Conclusions

The application of brick powder as a supplementary sorption material in a Portland cement paste was tested in this paper. Four mixes with 20 %, 30 %, 40 %, and 50% of the cement mass replaced by a brick powder were designed. The sorption capacity of the mixes were quite higher for Sr than for Cs, and so were the distribution coefficients. In comparison with the reference samples (with no brick powder), it turns out, however, that the addition of the brick powder did not significantly increased the sorption of neither Sr not Cs. In fact, a decrease in sorption was observed when for some mixes. Therefore, it should be concluded that the studied brick powder is not very suitable for the sorption of Sr and Cs elements in concrete.

Acknowledgment

This work was supported by the Czech Science Foundation, under project No. 17-11635S and by the project No. SGS19/143/OHK1/3T/11.

References

[1] Koťátková J, Zatloukal J, Reiterman P and Kolář K 2017 *J Environ Radioactiv* 147-55
[2] Ma B, Charlet L, Fernandez-Martinez A, Kang M and Madé B 2019 *Applied Geochemistry* **100** pp 414-31
[3] Abu-Khader MM 2009 Prog Nucl Ener **51** pp 225-35
[4] IAEA 2001 Performance of engineered barrier materials in near surface disposal facilities for

IOP Conf. Series: Materials Science and Engineering **549** (2019) 012046 doi:10.1088/1757-899X/549/1/012046

radioactive waste, results of a coordinated research project. Vienna: International Atomic Energy Agency (IAEA-TECDOC-1255)
[5] CEA 2009 Nuclear waste conditioning. A Nuclear Energy Division Monograph, Paris, France 78-82
[6] Ojovan MI 2011 Handbook of advanced radioactive waste conditioning technologies
[7] Elkamash Q, Elnaggar M and Eldessouky M 2006 *J Hazard Mat* **136** (2) pp 310-16
[8] Plecas I, Dimovic S and Smiciklas I 2006 *Prog Nucl Energ* **48** (6), pp 495-503
[9] Karakurt C, Kurama, H and Topçu IB 2010 *Cem Con Comp* **32** (1) pp 1-8
[10] Najimi M, Sobhani J, Ahmadi and Shekarchi M 2012 *Constr Build Mater* **35** pp 1023-33
[11] *ČSN EN 12390-3:* Testing of hardened concrete – Part 3: Compressive strength. 2002
[12] *ČSN EN 12390-5:* Testing of hardened concrete - Part5: Flexural strength. 2009

The ageing effect of raw materials on the final properties of high alumina refractory grog

D Zemánek[1] and L Nevřivová[1]

[1] Brno University of Technology, Faculty of Civil Engineering, Institute of Technology of Building Materials and Components, Veveří 331/95, 602 00 Brno, Czech Republic

E-mail: zemanek.d@fce.vutbr.cz

Abstract. High alumina ceramics is a part of the refractory ceramics in SiO_2-Al_2O_3 binary system. The aim of this paper is to characterize the effect of ageing raw materials and its influence on final properties of fired refractory grog. Refractory grog is very important component in production of refractory materials, such as fireclay, mullite and andalusite bricks. Refractory grog is manufactured generally from clay, claystone and kaolin, nowadays is mainly burnt in rotary and shaft kilns. The primary objective of this article is to determine an influence of ageing effect, which has impact on mineralogical composition of raw material which obviously results in different mineralogical composition of fired grog. Raw material with different age was used for the mixtures and properties of fired samples were investigated, considering the influence of the raw materials mixtures. Specifically, the apparent porosity, water absorption, bulk density, apparent porosity, mineralogical composition and pore distribution were investigated.

1. Introduction

Refractory grog is a basic component of ceramic products and has a major influence on the final properties. Very important component in refractory manufacturing is chamotte and high alumina grog. These grogs are manufactured from suitable sources, especially from clays, claystones and shales. Nowadays, the firing process is carried out in shaft and rotary kilns at temperatures between 1350 °C and 1550 °C. [1] High temperatures leads to changes of mineralogical composition. The minerals mullite, cristobalite and amorphous phase are formed during this process. Mullite is in these types of refractory grog a main carrier of refractory properties, therefore presence of mullite is required at the expense of other minerals. Chemical formula of mullite is $3Al_2O_3{\cdot}2SiO_2$, and its occurrence in nature is very rare.[2]

Current requirements on high quality of ceramic products leads to decreasing of natural sources. At the same time waste or secondary raw materials, which are not directly applicable in the required applications.[3] Ageing is spontaneous or intentionally controlled process by which the properties of raw materials are improved when are stored in stockpiles. In particularly, plasticity, viscosity and other rheological properties are improved. During the ageing, water, additives and chemical and biological processes in material are evenly distributed.

It is common practice in the ceramic industry to land freshly mined clays or their sieve-passing fractions and eventually expose them to external influences before entering the production process. In the past, clays and kaolins intended for the preparation of fine ceramics were left to rest in moist pits.

A small amount of biological plant waste was added to the surface of the deposited material. This activity aimed to create an environment that is natural for clay materials. The action of microorganisms on the clay material has a positive effect on flexural strength and other physical and mechanical properties. [4]

Natural raw materials are becoming rare, while millions of tons of inorganic wastes are produced daily in the mining and processing industries worldwide. However, some wastes with their composition resemble the original raw materials or are beneficial in other parts of the production process. Thus, the use of waste as an alternative to raw materials deserves particular emphasis in terms of economics and ecology. Processing wastes from the mining and treatment of natural raw materials has been used for landfill for decades or stockpiled near plants.

The current state of the mining industry, when new deposits opening is complicated and takes more than 20 years and the old ones are almost mined, leads to the need to deal with materials that were previously unwanted in manufacturing processes and were considered waste. Finding suitable ways to use them involves a reduction in production costs, alternative products or new products for other industrial applications. [5]

In Czech Republic, over the decades, the sieve-passing fractions of W claystone have been stockpiled. The reason was their difficult processing and the negative effect on firing process in the shaft kilns. There are several thousand tons of high-quality raw material available. The way that has been developed in recent years is the briquetting of mentioned sieve-passing fractions. By compressing the raw material, it is possible to form bodies corresponding to the appropriate claystone fraction and to utilize the material that was previously considered as waste. [6]

Because the material is stockpiled for different periods of time, its age is divided as follows:

- W claystone freshly mined F – without ageing (0)
- W claystone charcoal piles C – ageing 5 years (5)
- W claystone matured M – ageing 10 years (10)

The different ageing time may have an impact on the mineralogical properties of the sieve-passing fractions of claystone, but also on the resulting mineralogical composition of the fired ceramic body. In Table 1 are presented chemical compositions of materials used for the experiment.

Table 1. Chemical composition of input materials.

material	Al_2O_3 [%]	SiO_2 [%]	Fe_2O_3 [%]	TiO_2 [%]	CaO [%]	MgO [%]	Na_2O [%]	K_2O [%]	LOI [%]
W claystone F	42.15	53.19	1.03	1.68	0.38	0.17	0.45	0.85	14.24
W claystone C	41.71	54.02	1.23	1.67	0.28	0.14	0.10	0.75	14.63
W claystone M	41.72	53.51	1.58	1.80	0.34	0.17	0.10	0.68	14.77
Mullite MOTIM	76.04	23.00	0.26	0.05	0.32	0.05	0.10	0.12	0.19

2. Experimental

Three types of undersize W claystone fractions were used for specimen preparation. Physical and chemical properties are mentioned above. In the first phase, the claystone moisture was found to be 12 % for W claystone C and M and 9 % for W claystone F. Claystone was dried and then mixed with mullite MOTIM in proportion 90:10 to achieve content of alumina higher than 45 %.

Table 2. Designed mixtures with different type of W claystone.

Mixture ID	type of W claystone	W claystone [%]	mullite MOTIM [%]
0	F – freshly mined	90	10
5	C – charcoal piles	90	10
10	M - matured	90	10

Mixtures were homogenized in mull mixer. There were prepared three mixtures of high alumina grog, each contained same amount of mullite MOTIM and different type of W claystone, average designed chemical composition is shown in Table 3.

Table 3. Average designed chemical composition of high alumina grog.

material composition	Al_2O_3 [%]	SiO_2 [%]	Fe_2O_3 [%]	TiO_2 [%]	Na_2O+K_2O [%]
90 % W claystone + 10 % mullite MOTIM	46.20	49.31	1.81	1.82	0.99

Prepared mixtures were uniaxially pressed in a cylindrical steel mold using 20 MPa, 25 MPa and 30 MPa of pressure. The green bodies were dried in a laboratory drier at 110 °C and then fired at temperatures 1300 °C, 1400 °C and 1500 °C with a heating rate of 3,5 °C/min and soaking time of 2 hours at the maximum temperature in a laboratory furnace.

After firing the apparent porosity, bulk density, water absorption and apparent porosity were determined by the official standard EN 993-1:1995 [7] (vacuum water absorption). Pore structure was determined by mercury intrusion porosimetry test. Phases were identified by X-ray diffraction (PANALYTICAL Empyrean XRD) with Cu Kα as the radiation source, accelerating voltage 45 kV, beam current 40 mA in the range from 5° to 80° with a step scan of 0.01°. The obtained X-ray diffraction patterns were compared with the Crystallography Open Database (COD). [8]

3. Results and discussion

The most important properties of high alumina grog are apparent porosity and bulk density, along with final chemical and mineralogical composition suitable for the intended use. Higher linear shrinkage is appropriate but may be caused by higher content of fluxes. For example, in case of anode bake furnaces is suitable content of flux lower than 1 %.

Apparent porosity and bulk density of mixtures are presented in Table 4, 5 and 6. Increasing of firing temperature is associated with decreasing of apparent porosity, also with increasing bulk density. Increasing of pressure provides us diverse results.

Table 4. Apparent porosity and bulk density of Mixture 0 based on W claystone F at different firing temperatures and variable pressure.

firing temperature [°C]	pressure [MPa]	apparent porosity [%]	bulk density [$kg \cdot m^{-3}$]
	20	18.5	2230
1300	25	18.1	2220
	30	17.5	2200
	20	14.7	2300
1400	25	14.2	2300
	30	14.5	2290
	20	5.6	2440
1500	25	6.4	2410
	30	7.6	2390

IOP Conf. Series: Materials Science and Engineering **549** (2019) 012047 doi:10.1088/1757-899X/549/1/012047

Table 5. Apparent porosity and bulk density of Mixture 5 based on W claystone C at different firing temperatures and variable pressure.

firing temperature [°C]	pressure [MPa]	apparent porosity [%]	bulk density [kg·m^{-3}]
	20	17.8	2210
1300	25	19.2	2240
	30	18.4	2230
	20	12.8	2260
1400	25	13.3	2290
	30	14.4	2280
	20	9.7	2370
1500	25	10.9	2360
	30	9.4	2420

Table 6. Apparent porosity and water absorption of Mixture 10 based on W claystone M at different firing temperatures and variable pressure.

firing temperature [°C]	pressure [MPa]	apparent porosity [%]	bulk density [kg·m^{-3}]
	20	21.3	2160
1300	25	20.2	2160
	30	19.4	2180
	20	16.3	2190
1400	25	17.4	2250
	30	17.0	2230
	20	10.3	2340
1500	25	10.7	2360
	30	9.8	2380

The addition of mullite MOTIM affected the physical properties when fired. After firing the aluminium oxide content was higher than 45 % as designed.

The firing temperature has a strong impact on the density of the fired high alumina grog samples. Apparent porosity does not decrease at higher pressures for all samples and some results have opposite trend, but the results are in range lower than 1 %. Best results in case of apparent porosity at temperature 1500 °C are achieved with the Mixture 0, final apparent porosity is between 5,6 % and 7,4 % depending on applied pressure. Apparent porosity results of the Mixture 5 and Mixture 10 at this temperature are higher than Mixture 0 at pressure 20 MPa and 25 MPa.

XRD analysis did not proved significant differences among three types of raw W claystone in case of mineralogical composition.

IOP Conf. Series: Materials Science and Engineering **549** (2019) 012047 doi:10.1088/1757-899X/549/1/012047

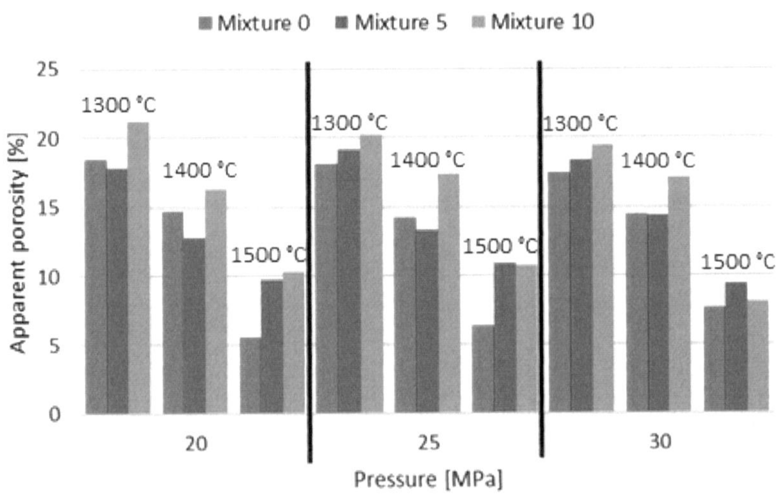

Figure 1. The influence of pressure and firing temperature on apparent porosity.

The result of XRD analysis of fired samples is shown in Figure 2. Enlarged section between 2Θ angle 25° and 27° is shown in Figure 3.

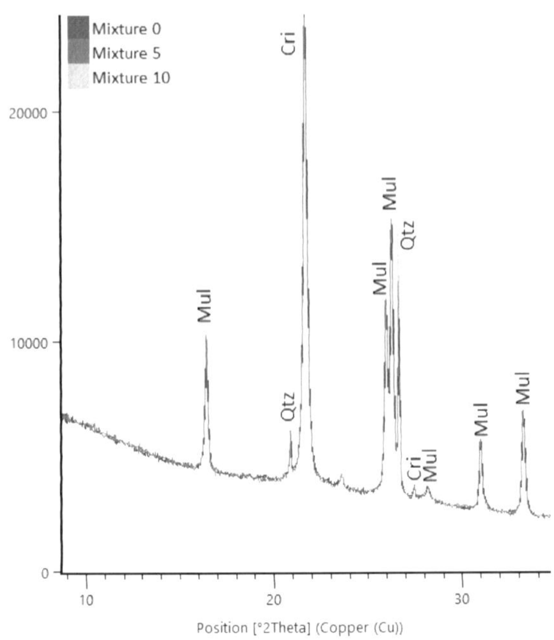

Figure 2. Influence of different W claystone age on mineralogy of high alumina grog at temperature 1300 °C.

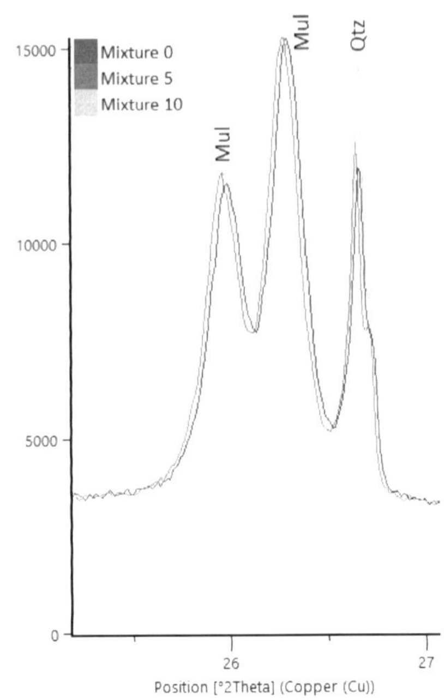

Figure 3. Influence of different W claystone age on residual quartz presence at temperature 1300 °C.

Legend: Mul-mullite, Qtz-quartz, Cri-cristobalite

IOP Conf. Series: Materials Science and Engineering **549** (2019) 012047 doi:10.1088/1757-899X/549/1/012047

Acknowledgments
This article was created within the project no. 2016TH02020321 with the title *"Production of Mullite Grog in a Shaft Kiln"* sponsored by the Technical Agency of the Czech Republic.

References
[1] Kutzendörfer J and Tomšů F 2008 *Žáruvzdorné materiály: díl 1.* 1st ed. Prague: Czech Ceramic Society. ISBN 978-80-02-02076-9
[2] Schneider H 2008 *Structure and properties of mullite—A review.* Journal of the European Ceramic Society. **28**, pp 329-344
[3] Brasileiro M, Rodrigues A and Menezes R 2012 *The Kaolin Residue and Its Use for Production of Mullite Bodies.* Energy, Engineering and Technologies
[4] Gaidzinski R, Osterreicher-Cunha P and Duailibi J 2009 *Modification of clay properties by aging: Role of indigenous microbiota and implications for ceramic processing.* Applied Clay Science. **43**, pp 98-102
[5] Menezes R, Brasileiro M, Gonçalves W and Santana L 2009 *Statistical Design for Recycling Kaolin Processing Waste in the Manufacturing of Mullite-Based Ceramics.* Mat. Res., **12**
[6] Pešová A, Andertová J, Machovič V and Gedeon O 2008 *Ageing of clay-based materials.* Advanced Materials Research, **39-40**, pp 359-362
[7] EN 993-1:1995 *Methods of test for dense shaped refractory products.* Part 1: Determination of bulk density, apparent porosity and true porosity
[8] *Crystallography Open Database: an open-access collection of crystal structures.* [online]. Available from: http://www.crystallography.net/
[9] Nevřivová L 2018 *Use of Undersize Fractions of Refractory Claystone for Grog Manufacturing in Shaft Kilns*
[10] Nevřivová L and Zemánek D 2019 *The Influence of Al2O3 Microparticles on the Pore Structure of Fired Clay*

IOP Conf. Series: Materials Science and Engineering **549** (2019) 012048 doi:10.1088/1757-899X/549/1/012048

Influence of type and composition of aggregate on mechanical parameters of concrete

M Tazky[1], L Osuska[1] and L Bodnarova[1]

[1] Brno University of Technology, Faculty of Civil Engineering, Veveri 95, 602 00 Brno, Czech Republic

E-mail: bodnarova.l@fce.vutbr.cz

Abstract. The paper deals with a thorough analysis of the influence of the type of aggregate and its resulting grain size curve on the basic strength parameters of concrete. From the point of view of the type, natural crushed aggregate of amphibolitic type, natural crushed aggregate of granodioritic type and natural mined sedimentary aggregate of the psammitic type were monitored. The same concrete strength class was proposed for the analysis of the impact of individual types of aggregates. In one case it was a concrete mixture with a maximum aggregate grain size D_{max} 16 mm with a continuous grain size curve and in the second case a concrete mixture with a maximum grain size D_{max} of 22 mm with a discontinuous grain size curve, in the absence of a 4-8 mm fiction. In both cases, the mixture of individual aggregate fractions was determined so that the resulting grain size curve of the mixture was the same for all aggregate types. Thanks to this approach it was possible to directly monitor the influence of the type of aggregate and the grain size curves on the resulting concrete parameters. The results of the experiment show that a significant impact while maintaining the same grain size curves of the resulting aggregate mixture can be seen mainly in the type of aggregate. The absence of a 4-8 mm fraction when designing a concrete mixture with a D_{max} of 22 mm proves to be completely problem-free from the point of view of concrete parameters in both fresh and hardened condition. A part of the research was also verification of the possibility of using natural crushed sand of fraction 0-4 mm for production of high-strength situ ready-mixed concretes.

1. Introduction

New trends in the production of high performance concretes provide a prerequisite for mixture designs based on precisely described and defined parameters entering the production process. In terms of concrete technology, it is mainly the creation of optimal grain size curves of aggregate, but also the optimal combination of individual binder components. Special requirements for high performance concretes are laid both in terms of strengths and, for example, high resistance against mechanical wear or wear by water flow. For the production of high performance concretes, it is often necessary to use different input components, such as quality dense natural aggregates, as opposed to conventional concrete.

This paper deals with the assessment of the possibility of producing concretes at the boundary of high-strength concretes with a discontinuous grain size curve. The aim of the experiment is to define the suitability of the resulting grain size curve of the aggregate mixture in terms of the maximum grain used and the absence of the 4-8 mm fraction. A large part of authors tends to the theory that with increasing volume of aggregate and increasing maximum aggregate grain, undesirable volume changes of the composite decrease. This fact often results in improvements in the mechanical and durability parameters of the concrete. [1, 2, 3] Generally, however, no consensus can be found across scientific publications and published results within the impact of the absence of any of the aggregate fractions on the resulting composite parameters. However, most authors tend to the theory that a discontinuous grain size curve has a positive impact on high-strength concrete (HSC) parameters. [4, 5].

The formation of a discontinuous grain size curve of the aggregate also brings savings from the concrete manufacturer's perspective, because the 4-8 mm fraction is generally the most omitted from the composition of the mixture, which is the most economically demanding in terms of crushed aggregates used for production of HSC. Further economic saving, but also the achievement of more suitable mechanical parameters, may mean the use of a 0-4 mm fraction of crushed aggregate. In general, however, this type of aggregate is mainly used in prefabricated production, mainly due to the negative impact on the consistency of the mixture. However, the impact of the use of this crushed aggregate fraction in connection with conventional sand for HSC production was monitored in the experiment.

A good precondition for using this aggregate for the production of HSC can be seen in higher doses of binder components which can provide enough cement slurry for the desired degree of consistency for ready-mixed concrete.

2. Experiment

Concrete mixtures of strength class C 50/60, i.e. concretes at the boundary of HSC, were designed within the experiment. All mixture designs matched the S4 degree of consistency by slump test according to EN 12350-2. [6] CEM I 42.5 R Portland cement was used for the production in consistent dosage and aggregates of various types in terms of both extraction technology and petrography. In two cases, the aggregates were crushed and in one case the aggregate was mined. Crushed aggregates were from the perspective of petrography classified as granodiorite and as amphibolite. In the case of mined aggregate, it was a psammitic sedimentary rock. Before the design of concrete mixtures, basic tests were carried out on the aggregates to determine grain size and density. The specific results, including a description of the aggregates used, are given in the following Table 1.

2.1. Aggregate properties

Grain size was determined in accordance with EN 933-1 [7], grain density in accordance with EN 1097-6 [8] and shape index in accordance with EN 933-4 [9].

Table 1. Density and shape index of used aggregates.

Aggregate origin/locality	Rock	Fraction [mm]	Density [$kg \cdot m^{-3}$]	Shape index [-]
Natural crushed / Zelesice	amphibolite	0 – 4	2930	-
		4 – 8	2960	18
		8 – 16	2980	22
		11 – 22	2940	13
Natural crushed / Olbramovice	granodiorite	4 – 8	2620	23
		8 – 16	2650	21
		11 – 22	2680	22
Natural mined / Zabcice	psammitic sediment	0 – 4	2520	-
		4 – 8	2540	15
		8 – 16	2530	18
		16 – 22	2550	16

IOP Conf. Series: Materials Science and Engineering **549** (2019) 012048 doi:10.1088/1757-899X/549/1/012048

2.2. Composition of concretes and results of mechanical parameters

Based on the results obtained on the aggregates used, a total of 6 concrete formulas were designed. The individual formulas differed only in the type of aggregate used. A total of 3 formulas were designed with a maximum aggregate grain of 16 mm with a composition of fractions 0-4, 4-8 and 8-16 mm and 3 recipes with a maximum aggregate grain of 22 mm with a composition 0-4, 8-16 and 11-22 mm. The grain size curves of each aggregate mixture were designed for each D_{max} as constant. Thanks to this approach it was possible to monitor the impact of individual types of aggregate, eventually the use of crushed sand, on the resulting basic mechanical parameters of the concrete. The following Table 2 and Figure 1 and 2 show the composition of individual concretes and the resulting grain size curves of aggregate mixtures.

Table 2. Composition of individual concrete formulas.

	Material	Formula designation					
		1	2	3	4	5	6
	CEM I 42.5 R [kg]	435	435	435	435	435	435
Aggregate [kg]	0-4 mm Zabcice	440	440	815	815	775	775
	0-4 mm Zelesice	445	445	-	-	-	-
	4-8 mm Zabcice	-	-	-	-	225	-
	4-8 mm Zelesice	230	-	-	-	-	-
	4-8 mm Olbramovice	-	-	200	-	-	-
	8-16 mm Zabcice	-	-	-	-	688	540
	8-16 mm Zelesice	730	555	-	-	-	-
	8-16 mm Olbramovice	-	-	713	535	-	-
	11-22 mm Zabcice	-	-	-	-	-	373
	11-22 mm Zelesice	-	405	-	-	-	-
	11-22 mm Olbramovice	-	-	-	378	-	-
	Superplasticizer [% m_c]	0.8	0.8	0.8	0.8	0.8	0.8
	Water-cement ratio [-]	0.39	0.39	0.39	0.39	0.39	0.39

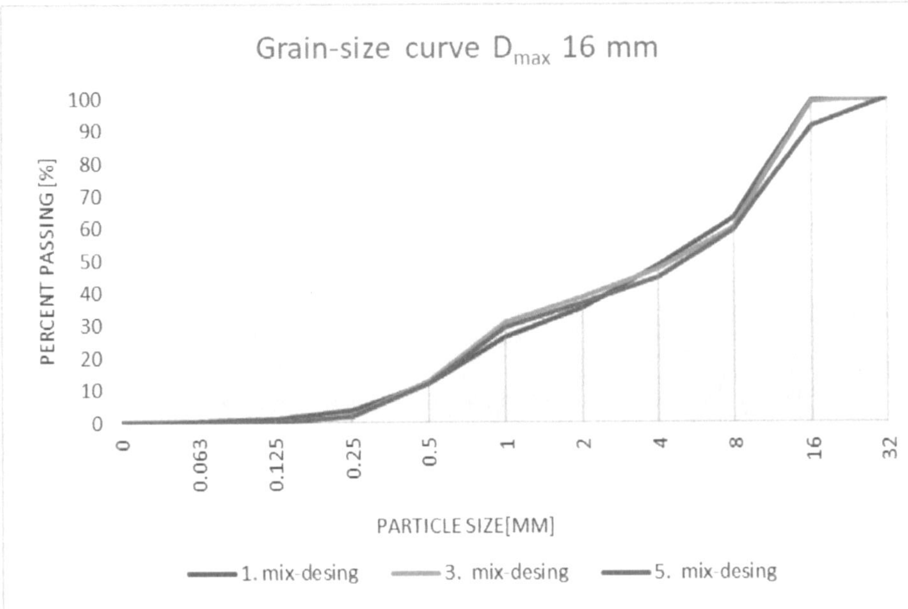

Figure 1. Grain size curves of aggregate mixtures for formulas with D_{max} 16 mm.

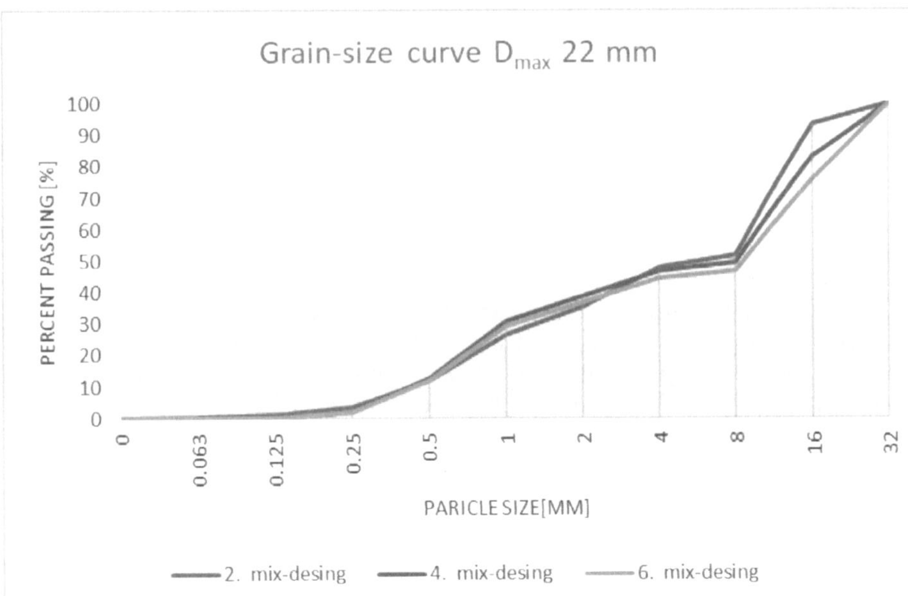

Figure 2. Grain size curves of aggregate mixtures for formulas with D_{max} 22 mm.

On produced concretes was determined the consistency by cone setting at time 5 minutes and further after 7, 28 and 90 days, compressive strength on cubes with 150 mm edge according to EN 12390-3 [10] and flexural strength on prisms with size 100 × 100 × 400 in 4-point distribution according to EN 12390-5 were determined [11]. Following Table 3 summarizes the results of all tests performed.

Table 3. Results of realized tests.

Parameter		Formula designation					
		1	2	3	4	5	6
Slump-test [mm]		160	170	160	170	170	170
Density $[kg \cdot m^{-3}]$	7-day	2500	2510	2350	2370	2320	2360
	28-day	2520	2530	2380	2360	2340	2390
	90-day	2520	2520	2360	2350	2340	2370
Flexural strength [MPa]	7-day	6.6	7.4	6.6	6.0	5.3	5.6
	28-day	7.8	8.0	7.0	7.0	5.6	6.0
	90-day	8.0	7.9	7.0	7.2	5.8	6.5
Compressive strength [MPa]	7-day	59.7	69.9	57.3	60.4	52.0	56.4
	28-day	73.6	76.5	73.8	76.3	61.5	67.2
	90-day	73.3	74.5	76.2	79.3	63.2	68.1

From the results of consistency determination by cone setting method it is evident that crushed sand of fraction 0-4 mm can be easily used for the production of high-strength ready mixed concrete of the S4 degree of consistency with a suitable combination with 0-4 mm fraction of dredged sand. The use of this crushed sand can then be particularly suitable for structures with high abrasion resistance. According to the assumption it has been proven that if concretes with a higher content of binder components are produced, the crushed sand is also not an obstacle for their production in an appropriate ratio with the mined sand. It has also been proven that the maximum grain size of the aggregate has practically no effect on the consistency of the mixture.

The results of the compressive strength show significant differences, especially in terms of the type of aggregate used. The experiment clearly demonstrates the suitability of using quality crushed aggregates for the production of high strength concretes. It has also been demonstrated that the use of a discontinuous grain size curve and a larger maximum grain appears to be more appropriate for high strength concrete production. This trend is also evident in the determination of flexural strength of concrete. In this case, the quality of the aggregate is a decisive parameter and the use of quality magmatic rock appears to be the most suitable based on the results.

3. Conclusion

The realized experiment was focused on determining the influence of the maximum aggregate grain and the type of aggregate on the resulting mechanical parameters of high-strength concretes. Several concretes were produced using different types of aggregates (crushed × mined) and various petrographies (amphibolite × granodiorite × sedimentary psammite). All the concretes produced were designed with a constant dose of cement, superplasticizer and a water cement ratio. Mixtures were produced with a maximum aggregate grain size of 16 and 22 mm, while the individual grain size curves were also designed as constant. Thanks to this approach and the results obtained, the impact of different aggregates on the consistency of the concrete and on the basic strength parameters can be assessed.

The results of the experiment proved the suitability of using quality crushed aggregate for the production of concretes of higher strength classes. The type and mechanical parameters of the aggregate itself seem to be absolutely deciding in the production of HSC. This experiment also pointed to the suitability of using aggregates with a greater maximum grain size, wherein the composition is particularly suitable in terms of the achieved flexural strengths.

A part of the research was also an assessment of the possibility of using crushed sand of fraction 0-4 mm for ready-mixed concretes by consistency of slump test to S4. The results clearly prove that this aggregate can be used without difficulty in the production of concretes with a larger amount of binder component with a suitable combination with conventional mined or dredged sand with fraction 0-4 mm. This larger amount of binder component together in combination with conventional sand is necessary to achieve the desired plasticity of the mixture. The results of the strength of concretes with this aggregate show very good preconditions for its use especially in concretes exposed to increased abrasion, where the sand coming from crushed aggregate contains noticeably more resistant grains than the classic sand.

The results obtained will be used for further development of high-strength concretes with high strength, high resistance to mechanical wear and wear by water flow and minimal volume changes.

Acknowledgments
Theoretical background for the presented results was obtained with financial support of the project GA 18-25035S „Study and subsequent modelling of mechanical corrosion of cementitious composites exposed to flowing fluids".

References

[1] Gonzalez-Corominas A and Etxeberria M 2016 Effects of using recycled concrete aggregates on the shrinkage of high performance concrete. *Construction and Building Materials* **115** 32-41
[2] Grassl P, Wong H S and Buenfeld N R 2010 Influence of aggregate size and volume fraction on shrinkage induced micro-cracking of concrete and mortar. *Cement and Concrete Research* **40** 85-93
[3] Chen Y, Wie J, Huang H, Jin W and Yu Q 2018 Application of 3D-DIC to characterize the effect of aggregate size and volume on non-uniform shrinkage strain distribution in concrete. *Cement and Concrete Research* **86** 178-189
[4] Huynh Trong-Phuoc, Si-Huy Ngo and Chao-Lung Hwang 2018 Fresh and Hardened Properties of Concrete Produced with Different Particle Sizes of Coarse Aggregate. *Advanced Materials Research* **1147** 18-23
[5] Meddah M S, Zitouni S and Belâabes S 2010 Effect of content and particle size distribution of coarse aggregate on the compressive strength of concrete. *Construction and Building Materials* **24** 505–512
[6] EN 12350–2:2009 Testing fresh concrete - Part 2: Slump-test
[7] EN 933 – 1:2012 Test for geometrical properties of aggregates - Part 1: Determination of particle size distribution– Sieving method
[8] EN 1097–6:2014 Test for mechanical and physial properties of aggregates - Part 6: Determination of particle density and water absorption
[9] EN 933–4:2008 Test for geometrical properties of aggregates - Part 4: Determination of particle shape – Shape index
[10] EN 12390–3:2009 Testing hardened concrete - Part 3: Compressive stregth of test specimens
[11] EN 12390–5:2006 Testing hardened concrete - Part 5: Flexural strength of test specimens

Experimental study of the modulus of elasticity of concrete at different ambient temperature

P Bujnakova[1], J Jost[1], M Farbak[1] and P Kotes[1]

[1] University of Žilina, Faculty of Civil Engineering, Univerzitna 1, Žilina, Slovakia

E-mail: petra.bujnakova@fstav.uniza.sk

Abstract. The structural performance of concrete construction and bridges is directly influenced by the strength and the modulus of elasticity of concrete. The modulus of elasticity is an input parameter to the creep compliance important especially in design calculation of large concrete bridges. The development of strength and modulus of elasticity at the early ages depends also on the temperature of the environment during casting and curing of concrete. The effect of ambient temperature on modulus of elasticity is however not assessed by the EN 1992-1-1 [1]. The experimental program was focused on determination of static modulus of elasticity and compressive strength cast and cured at ambient temperature +20°C and temperature -5°C at different time. The paper also formulates preliminary design recommendations applicable in practice.

1. Introduction

The proper execution of concrete is controlled by using adequate recipes and technology process specified by relevant material standards when casting concrete on site [4-6]. The mechanical properties of concrete strength class can achieve different values depending on the concrete composition such as the type of aggregate, type of cement, the w/c ratio and admixtures [10-12]. The various results are obtained by using different testing methods (quasistatic, ultrasonic and resonant frequency) especially at early ages [13, 14]. Ultrasonic measurements provide values of modulus of elasticity that are higher than the values provided by the quasi-static test [12], [15]. The test procedure is not universal in several country [3], [7, 8]. Current knowledge [18-20] indicates that besides the above-mentioned factors, the development of strength and elasticity modulus during the first 28 days depends also on the temperature of the environment during casting and curing of concrete [9], [21]. It is usual that concrete structures and bridges are cast at temperatures below zero degrees Celsius. The hydration of cement requires water to be in liquid state. With low temperatures (or temperatures below zero), hydration might slow down or be stopped completely. This might directly impact the development of concrete properties during the first 28 days. The experimental studies [22, 23] observed that the mechanical properties of concrete are affected by hydration even before the setting time. The available references show [16, 17], [20] that the effect of low temperatures on the development of concrete strength is relatively small, but if the concrete temperature reaches freezing before a certain strength, the concrete will have reduced overall strength. Also the low temperature will affect the modulus of elasticity of concrete. The possible negative effect might be critical especially for structures where the design is governed by SLS criterions (e.g. large span bridges),

IOP Conf. Series: Materials Science and Engineering **549** (2019) 012049 doi:10.1088/1757-899X/549/1/012049

especially given that EN 1992-1-1 might overestimate the modulus of elasticity of concrete even when concrete is cast in ambient temperatures [24, 25].

2. Experimental program

A series of laboratory tests focused on demonstrating the static modulus of elasticity of concrete cast at different temperatures has been performed at the testing laboratory of the University of Žilina in Slovakia. A concrete with a recipe corresponding to concrete class C 30/37 was used.

The test procedure for determination of the static modulus of elasticity in compression has been defined in the Slovak standard STN EN 12390-13 [2] since 2014. The standard code STN ISO 6784 (1993) [3] is valid simultaneously (replaced the STN 73 1319, from 1968). The test of modulus of elasticity according to this standard is recommended for the structures with large expected deflection. This procedure was used for the determination of static modulus of elasticity for cylindrical concrete samples. A series of compressive stress cycles loaded up to approximately 1/3 of the concrete compressive strength measured immediately before the tests of modulus of elasticity. Than the modulus of elasticity is calculated as follows:

$$E_c = \frac{\Delta\sigma}{\Delta\varepsilon} = \frac{\sigma_a - \sigma_b}{\varepsilon_a - \varepsilon_b} \qquad (1)$$

where:
σ_a – the upper stress at loading [N/mm^2],
σ_b – the basic stress [N/mm^2],
ε_a – the average strain at loading [-],
ε_b – the average basic strain [-].

The tests were carried out on 44 cylinders with a diameter of 150 mm and a height of 300 mm, Fig.1. The specimens were cast from the same concrete composition corresponding to the concrete class C 30/37 used commonly in bridge superstructures. After casting, the specimens were stored in a mold and curing in a room with ambient conditions at (20.0±2) °C during 24 hours. After 24 hours, all specimens were removed from the moulds and stored in freezing and climate chamber, Fig. 2. The 24 concrete cylinder specimens were stored in freezing chamber with a temperature of -5.0 °C. The others 20 specimens were stored in climate chamber with temperature at +20.0 °C.

Figure 1. Casting of specimens.

IOP Conf. Series: Materials Science and Engineering **549** (2019) 012049 doi:10.1088/1757-899X/549/1/012049

Thereafter, the four cylinder specimens were tested for each series. The one reference specimen was used for determination of compressive strength and three specimens were tested for secant modulus of elasticity and cylindrical compressive strength.

Figure 2. Storage of specimens in a) climate chamber and b) freezing chamber.

The specimens were measured at the temperature of +20.0 °C after 3, 7, 14, 21, 28, 90 days and at the temperature of -5.0 °C after 7, 14, 21, 28, 90 days. Total 11 series have been tested.

The top of the specimen has been ground before testing to achieve a parallel surface of both compressive ends. Before the test, the cylindrical compressive strength $f_{c,cyl}$ was determined for each series. The upper stress level at loading was obtained as approximately 1/3 of cylindrical compressive strength ($\sigma_a = f_{c,cyl}/3$). The basic stress was defined with the value of 0.5 N/mm^2 (σ_b). The special devices- extensometers were used for measuring the changes of the length. The two strain gauges were installed circumferentially at opposite points at the height of the specimen to measure transverse deformation, Fig.3. The modulus of elasticity of the concrete corresponds to the average slope of stress-strain responses captured during cyclic loading.

In the first loading cycle, the specimen was loaded up to σ_b = 0.50 N/mm^2 and recorded the longitudinal strain ε_b on both extensometers. Than load was applied at a continuous rate of movement corresponding to a stress 0.60 N/mm^2 per second until a stress level reached the 1/3 of measured compressive strength. The load was kept at this level for 60 seconds and the strain ε_a was read. The two additional loading cycle was repeated. The load was kept constant for 60 seconds again and the linear-elastic response of the specimen ε_a and ε_b was recorded. Then the specimens were loaded to failure and the static modulus of elasticity was calculated, Fig. 3.

Figure 3. Testing arrangement of a) modulus of elasticity and b) cylindrical compressive strength.

IOP Conf. Series: Materials Science and Engineering **549** (2019) 012049 doi:10.1088/1757-899X/549/1/012049

3. Results and discussion

The first tests were performed at 3 days after casting of specimens at temperature +20°C. The results of the static modulus of elasticity and compressive strength are listed in Table 1. The results of tests carried out on specimens at temperature (-5°C) are shown in Table 2. The average compressive strength of specimen tested cast low temperature was about 27 percent lower that the strength of specimen cast at ambient temperature while the modulus of elasticity was only about 17 percent lower. The negative effect of an ambient temperature (below zero) is more visible in the cylinder compressive strength. Development of the compressive strength of concrete over the first 90 days at different ambient temperature is shown in Fig. 4. The comparison of the development of the modulus of elasticity of concrete over the first 90 days at different ambient temperature is shown in Fig. 5.

Table 1. Mechanical properties of concrete cylinders at the temperature -5 °C.

Series	No. of specimen	Density [kg/m^3]	Compressive strength f_{ci} [N/mm^2]	Compressive strength $f_{cm,i}$ [N/mm^2]	Modulus of elasticity E_{ci} [N/mm^2]	Modulus of elasticity $E_{cm,i}$ [N/mm^2]
B-7	5*	2345	31.0	30.7		23 600
	6*	2348	30.6		24 100	
	7*	2371	31.3		24 100	
	8*	2330	29.8		22 700	
D-14	13*	2328	33.6	32.5		25 100
	14*	2357	29.8		25 100	
	15*	2351	31.6		24 800	
	16*	2350	35.0		25 500	
F-21	21*	2321	34.2	34.6		25 600
	22*	2385	35.6		25 600	
	23*	2332	34.0		26 500	
	24*	2325	34.8		24 800	
H-28	29*	2365	33.9	37.0		26 000
	30*	2342	38.0		26 800	
	31*	2320	37.2		24 800	
	32*	2363	38.8		26 300	
J-90	37*	2405	37.9	39.2		26 600
	38*	2339	40.9		27 100	
	39*	2340	40.7		26 000	
	40*	2305	37.1		26 700	

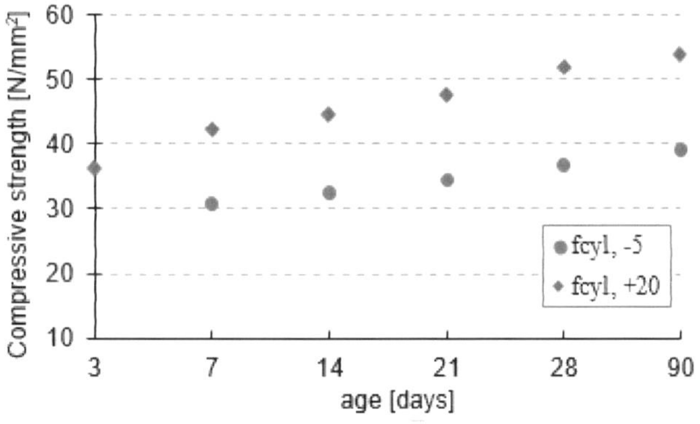

Figure 4. Measured compressive strength of concrete.

Table 2. Mechanical properties of concrete cylinders at the temperature +20 °C.

Series	No. of specimen	Density [kg/m³]	Compressive strength f_{ci} [N/mm²]	Compressive strength $f_{cm,i}$ [N/mm²]	Modulus of elasticity E_{ci} [N/mm²]	Modulus of elasticity $E_{cm,i}$ [N/mm²]
	1	2312	35.7			
A-3	2	2349	35.4	36.4	25 400	25 600
	3	2348	37.3		25 400	
	4	2344	37.4		25 900	
	9	2345	41.6			
C-7	10	2343	43.5	42.2	28 500	28 500
	11	2325	42.1		27 600	
	12	2354	41.7		29 400	
	17	2336	36.9			
E-14	18	2355	46.1	44.6	29 600	29 900
	19	2321	46.3		30 100	
	20	2335	49.0		30 100	
	25	2322	47.4			
G-21	26	2339	48.9	47.6	29 300	31 000
	27	2309	45.4		33 400	
	28	2346	48.6		30 400	
I-28	33	2327	51.4			
	34	2335	53.5	51.9	31 500	30 900
	35	2332	52.6		30 900	
	36	2321	50.2		30 300	
K-90	41	2310	55.4			
	42	2318	56.1	54.1	32 000	32 100
	43	2336	48.9		33 300	
	44	2323	56.1		31 100	

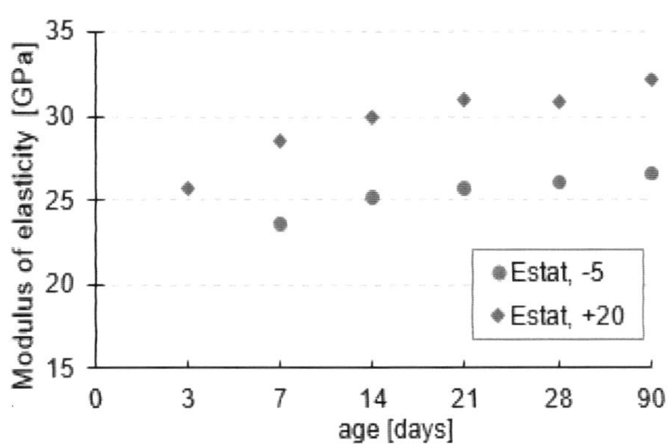

Figure 5. Measured static modulus of elasticity of concrete.

The characteristic value of compressive strength of concrete is estimated in accordance with Table 3.12 of EN 1992-1-1 as:

$$f_{cm} = f_{ck} + 8 \quad (2)$$

The modulus of elasticity is estimated as:

$$E_{\mathrm{cm}} = 22 \cdot [(f_{\mathrm{cm}})/10]^{0,3} \qquad (3)$$

Then the values of modulus of elasticity of the tested concrete specimens calculated in accordance with STN EN 1992-1-1 [1] are plotted in Table 3.

Table 3. Compression test and elastic modulus test results.

days	$f_{cm,20}$ [MPa]	$f_{cm,-5}$ [MPa]	$f_{ck,20}$ [MPa]	$f_{ck,-5}$ [MPa]	$E_{cm,20}$ [GPa]	$E_{cm,-5}$ [GPa]
7	42.2	30.7	34.2	22.7	31.81	28.13
14	44.6	32.5	36.6	24.5	32.47	28.79
21	47.6	34.6	39.6	26.6	33.25	29.50
28	**51.9**	**37.0**	**43.9**	**29.0**	**34.29**	**30.28**
90	54.1	39.2	46.1	31.2	34.80	30.95

The measured values were compared with values of modulus of elasticity specified in EN 1992-1-1 [1]. The concrete strength class specified to the manufacturer was C 30/37. While the compressive strength of concrete measured at ambient temperature by tests at 28 days is largely sufficient to classify the concrete in that class (f_{ck}= 43.9 MPa), the measured modulus (30.9 GPa) is slightly inferior to the value specified for C 30/37 in Table 3.12 of EN 1992-1-1 (32 GPa). The value E_{cm} determined using Eq. (3) overestimates the elasticity modulus in comparison to the tested value. This is an observation that confirms findings available in earlier references [8]. The elasticity modulus of specimen cast at -5 degrees determined by tests (26 GPa) is 20 percent lower than the elasticity modulus specified for class C 30/37 in Table 3.12 of EN 1992-1-1 [1]. The value calculated using the characteristic compressive strength $f_{ck,-5}$ in Eq. (2) provides the non-conservative estimate of the modulus of elasticity.

4. Conclusions

The experimental research allowed to demonstrate the influence of the low temperature at casting and curing on the elasticity modulus of concrete. While the modulus of elasticity determined by tests is relatively close to the value of the modulus specified by EN 1992-1-1 for concrete C 30/37 for the concrete cast and cured at ambient temperature, the modulus of specimen cast and cured at -5 degrees is significantly lower. In absence of more accurate design recommendations, the authors recommend to design structures expected to be constructed at low temperatures assuming a reduced value 0.8.Ecm. The design recommendations will be further developed following the completion of a research program. Currently, the testes of dynamic modulus of elasticity are verified. Based on the experimental results, the numerical model will be created for analysis of segmentally cast cantilever bridges considering additional mechanical properties (creep, shrinkage).

Acknowledgments

This work was supported by the Slovak Research and Development Agency under contract APVV-14-0772, by Research project No. 1/0413/18, No. 1/0343/18 and No. 1/0045/19 of Slovak Grant Agency.

References

[1] STN EN 1992-1-1 2015 *Eurocode 2: Design of concrete structures- Part 1-1. General rules and rules for building* (Bratislava: Office of Standards, Metrology and Testing)
[2] STN EN 12390-13 2014 *Testing hardened concrete. Part 13: Determination of secant modulus of elasticity in compression.* (Bratislava: Office of Standards, Metrology and Testing)
[3] ISO 6784 1993 *Concrete. Determination of static modulus of elasticity in compression* (Bratislava: Office of Standards, Metrology and Testing)

[4] STN EN 13670 2012 *Execution of concrete structures.* (Bratislava: Office of Standards, Metrology and Testing)
[5] STN EN 206 +A1 2017 *Concrete. Specification, performance, production and conformity.* (Bratislava: Office of Standards, Metrology and Testing)
[6] ACI 306R-88 2002 *Cold Weather Concreting* Reported by ACI Committee 306, USA
[7] ASTM Standard C469 1994 *Standard Test Method for Static Modulus of Elasticity and Poisson's ratio of Concrete in Compression (Pennsylvania, USA)*
[8] Huňka P *et al* 2012 *Beton TKS* vol. **4** pp 62-67
[9] Kocab D at al 2017 *Experimental Analysis of the development of Elastic properties and Strength under Different Ambient temperature during the hardening of concrete Proc. Eng.* **195** pp 102-107
[10] Nikbin I *et al* 2014 *Constr. and Build. Mater.* **57** pp 69-80
[11] Liu B at al 2014 *Constr. and Build. Mater.* **69** pp 133-142
[12] Benboudjema F et al 2019 Thermal cracking of Massive Concrete Structures *Mechanical properties* chapter 4 pp 69-114
[13] Boulay C *et al* 2014 *Mater. and Struc.* **47** (1) pp 141-155
[14] Delasuate et al 2016 *Strain* vol **52** issue 2
[15] Han S H et al 2004 *Cem. And Concr. Res.* **34** pp 1219-1227
[16] Zandi H K et al 2011*Cem. and Concr. Res.* **41** pp 244-254
[17] Petersen L *et al* 2007 *ACI Mater. J.* **104** 4 pp 369–378
[18] Jiao Y et al 2014 *Adv. in Mater. Sci. and Eng.* ID **191360** 10p.
[19] Jacobsen S at el 1996 *Cem. and Concr. Res.* **26** 6 pp 919-931
[20] Madhavi T Ch et al 2016 ARPN J. of Eng. and App. Scien. **11** 9 pp 5990-5992
[21] Kim J at al 2002 *Cem. and Concr. Res.* **32** pp 1087-1094
[22] Carette J et al 2016 *Cem. And Concr. Compos.* **73** pp 1-9
[23] De Schutter G et al 1996 *Mater. and Struc.* Vol **29** pp 335-344
[24] Moravcik M et al 2014 *J of Civ. Struc. And Transpor. Eng.* **1** pp 1-6
[25] Holy M et al 2015 *Beton TKS* **4** pp 48-51

Institute of Physics
Dirac House, Temple Back
Bristol BS1 6BE UK

ISSN: 1757-8981
ISBN 978-1-5108-9089-3